THE INTERNATIONAL BOOK OF

WOOD

THE INTERNATIONAL BOOK OF
WOOD

MITCHELL BEAZLEY PUBLISHERS LIMITED
LONDON

Editor	Martyn Bramwell
Art Editor	Janette Palmer
Editorial Advisor	Bryan Haynes
Assistant Editors	Sue Farr, Paul Holberton
Designers	Mike Blore, Celia Welcomme,
	Len Roberts, John Ridgeway
Researchers	Gillian Abrahams, Helen Varley,
	Jinny Johnson, Yvonne McFarlane,
	Mike Janson, Karen de Groot,
	Tristan Allsop
Photo-research Editor	Susan Pinkus
Picture Researchers	Jackum Brown, Juliet Scott
Production	Elsie Day
Publisher	Bruce Marshall

The International Book of Wood was edited and
designed by Mitchell Beazley Publishers Limited,
87–89 Shaftesbury Avenue, London W1V 7AD

© Mitchell Beazley Publishers Limited 1976
Reprinted 1979
All rights reserved
ISBN 0 85533 081 3 (Hardbound Edition)
ISBN 0 85533 182 8 (Paperback Edition)

Photoset in Great Britain by Servis Filmsetting Limited
Printed and bound in Hong Kong
by Mandarin Publishers Ltd.

Wood is Man's oldest natural resource. It has provided him with fuel, tools, food and shelter ever since he started his long journey. Yet the properties of wood – its unique range of qualities and capabilities – are esoteric secrets today. It is easy to appreciate wood for its natural beauty and for its usefulness; far from easy to learn its lore: what wood makes the best boats, the strongest spokes or gunstocks; how the craftsmen of the eighteenth century achieved such delicate work in veneering and inlay, or how this remarkable material has taken its place, in countless ways, in the expanding technology of our present day. This hidden body of knowledge is brought to light in print for the first time in the pages of this book.

Much of the technology of wood clings to life in the form of ancient crafts held in precariously few hands. Much more has been irretrievably lost – superseded (for our era; not necessarily for all time) – by new materials and new methods.

We are foolish to treat it so casually. Wood has the priceless, indeed the unique, value of being the one basic resource that Man can renew. Oil wells run dry. Mines are depleted of coal and ore. But a well-managed forest (or indeed a forest not managed at all) goes on creating wood for ever.

Perhaps one day Man will harness the sun's energy to produce his needs, imitating, by costly and elaborate mechanisms, the miracle of organic growth. So far there is no sign of it. We do not even fully understand how a tree works. What force pumps the sap from the roots to the crown when there are no leaves to transpire moisture? What enables some trees to achieve six feet of growth in a single season, multiplying their bulk many times over yet consuming nothing but sunlight and ground-water?

In my own *International Book of Trees* I dwelt on these things, and vastly enjoyed enumerating the beauties of the trees we grow.

The *International Book of Wood* is the natural successor to that work and I am delighted to see it published with the same skill, care and imagination.

To those with a capacity for wonder, with respect for artistry and a curiosity for ingenious techniques, what follows will be sheer pleasure. But, even more important, this book puts on record for all to see what Man owes to trees and their wood. He has no older or deeper debt.

The Anatomy of Wood

After standing in the Alice Holt Forest
in southern England for more than
150 years, this straight-stemmed
forest oak crashes to the ground –
felled as part of a thinning
programme to give neighbouring
trees more room to develop

Wood in the Making

Like all green plants, trees make the materials for growth in their leaves by a process known as photosynthesis. Deriving its energy from sunlight, it is a complex chemical reaction in which carbon dioxide from the air combines with water taken from the ground to form sugars. The reaction takes place in the presence of chlorophyll – the green substance which gives leaves their characteristic colour.

Carbon dioxide passes directly into a leaf through tiny openings called stomata, but the water has a long journey from the ground to the chemical factory in the leaf. It passes into the roots through the root hairs by osmosis – the flow of water from a solution of low salt concentration, as is normal in the ground, to one of high salt concentration, as in the hair cell. The sap then flows through the xylem, or sapwood, to the crown of the tree.

But wood has other important functions in addition to the conduction of sap. It provides mechanical strength to support the weight of the crown of the tree, and stores food created by the leaves. Food is moved in solution from the leaves to all parts of the tree through the inner bark, or phloem, and is used either immediately or after a period of storage, for the generation of new growth.

New wood is produced by a specialized cell layer called the cambium, lying between the wood and the phloem. The wood cambium completely encloses the living parts of the tree and during periods of active growth the cambial cells divide to produce new wood cells on the inside and phloem cells on the outside; thus new wood is laid down on a core of existing wood. If part of the year is unfavourable for growth because of cold or drought, wood is laid down as seasonal increments which are seen as growth rings. If growth is continuous, as in tropical regions, no growth rings are produced in the wood.

Two important functions, sap flow and food storage, occur in the most recently formed wood, called the sapwood. However, there comes a time when the innermost sapwood is so far removed from the active growth region that it dies and the cell contents undergo chemical change. The new substances produced may colour the wood to form a distinctive heartwood.

MAJESTIC MATURITY
The tree has the longest life-span of all Earth's higher forms of life and during its life may produce literally millions of progeny. To maintain this remarkable existence it has highly specialized organs. The roots anchor the tree in the earth and absorb water and mineral salts. The stem conducts the nutrient sap to the leaves, where food is manufactured, and then carries the food to all parts of the tree. Fruits produced each year ensure the healthy survival of the species

LEAF ANATOMY

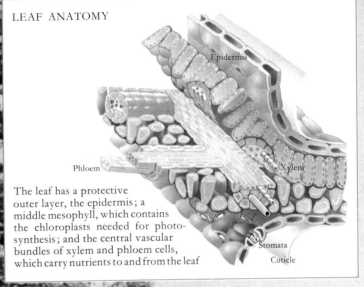

The leaf has a protective outer layer, the epidermis; a middle mesophyll, which contains the chloroplasts needed for photosynthesis; and the central vascular bundles of xylem and phloem cells, which carry nutrients to and from the leaf

STEM ANATOMY

A tree grows in thickness by the activity of a single layer of cells called the cambium. This produces sapwood, or xylem, on the inside and bark, or phloem, on the outside. As the cambium continually divides, the first-formed xylem cells become progressively detached from the cambium, undergoing chemical and physical changes to form the distinctive heartwood

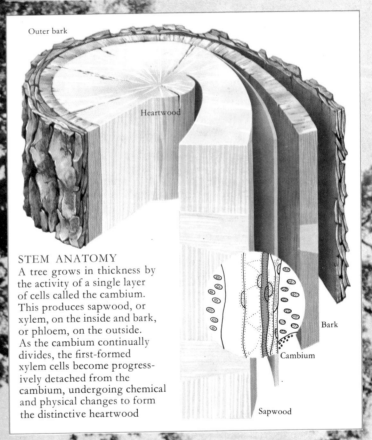

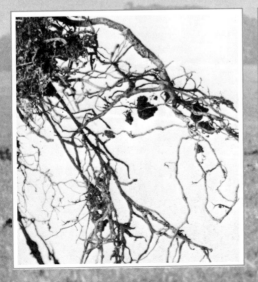

ROOT ANATOMY

Small hairs on the roots extend into soil and take up moisture by the process of osmosis. Mineral salts essential to growth are also absorbed and pass into the water stream. The solution enters the xylem vessels of the roots and is distributed throughout the tree. Ninety-nine per cent of the water is lost through the stomata of the leaves by evaporation. There is, therefore, a continuous movement of water through the tree and on a summer day an oak may take up more than 100 gallons

The Tree's Recorded History

The life history of a tree is recorded in the structure of its wood, and can be seen most clearly, if growth is seasonal, in the pattern of the growth rings.

Timbers differ greatly in ring width. Some, like box and yew, grow slowly with narrow rings while others, like poplar and some of the pines, grow vigorously and may produce rings more than half an inch in width. Ring width may, however, vary within a single tree due to variations in the conditions of growth from year to year.

Growth conditions are particularly important. Trees on a fertile site will produce wider rings than those on infertile soil; parkland trees with space for full crown and root development have wider rings than forest trees, where there is fierce competition for space; trees with a short growing season, as in arctic or mountain conditions near the snow-line, generally have very fine rings.

The width of a ring, when traced round the cross-section of a log, is rarely constant, and in a tree grown on a slope or exposed to a strong prevailing wind, the trunk is often oval in cross-section with growth rings differing appreciably in width on either side of the pith. In such trees the vigorous growth is known as reaction wood; it has a special structure and chemical composition that enable the tree to withstand the unusual conditions of growth and help it return to a normal upright position. However, when sawn into lumber and dried for use, reaction wood has an unusually high longitudinal shrinkage and tends to split and distort.

Seasonal conditions affect growth and are reflected in the width of the rings. Abnormal conditions such as prolonged drought can cause a cessation of growth and the formation of "false rings", while insect attack, fire damage and frost damage can be seen as wound responses in the wood. Recognition that seasonal effects on ring width often occur over a wide geographical range has enabled ring patterns to be dated by sampling very old trees and timbers of known age; these patterns provide information on seasonal weather conditions in past centuries and make it possible, by comparison, to date timber samples from archaeological sites.

BARK AND PHLOEM
The protective bark and nutrient-carrying phloem lie outside the cambium – the single cell layer responsible for growth. In tree-ring studies, the only value of these parts is to confirm that all the xylem is present

SAPWOOD
Fluid-conducting, and food-storage, tissue often makes the sapwood paler than the heartwood. Wide rings indicate vigorous growth in recent years

POOR GROWTH
Five years of poor growth are indicated by narrow rings; possibly due to repeated defoliation by insect pests

HEALING GROWTH
During each seasonal growth period, new wood extended farther across the wound, leaving a clear pattern in the growth rings

FIRE DAMAGE
Fire, driven through the forest by its own self-generated winds, has badly burned one quadrant of the trunk. The healing process continued for ten years – finally covering the scar with sound wood

YOUNG VIGOUR
The young tree, seeded in a beneficial environment, grew steadily, adding even annual increments

PITH
Often difficult, or even impossible, to see on the section of a mature tree, the pith may sometimes be seen as a pale area a few millimetres across

DROUGHT
A drastic shortage of water lasting for several successive years restricted growth and only extremely narrow rings were formed

NORMAL GROWTH
While the far side of the tree was repairing damage inflicted by the forest fire, the rest of the tree's girth increased normally. Growth rings are even and of moderate width

TENSION WOOD
Laid down in fairly wide increments, tension wood is a growth response to stress. The tree may have developed a lean due to soil slip on a slope, or through exposure to a prevailing strong wind

DISEASE
Growth may be severely inhibited by diseases or by insect or fungal damage to the roots or leaves. Successful competition, for water and nutrients, by neighbouring trees may also produce this pattern

LABORATORY SAMPLES
In studying growth patterns and variations in timbers, disc samples are taken from felled trees and retained for analysis. Most of these discs, dried rapidly, split along a radius as shrinkage is greater tangentially than across the growth rings

SLOW GROWTH PATTERN
In the sample of fir below, the annual increments, or growth rings, are narrow throughout much of the tree's life. This pattern suggests that the tree grew in far from ideal circumstances – perhaps on a bare hillside with thin, poor soil

FAST GROWTH PATTERN
Contrasting with the sample of slow-grown fir (left), the sample of spruce below has developed vigorously – adding wide annual increments, notably during its first twelve years. The tree apparently had an abundant water supply and good soil

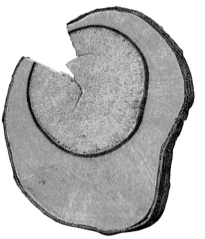

PROTECTIVE BARRIER
In the cross-section above, a plum tree, Prunus sp, has responded to a fungal attack by laying down a barrier of resin. Originally secreted just below the bark, this defensive ring has been totally enclosed by subsequent wood growth

COMPRESSION WOOD (SPRUCE)
In softwoods, a modified form of tissue is laid down on the "down-hill" side of a leaning trunk. This compression wood, a response similar to that producing tension wood in hardwoods, is abnormally dense, with a high lignin content, but the wood lacks toughness

TENSION WOOD (HARDWOOD)
This reaction wood, produced on the "uphill" side of the tree to counter-act a tendency to lean, has exceptionally high longitudinal shrinkage. Tension wood in beech, below, has split along the grain and torn fibres are visible on the inner surfaces

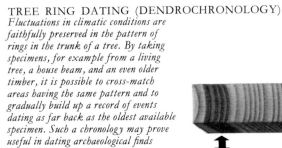

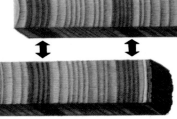

TREE RING DATING (DENDROCHRONOLOGY)
Fluctuations in climatic conditions are faithfully preserved in the pattern of rings in the trunk of a tree. By taking specimens, for example from a living tree, a house beam, and an even older timber, it is possible to cross-match areas having the same pattern and to gradually build up a record of events dating as far back as the oldest available specimen. Such a chronology may prove useful in dating archaeological finds

A sample core from a living tree shows growth rings over a period of nearly 40 years

By comparing this sample from a house beam with that from the living tree, above, the beam is shown to have been cut from a tree felled nineteen years ago

Adding a third sample from an even older timber gives a correlated sequence covering nearly 90 years

15

Patterns of Growth

Events during the life of the tree are recorded by other features of growth in addition to the growth rings. Branch development is revealed by the patterns of knots within the wood and sometimes even the arrangement of the leaf shoots on the young stem may be seen preserved in the wood. This occurs in some softwoods with leaves which last for several years. The tissues at the base of each leaf are enclosed by the growth of the stem and, when the wood is cut, the regular pattern of leaf insertions is clearly seen.

While a shoot produces leaves, its tissues are continuous with those in the stem; when a branch dies, though it continues to be enclosed by the main stem, there is no further tissue connection. If sawn through at a later date, a branch with cells which are continuous with those of the main stem produces a knot which stays in place – called a live knot. If the branch was dead there is no tissue connection and, on drying, the knot often shrinks and falls out – a dead knot and a feature which reduces the value of the timber. When a branch is pruned, the stem grows over the cut end to repair the damage and, once the wound is completely covered, further growth is of sound, clear wood. The covering, or occlusion, of large branch stubs may sometimes be seen as swellings on the stem.

Near large limbs, or on trees with fluted stems, bark may become enclosed in the growing wood, while cones have been found embedded in the wood of some pines. Another type of inclusion occurs where splits or cavities in the wood are filled – often with gum or resin, but sometimes with mineral-like accumulations deposited from the sap and so hard that they damage saw blades. Gums, resins and oils occur in many woods, often giving a characteristic smell, and crystalline deposits are common, though typically they are so small that they can only be seen under the microscope. In some woods, fine silica grains may be deposited, making the wood abrasive to saws and knives.

Objects near to, or attached to, a tree may become completely overgrown; they are often difficult to detect and may cause extensive damage to saw-mill equipment as well as spoiling the timber for commercial use.

CONSUMING GROWTH
While the boys who sat on it grew into middle age, a wrought-iron seat, leaning against an oak tree beside the cricket field of an English public school, was gradually engulfed by the remorseless growth of the sheltering tree

SPIRAL GRAIN
Normally, wood fibres run straight and parallel to the tree's axis, but spiral grain, twisting about the vertical axis, has been found in more than 200 species. Shown below in wood of the plum tree, it causes the timber to split

KNOT FORMATION
Knots in timber mark the "roots" of branches in the stem of the tree. Living branches maintain organic continuity with the stem, and the "live" knots they form distort its grain but do not much affect the timber's strength

LEAF SCARS
Leaf scars are a form of knot marking the emplace-ment of the leaf shoots on the stem of a young sapling. Shown below as a young sapling and, right, on the surface of a flat-sawn plank of pine, the scars form a distinct spiral pattern in the timber

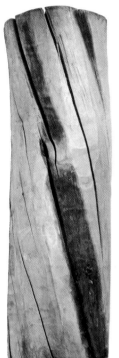

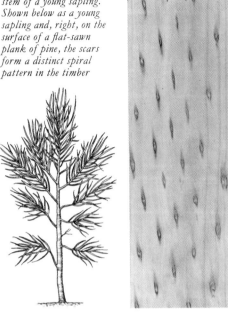

"DEAD" KNOT *(above)*
A "dead" knot marks the base of a branch which has died. From the point of death, it interrupts the growth rings like a foreign body. Dead knots are loosely enclosed and may easily fall out when the sawn timber dries

STONY DEPOSITS
The roots of a tree are continuously extracting water and mineral salts from the soil. The salts, which were dissolved in the water, are frequently precipitated from the sap to form crystalline deposits. This calcareous deposit in afzelia is a serious defect in the wood and a danger in the mill

BARK INCLUSION
An area of bark may die and be absorbed into the trunk as new wood grows round it. But shown here is bark enclosed by two limbs of an East African muninga which have grown into contact as new wood is laid down by each

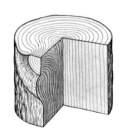

HEALTHY PRUNING
Properly pruned branches are cut close in at the base, so that a minimum of dead wood is enclosed. Though a bulge will form, and the cut surface will not be continuous with the stem, subsequent wood laid down will be sound

The Form and Function of Cells

Within the tree, the functions of sap conduction, physical support and food storage are carried out by cells specially adapted for each purpose, though the type of cell involved sometimes differs between softwoods and hardwoods. Conduction in softwoods is by thin-walled tracheids; in hardwoods by vessels: the support in softwoods is provided by thick-walled tracheids; in hardwoods by fibres: in both softwoods and hardwoods, food storage occurs in parenchyma cells.

Softwood tracheids specialized for fluid conduction are long, thin-walled, tubular cells with large internal cavities, or lumen, and with many valves, called pits, in their walls to control the flow of sap; tracheids specialized for the support role have thick walls and a small lumen. In temperate softwoods, bands of thin-walled tracheids laid down when seasonal growth starts, and called early or spring wood, alternate with bands of thick-walled summer wood cells.

In hardwoods, different types of cell are responsible for conduction and support. Sap flow occurs in vessels which form a pipe-like system from roots to leaves; on a cleanly cut end surface they appear as small openings or pores and their presence distinguishes hardwoods from softwoods. Hardwoods with a uniform distribution of pores, as seen on the end grain, are called diffuse-porous woods; more rarely the pores occur in zones and, when those in the earlywood form a distinct line and are larger than those in the latewood, the wood is called ring-porous.

The main supporting cells in hardwoods are fibres, generally needle-shaped cells that are more or less thick walled according to the species of timber. The thickness of the fibre wall, more than any other feature, determines the density, and hence many of the physical properties, of a hardwood.

In both softwoods and hardwoods, storage of food occurs in the parenchyma – small, box-like cells which are sometimes arranged axially, especially in hardwoods, and always in horizontal bands called rays which run across the grain in a radial pattern. Where the rays are large, as in oak and beech, they give a characteristic "figure" on cut surfaces.

SOFTWOODS AND HARDWOODS

The terms softwood and hardwood refer to the botanical origins of woods and not to their density or physical hardness. Softwoods come from cone-bearing trees, often with evergreen needle-like leaves, belonging to the botanical group gymnosperms – plants with naked seeds. Hardwoods come from broad-leaved trees, either evergreen or deciduous, belonging to the angiosperm group.

THE FORMATION OF CELLS

New wood is formed by the cambium, a single cell layer between the bark and the wood. Cambial cells divide when conditions are favourable for growth, one cell remaining in the cambium, the other developing into a xylem, or wood, cell if on the inside of the cambial cell, or into a phloem, or inner bark, cell if on the outside. The thin-walled new cells grow rapidly as additional layers are deposited from inside the cell, and specialized features, like wall pits, form

SPECIALIZED CELL TYPES

Softwood tracheids are long cells, often about 3mm; some have thin walls and abundant pits which control the flow of sap; others, the main support cells, have thicker walls, pointed ends and few pits. Hardwood fibres are shorter, about 1mm, with thick walls and pointed ends. Vessels consist of short, pipe-like cells with open ends Parenchyma, or storage cells, are small, thin-walled and box-like in structure

Scots pine
Pinus sylvestris

European oak
Quercus robur

Common beech
Fagus sylvatica

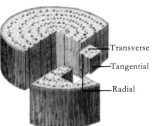

Transverse
Tangential
Radial

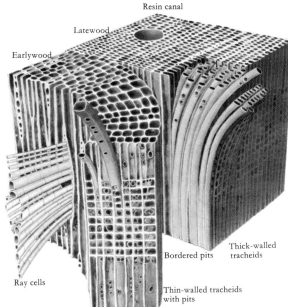

Resin canal
Latewood
Earlywood
Bordered pits
Thick-walled tracheids
Thin-walled tracheids with pits
Ray cells

Latewood consists of thick-walled tracheids and is the main tissue giving the wood its strength. Resin canals, often clearly visible on transverse sections, are a common feature of many softwoods

The earlywood consists of thin-walled tracheids, which give this region of the annual growth ring a characteristically open structure. Thin-walled tracheids have an abundance of bordered pits, which allow the flow of sap

Storage tissue in the majority of softwoods is confined to the rays, which are always very fine. Occasionally rays have a central cavity which allows the flow of resin; similar channels occur among the tracheids

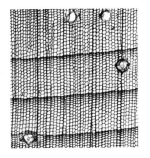
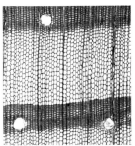

SOFTWOOD VARIETY
The most significant structural difference between softwoods is in growth-ring pattern, which influences density and texture. Density is largely determined by the proportions of thin-walled earlywood tissue and thick-walled latewood, which is dense. This contrast is seen *above in the sections of light-weight yellow pine, left, and pitch pine. Texture, too, is influenced by the earlywood/latewood contrast, and also by the vigour of growth. In fast-grown timber, the texture tends to be coarse. Where growth is continuous, as in Parana pine, texture is fine and even*

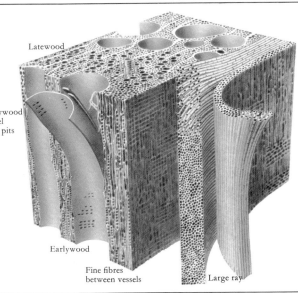

Latewood
wood el pits
Earlywood
Fine fibres between vessels
Large ray

In the ring-porous hardwood, a conspicuous line of very large earlywood vessels contrasts with the numerous small vessels found in the latewood zone. Together, the two zones constitute an annual ring

Latewood is made up very largely of fibres, with some associated storage cells, called parenchyma. The fibres give the wood its strength and contribute to other physical characteristics

Much of the wood's storage tissue is in the form of rays – clearly visible as strands running at right angles to the growth rings. In most ring-porous woods the rays are rather fine, but in oak (illustrated) some rays are very large and give the wood an attractive silver grain figure on radial surfaces

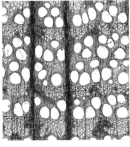
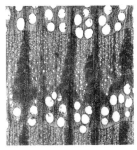

CONTRASTING OAKS
The structure and properties of ring-porous woods are markedly affected by their vigour of growth. As growth slows, the ring width decreases, mainly as a result of a reduction in the amount of dense latewood. The proportion of the wood occupied by large earlywood cells is *increased and, as this tissue is light, overall density of the wood is reduced. The contrast can be seen above in slow-grown oak, left, and fast-grown oak. The lighter, slow-grown timber is low in strength but easily worked: more vigorous wood is heavier and stronger, but can prove difficult to process*

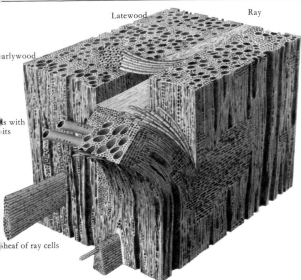

Latewood
Ray
arlywood
s with its
heaf of ray cells

The diffuse-porous hardwood, here represented by beech, has numerous small vessels evenly distributed and showing only slight gradation in size across the annual growth ring. The vessels are surrounded by a mass of fine fibres constituting the bulk of the wood

Storage tissue in beech occurs almost entirely in the form of rays. These are very variable in size but are conspicuous on all surfaces of the wood: surfaces cut tangentially expose the characteristic cigar shape of the beech ray

The vessel walls are provided with groups of small pits, but sap-flow in hardwoods takes place through the "open" ends of the cells rather than across the grain of the wood

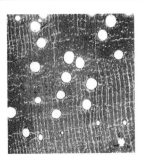
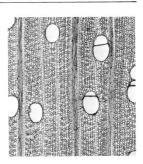

RANGE OF CHARACTER
Most of the world's hardwoods have a diffuse-porous structure and differ greatly in character due to differences in the size, number and distribution of their cell types. Cross-sections of lignum vitae (above left) and balsa (right) show a difference in texture – balsa being much *coarser than the heavier lignum vitae. More significant, however, is their fibre structure; in balsa the fibres are thin walled and enclose a high proportion of air, making the wood very light in weight; in lignum vitae, the fibres are very thick walled producing a very heavy wood*

THE ANATOMY OF WOOD

The Inside Story

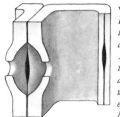

For many years it has been possible to examine the structure of wood by studying thin sections under an optical microscope, but the amount of detail that can be observed is limited. Because an electron beam has a much shorter wavelength than that of light, the transmission electron microscope allows much smaller items to be seen. But a more recent, and far more dramatic, advance has been made with the electron scanning microscope, which can focus through such a deep field of view that remarkable three-dimensional images may be achieved – revealing the internal "architecture" of wood.

This remarkable technological tool has revealed visually the cell-wall structure that had been deduced by optical and other laboratory methods. Wood consists of cellulose molecules aggregated to form long string-like units called microfibrils. These microfibrils are encrusted with lignin and hemicelluloses; the microfibrils giving the wood its tensile and bending strength while the lignin provides stiffness. The microfibrils are visible under the electron microscope and their orientation in the complex layered wall structure is clearly seen.

The cell, when first derived from the cambium, has a primary wall composed of loosely aggregated, irregularly arranged microfibrils, which allow the cell to continue to enlarge. With growth, a second and much thicker cell wall is laid down inside the primary wall; this wall usually has three layers. The outer and inner layers are thin; the middle layer is much thicker with its constituent microfibrils closely aligned with the long axis of the cell. To a large extent, the physical and mechanical properties of the wood depend on this thick middle layer of the secondary cell wall.

In hardwoods, sap flows from cell to cell through openings, or perforations, in the membrane between abutting vessels. In softwoods, the sap flow is controlled by movable valves, or pits, in the walls of conducting tracheids. With these fine details of the cellular structure and the sculpturing of cell walls, the scanning electron microscope has quite literally opened our eyes to the incredible wealth of detail and form within wood.

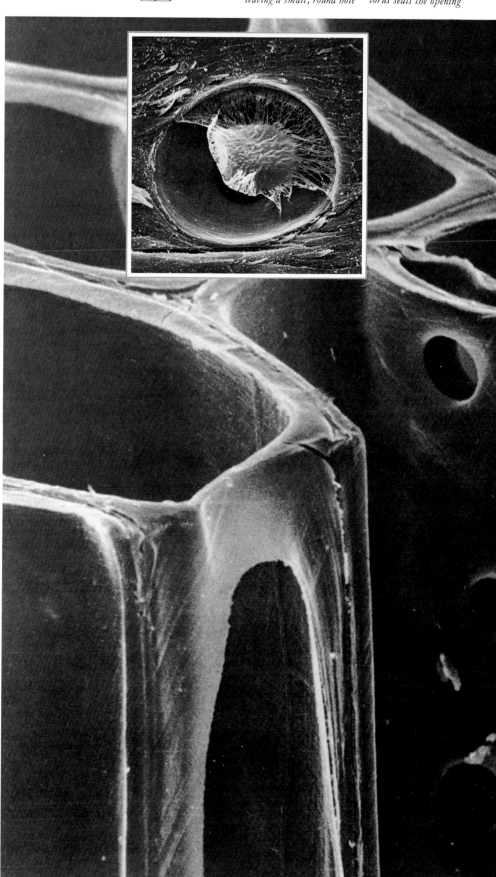

CELL WALLS AND PITS
The micrograph shows pit-openings in the cell walls, and a pit chamber where the cells are cut across. An inset shows a torus and its supporting net, torn to reveal the underlying pit chamber and the pit-opening into the cell

PERFORATION PLATE
*The vessels of hardwoods
have densely pitted walls,
but sap flows from cell
to cell through perforations
formed by the dissolution
of an area of wall common
to adjoining vessels. This
is one such opening with
an overarching border*

MULTILAYERED WALL
*The thickness of the cell
wall is largely made up
of the middle layer of the
secondary cell wall. This
can be seen in the micro-
graph, separating the thin
inner lining of the wall
from the several layers
between the two cells*

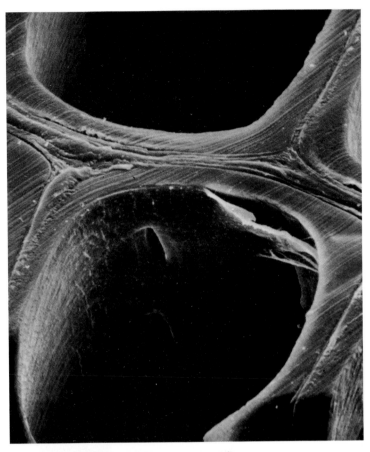

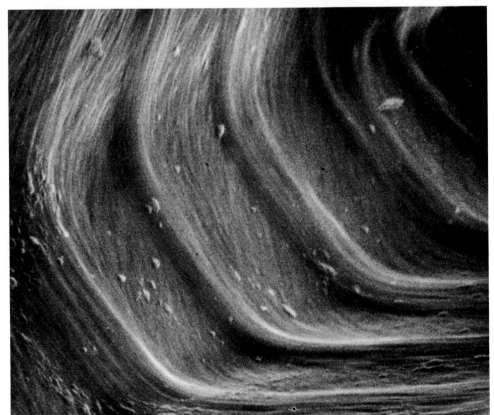

MODIFICATIONS
*In some woods the normal
pattern of wall structure
is modified. Development
of helical thickenings to
the secondary cell wall
may occur. The micrograph
shows helical thickenings
strengthening the inside of
a Douglas fir cell – just
as an engineer strengthens
tubular metal structures*

Earth's Petrified Forests

At best fragmentary, our knowledge of plant evolution is based solely on fossils. Some light has been shed by the painstaking study of the fossil evidence, but most fossils are no more than impressions of stems or leaves of soft, perishable plants which have been trapped in sedimentary rock. Trees constitute an exception. Silica, quartz or other minerals may fill the centres of tree tissue cells and in so doing preserve the cellulose wall. Vast petrified forests formed in this manner survive in many parts of the world with all their anatomical detail intact.

The first thousand million years of plant evolution were slow. Single-celled plants languished in the primeval seas, while a layer of life-supporting oxygen gradually formed above them. Once sufficient oxygen screened the land from the harmful rays of the sun, an explosion of terrestrial plant life began.

The first true land plants appeared some 450 million years ago. They had no true leaves or roots, but from them the first species large enough to be described as trees emerged at the middle of the Devonian period. They were small by comparison with modern forest trees, but by the late Devonian period giant mosses and ferns had evolved more than a hundred feet high.

The development of male and female gametes and reproduction by seeds, instead of spores, were two important evolutionary steps which brought plant life close to its modern form. During the Carboniferous period, or Coal Age, the gymnosperms, early relatives of the conifers, the cycads and the ginkgos began to appear, although lower vascular plants dominated the swamp forests until the mass extinctions of the Permian period, 280 million years ago.

The dominance of gymnosperms was complete by the Cretaceous period – by which time angiosperms, the flowering plants which dominate the world's flora today, were coming into being. Subsequent large climatic and geological shifts brought about angiosperm dissemination and the decline of the gymnosperms – their empire reduced to the colder temperate and northern regions of the world, where they reign today.

RINGS OF STONE
The growth structures of this fossilized wood have been perfectly preserved by replacement minerals, and the even, well-spaced growth rings provide a clear guide to the good growth conditions prevailing during the tree's life

FOSSILIZED STEM
This polished section shows the internal structure of an American oak tree that lived and flourished more than twenty-six million years ago. The cell lumen have been filled with crystalline silica, coloured by a variety of mineral pigments

EVOLUTION OF PLANTS

The first vascular land plants appeared about 430 million years ago. The next major evolutionary step saw the appearance of the first seed-bearing plants, the gymnosperms, which continued to develop until, by the Cretaceous period, they were the dominant form of plant life. Their reign was, however, short-lived as a number of climatological and geological changes wiped out whole genera and paved the way for the spectacular rise of the angiosperms, or flowering plants, during the last 135 million years

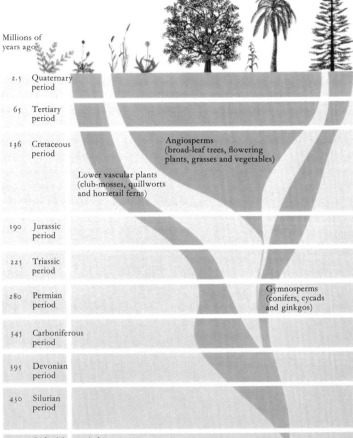

Millions of years ago		
2.5	Quaternary period	
65	Tertiary period	
136	Cretaceous period	Angiosperms (broad-leaf trees, flowering plants, grasses and vegetables)
		Lower vascular plants (club-mosses, quillworts and horsetail ferns)
190	Jurassic period	
225	Triassic period	
280	Permian period	Gymnosperms (conifers, cycads and ginkgos)
345	Carboniferous period	
395	Devonian period	
430	Silurian period	
500	Ordovician period	

THE FIRST TREES

Sigillaria "trees" were giant club-mosses common in the Carboniferous, or Coal Age. The trunk was unbranched and pitted with the scars of old leaves. These leaf-scar impressions are among the commonest fossils found in coal seams

COAL AGE GIANTS

Growing more than 100ft tall, the Lepidodendron trees were a dominant plant species in the Coal Age swamp forests. Leaf scars show the spiral arrangement of the leaf stems on the branches and the main stem of this primitive tree

COAL-FORMING FORESTS

During the Carboniferous period (345–280 million years ago), vast areas of low-lying land were clothed in dense, humid forests, repeatedly subjected to inundation by the sea. In these forests, giant ferns, horsetails and primitive trees grew in profusion — dead and decaying vegetable matter accumulating in a dense mat on the water-logged ground. Interleaved with layers of silt and clay, this carbon-rich material became compressed and changed chemically into the coal seams mined today

FOSSILIZED FERN

Because most plants are soft and decay rapidly, few really clear fossils are formed, and most are simply the impressions of stems and leaves left in muddy or silty deposits. Ferns, however, may be preserved in detail in coal deposits

23

The Renewable Resource

Recently felled larches bear witness to
Man's constant demands on Earth's
resources. Forestry practice balances
extraction and regeneration to ensure the
healthy survival of woodlands as a
resource, as a wildlife habitat, and as a
priceless amenity

The Living Forest

The forest from which Man takes his timber is the tallest and most impressive plant community on Earth. In terms of Man's brief life it appears permanent and unchanging, save for the seasonal growth and fall of the leaves, but to the forester it represents the climax of a long succession of events.

No wooded landscape we see today has been forest for all time. Plants have minimum requirements of temperature and moisture and, in ages past, virtually every part of Earth's surface has at some time been either too dry or too cold for plants to survive. However, as soon as climatic conditions change in favour of plant life, a fascinating sequence of changes occurs, called a primary succession.

First to colonize the barren land are the lowly lichens, surviving on bare rock. Slowly, the acids produced by these organisms pit the rock surface, plant debris accumulates, and mosses establish a shallow root-hold. Ferns may follow and, with short grasses and shrubs, gradually form a mantle of plant life. Roots probe ever deeper into the developing soil and eventually large shrubs give way to the first trees. These grow rapidly, cutting off sunlight from the smaller plants, and soon establish complete domination – closing their ranks and forming a climax community which may endure for thousands of years.

Yet even this community is not everlasting. Fire may destroy it outright and settlers may cut it down to gain land for pasture or cultivation. If the land is then abandoned, a secondary succession will take over, developing much faster on the more hospitable soil. Shrubs and trees are among the early invaders, their seeds carried by the wind, by birds and lodged in the pelts of mammals.

For as long as it stands and thrives, the forest is a vast machine storing energy and the many elements essential for life. Carbon products are stored in the wood of the tree and in forest-floor litter; the water-balance is maintained by rain and snowfall, held in the tree's tissues and in the soil, while a vast store of mineral nutrients derived from the soil is held in the leaves, only to be returned to the soil as the leaves fall and decay in the annual cycle of regeneration.

AGED SURVIVOR
This gnarled and twisted oak stands near the edge of the ancient Wistman's Wood on the bleak expanse of Dartmoor. Its deformed trunk and branches bear witness to the poor moorland soil and the constant high winds of the habitat

The primary succession, by which an area of barren ground is colonized by a series of plant regimes, reflects in just a few tens of years the whole story of plant evolution. First to establish a foothold are the simple lichens, which survive on bare rock and initiate the processes of soil production. These lowly pioneers are then succeeded by progressively higher plant groups, each contributing to the soil and preparing the way for the next group. The climax is reached in the fully mature forest

PRIMARY SUCCESSION

Stage 1: The barren rock is colonized by pioneer lichens and mosses

Stage 2: Ferns and small shrubs take root in the developing soil

Stage 3: Larger shrubs and small trees are dominant

Stage 4: Mature forest with trees and understorey

NATURE RECLAIMS THE LAND

On abandoned farmland, a secondary plant succession immediately begins the task of reclamation. The sequence is much faster than a primary succession as a healthy soil exists and the climate is favourable

Weeds, grasses and small shrubs immediately move in to colonize the abandoned farmland

As the ground-cover becomes more dense, shrubs, bushes and pine saplings take root

Pine-wood dominates for a while, but the trees cannot reproduce in their own shade

Eventually, broad-leaf (oak-hickory) forest becomes the dominant plant mix at climax

ENERGY BALANCE

BIOMASS

Total energy input

Lost through transpiration

Used by birds and animals

Transpiration from litter and ground

Stored energy

Production of new plant material

Retained in litter

LEAF LITTER

Retained by roots

HUMUS

The world's forests act as great collectors, converters and storehouses of energy. Solar energy is utilized in the chemical factory of the leaves to produce new plant material – a form of stored energy. As the plant material dies and decays its stored energy is taken up again in the form of dissolved nutrients. The gas exchanges that take place in the leaves renew the atmosphere, while water, lost into the air by transpiration, is released into the system to fall again elsewhere as rain or snow

THE RENEWABLE RESOURCE
A System in Balance

SEED-EATING CHIPMUNK
Chipmunks regularly store nuts in hoards, or caches, to tide them through the winter. Seeds they forget, or drop by accident, will sprout into young trees. Forest trees seed so freely that it is sufficient if one in a thousand survives

PREDATORY LADYBIRD
Seven-spot ladybirds prey in both their larval and adult stages on aphids, or greenflies, that suck the sap from leaves, damaging the tree's growth. But if aphids become too scarce, the ladybird population also falls, disastrously

Every old-established forest provides a home for vast numbers of animals, birds and insects which live in a delicate state of balance with each other and with their environment. Seed-eaters might, in theory, destroy the forest by devouring all the nuts destined to produce seedling trees, but in practice this rarely happens. Predatory owls, hawks, foxes and pine martens limit the numbers of seed-eating birds, mice and squirrels and allow sufficient seeds to survive to ensure the replacement of old and diseased trees. If the predators become too successful and destroy their prey, they in turn will decline in numbers through shortage of food.

The wildlife population is markedly different in deciduous and coniferous woodlands. Broad-leaf forests support a great variety of life. More than 200 insect species have been found on oak trees alone, including leaf-roller moths, gall-makers, predators and vegetarians. Insect-eating birds like thrushes, blackbirds and warblers abound, sharing their habitat with seed-eating wood-pigeons, pheasant, dormice and squirrels. Coniferous forests have a more restricted fauna comprising species better equipped to deal with their resinous bark and foliage and their tough leaves and seeds.

This delicate balance may, however, be upset by the introduction of a species into a new country where it has no natural enemy. Today, quarantine laws attempt to safeguard each country's woodlands, but without complete success. The bark beetles that carried a virulent strain of Dutch elm disease to Britain in 1970 made the crossing from Canada in consignments of unsawn rock elm used in the boat-building industry. Maps plotting the spread of the disease focused on ports at which the timber was landed.

Poisons are the simplest large-scale means of attack, but these need repeated application and may also harm innocent species. Biological control offers a better remedy. American Monterey pines, planted in Australia, suffered great damage from *Sirex* wood wasps accidentally imported from Europe. The balance was restored by the introduction of the European ichneumon fly – a natural, and destructive, parasite of the wood wasp.

FAMINE AND PLENTY
A hungry dormouse has made a clean sweep of these acorns, biting into each in turn to extract its nutritious kernel through a neat round hole. But next autumn the oaks will bear no seed, and few dormice, pigeons or squirrels will survive the ensuing winter. Then comes the bumper crop of a "mast" year, with far more acorns falling than the survivors can consume. Periodic seeding, between years of scarcity, is the oak tree's ruthless but effective answer to the destroyers of its seed

HUMMING-BIRD
Hovering motionless before a flower, the humming-bird sucks out the nectar with its long, curved beak, and, flying from tree to tree, cross-pollinates them. But to obtain its services, the tree must secrete a large quantity of nectar

BRACKET FUNGUS
The wind-borne spores of bracket fungi, alighting on wounds unprotected by bark, spread their threads relentlessly throughout the tree, eventually causing its death. However, by aiding the breakdown of dead wood, they assist the nutrient cycle

CONE-EATING CROSSBILL
The crossbill's crossed upper and lower bills can tear back the scales of cones to obtain the seed beneath. However, the soft fruits of mountain ash and holly contain tough seeds, which pass through its alimentary canal unharmed

LEAF-EATING MOTH
Here feeding greedily on cork-oak leaves, hairy caterpillars of the gipsy moth, Lymantria dispar, may attack a wide range of host trees. In Europe, its original home, the moth is checked by many enemies. Introduced into New England in 1869 by a mathematician who ingeniously planned to cross it with silk moths, it escaped and became a major pest. Millions of dollars have been spent on spraying infested forests with DDT in an attempt to stem its spread

BROWSING DEER
Deer such as the American elk, or wapiti, check tree growth by browsing back young shoots, especially in winter, when grass lies under snow. But they also trample seeds deep into the soil, safe from the attentions of woodmice

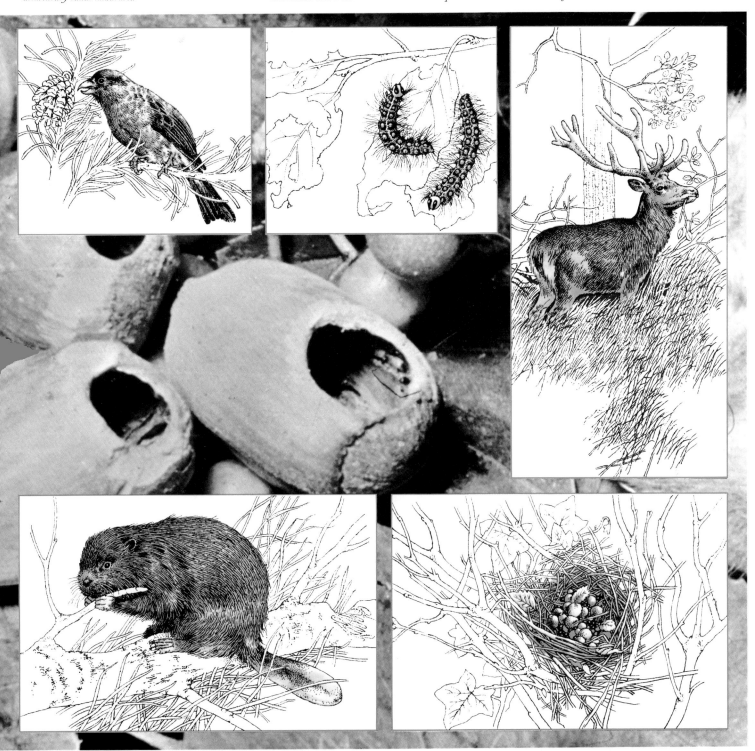

NATURE'S WOODSMAN
Beavers deliberately fell large trees for the logs and branchwood with which they build their dams and lodges. Inner bark, tender shoots, nuts and fruits are cached for winter food. The dams create a habitat for insects and waterlife

SPREADING THE SEED
The woodmouse's store of nuts and berries in an old bird's nest means movement of seed from established forest, often across or on to bare land. Relatively large, heavy seeds would not be carried far unaided by the beasts and birds

The Rainforest Hardwoods

The tropical rainforests of equatorial Central and South America, Asia and Africa form a world resource of vast extent – but uncertain future. Such forests are the only source of valuable and unique tropical hardwoods such as mahogany, teak, greenheart, obeche, iroko and padauk. But the great stems that repay exploitation occur only as scattered individuals, "emergent trees" soaring upwards from a host of smaller neighbours in a complex, multilayered plant society. The traditional practice has been to skim off the cream – to take only mature, desirable stems above a minimum size laid down by law. Inevitably the less desirable trees – and there may be two hundred different sorts on one square mile of tropical forest – colonized the gaps first, and high-grade timber trees have become progressively scarcer.

Today, full utilization of all available timber is advocated. High-powered mechanical harvesting will replace selective felling and elephant or buffalo haulage. Though prime stems of teak or mahogany will still be set aside, logs of other kinds will be converted in modern installations to sawn lumber for local use, to chipped fragments for particleboard, or to pulp for low-grade but serviceable paper.

In the wake of such clear-cutting, plantations may be established with fast-growing trees such as gmelina, which make full use of tropical heat and rainfall to achieve amazingly rapid growth. Sub-tropical pines also are proving adaptable to many tropical situations and their successful development has eased the volume of softwoods, and softwood products, previously imported from the north temperate forest regions.

New plantations, and the encroachments of land-hungry settlers, especially in Asian and African countries with ever-increasing populations, erode the tropical forest unremittingly. The true rainforest, with its two-hundred-foot-high emergent trees, its fascinating pattern of tree canopies and understoreys, its climbing lianas, strange ferns and orchids flourishing as epiphytes on tree trunks far above ground level, may eventually survive only in national parks and wildlife reserves.

BUTTRESSED GIANT
Many tropical broad-leaved trees develop huge woody buttresses around the base of the trunk. These help support the weight of the tree by spreading it over a larger area, and also brace the tree against the pressure of gusting winds

DANGEROUS PRACTICE
In Borneo and many other tropical areas, the practice of shifting cultivation poses a threat to forest resources. Semi-nomadic tribesmen clear-cut areas for cultivation but later move on – leaving the hill-side exposed to erosion

CONTROLLED POWER
In Thailand, elephants still play a major part in the valuable teak operations. Caught when young, they are trained to drag big logs in harness, raise smaller ones with trunk and tusks, and, shown above, to manoeuvre the largest using their weight

PRIMITIVE EQUIPMENT
Despite the great size and value of tropical hardwoods, the equipment used in the producer countries is often still small and inefficient. This diesel-powered mill in Cambodia has only recently replaced hand-sawing as a method of cutting planks

MARCH OF PROGRESS
A modern timber-felling operation in Sarawak, East Malaysia, is shown above. Concessions are leased by the government to lumber firms who pay a royalty, or "stumpage", on the crop. Only the largest trunks are commercially attractive

FALSE-COLOUR MONITORS
One overflight by an aircraft equipped with modern cameras can reveal more than months of survey on the ground. Infra-red, or "false-colour", photography brings out features that are normally hidden by the forest's natural greenery.

Here, healthy forest shows red while clear-cut areas appear in shades of blue. Satellite monitoring can reveal large-scale seasonal changes in the state of the forest and may also give advance warning of areas suffering from the ravages of pests or disease

31

Man-made Forests

Man-made forests, plantations and tree-farms should effectively answer civilization's ever-increasing demand for timber and wood-based products. The nineteenth century was an era of unrestrained exploitation of the forest resource, particularly in North America, where supply seemed inexhaustible. Nature had created the forests for nothing; Man could harvest them for profit; future timber-users were hardly considered. At the last moment conservationists led by Gifford Pinchot and President Theodore Roosevelt declared that outright exploitation could lead only to economic and environmental disaster. Man must not only conserve the remaining forests, he must urgently rebuild the reserves already destroyed and plan anew for the future of the forests.

Fundamental to the planning of man-made forests is the concept of "sustained yield". It is assumed that the market to be met – a sawmill, paper-mill or general demand from local timber-users – is constant, or expanding steadily. Output from the forests must match the demand, both present and future, within the limits of the natural cycle of the timber crop. In the simplest terms this means that if

the trees take fifty years to mature, one fiftieth part of the forest should be felled each year and a like proportion replanted. The forester in charge may select one large plot or many small ones, or he may cut out one individual tree in fifty; but he must never exceed the "allowable cut". This allowable cut, which corresponds to the forest's annual overall growth increment, is carefully assessed by sample measurements taken over a period of years and is adjusted regularly.

Tree-farming involves the injection of capital into the forest for future financial return, with the calculation of real, or assumed, interest charges on the money invested. Land, labour and trees must all be paid for, but there can be no return on outlay until the crop reaches marketable size. Foresters therefore apply the latest scientific techniques to achieve rapid growth at lowest cost, and to minimize losses among their young trees. Mechanization has accordingly supplanted many traditional methods, especially in the raising of plantation stock from seed in nurseries.

Where hardwood trees, which take several decades to mature, are being planted, methods

of multiple cropping have been adopted – as for instance in the southern United States, where soya beans and bluegrass have been planted between widely spaced black walnut seedlings. First the agricultural crop is harvested and later the valuable harvest of nuts provides an additional interim return while the trees mature. Though it takes many years to reach economic size, the walnut, used widely as a veneer, is very valuable.

Practical experience has taught that it pays best to plant far more trees at the outset than will be harvested fully grown. Roughly ten times as many may be planted, for a variable proportion will fail to survive the younger stages, being smothered by weeds, killed by browsing deer, eaten by rabbits or insects, or wiped out by disease. Another large proportion will be harvested as the thinnings, or selected trees cut out periodically to give the maturing crop more room. Thinnings, though small in size, are readily marketed as fence-posts, mine props, pulpwood and small saw-logs for lumber; they therefore represent a valuable return on initial investment before the final, fully mature crop can be harvested and the land replanted.

CLEAR-CUTTING METHOD
To ensure sustained yield, the forest is divided into blocks, cleared in strict rotation so that after, say, 50 years, it will contain stands of all ages from seedlings to 50-year-old trees. Soil erosion is prevented by standing timber farther up-slope and the clear-cut area provides excellent browsing for a variety of wildlife. New strains may be introduced on replanting

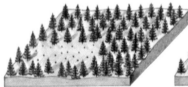

The chosen block is cleared completely; adjacent blocks provide soil protection and life habitats

The clear-cut area may be planted out with nursery stock, or be left to reseed from adjacent stands

Once the cleared block re-establishes itself, one of the adjacent areas may be worked for timber

SEED-TREE METHOD
Where trees grow readily from self-sown seeds, the area may be cut and suitably spaced seed-trees ("mother trees") left standing. Replacement costs are low, but there is no opportunity to introduce new strains. Successful regeneration may produce densely overcrowded stands, which require expensive thinning to avoid retarding the growth of young trees

A large area is cleared leaving well-spaced seed-trees; regrowth may be assisted by fertilizers

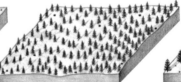

When cover is re-established, the seed-trees may be cut out along with any weak or deformed trees

To promote vigorous growth of young stock, one or two phases of thinning may be necessary

SINGLE TREE SELECTION
Selected trees of all sizes, from all parts of the forest, are felled, hauled out and graded for end-use by size and quality. The harvesting costs are high, but forest and soil are well protected. In mountainous areas, particularly those with tourist industries, the retained tree cover maintains the area's scenic attractions and protects against avalanches

Selected trees of various ages have been extracted leaving a thinner, but even, forest cover

Natural re-seeding from standing stock ensures regeneration, but additional planting may be needed

The young growth is well established and the intervening seed stock is available for cutting

NURSERY BREEDING
*Tubed seedlings, shown here
in the nursery, are raised
by sowing the seeds individually in tubes packed with
sand, compost and fertilizer.
As they sprout, the young
trees are planted out and
the split-sided tubes open
easily to release the roots*

TRANSPLANTER
*This American machine, first
developed to plant cabbages,
has a six-man crew and can
plant up to 30,000 seedlings
in a day. Revolving wheels
place the young trees in a
series of slots cut by a
plough, and rollers compress
the soil around the stems*

MECHANICAL WEEDER
*Well-spaced rows allow the
tractor to straddle lines
of seedlings, drawing tines,
or metal teeth, that uproot
most of the weeds. Survivors are killed by selective
weed-killers, applied as
sprays by tractor-operated
machines or light aircraft*

CROP CONTROL
*Use of aircraft enables the
forester to apply chemical
treatment to plantations too
dense for tractors to traverse. They can fertilize the
crop to promote growth, drop
insecticides to fight insect
pests, or apply fungicides
to control tree disease*

FOREST MANAGEMENT
*In order to harvest thin-
nings without damaging
the rest of the crop, gaps
called rackways are first
cut through the plant-
ation by removing several
adjacent rows of trees.
Lorries fitted with grabs,
and harvesting machines*

*called forwarders, can then
penetrate deep into the
forest to pick up their
loads of bucked poles and
pulpwood. Underplanting
techniques (inset) provide
an effective way of increasing yields. Here, European
birch of low economic value
has been thinned and under-*

*planted with giant silver
fir, Abies grandis, a high-
yield conifer from north-
western America. Birches
promote early growth by
shielding the firs from the
wind, sun and snow and by
creating the semi-shade in
which this particular tree
grows most vigorously*

33

In Search of the Supertree

Farmers have cultivated selected strains of maize, sugar-beet, potatoes and apples for thousands of years, but until the present century foresters grew only "unimproved" trees – almost exactly as they had evolved in the wild. Little or no attempt was made at breeding or selection: very little was known about the properties which might lead to faster timber production, and the long life-span of trees appeared to make experimentation impracticable. The search for the supertree began late – and was initiated only by the fortunate discovery of chance hybrids between native and imported species growing together in botanical collections.

Nevertheless, a whole chain of new methods had still to be evolved before trees could be bred scientifically and multiplied in sufficient quantities to make large-scale afforestation practical. First the forests must be scoured for outstanding parent trees and living shoots obtained either by climbing or by shooting off small branches. These shoots, called scions, are then grafted on to root-stocks of related trees of no special merit. Great skill is needed to make one resinous conifer stem unite with another.

Within just a few years, the grafted scions bear male and female flowers and their breeding must be carefully controlled. The female flowers are at once enclosed in plastic bags to prevent chance pollination and pollen from the male flowers of a selected tree is introduced into the bag with a syringe. Pine seed takes a further two years to mature, after which it may be sown in the nursery. Some seedlings quickly show exceptional vigour and good form; others, of inferior growth, are rejected. The winners are multiplied in seed orchards as rapidly as possible and scions cut from them are then grafted on to stocks of well-spaced, normal trees. Within a few years, these in turn will start to flower and pollination can safely be left to nature since all the trees in the orchard are now of selected stock.

The high prices fetched by each year's seed crop easily recover development costs and the supertrees, which cost no more to plant than ordinary trees, give improved timber yields and better land utilization.

THE FOREST ELITE
In this experimental plantation in the southern USA, a line of specially bred supertrees towers over ranks of unselected stock. (The trial required that both receive identical conditions of growth.) For the same cost of planting and maintenance, supertrees yield comparable volumes of timber in less than half the time taken by normal stock. The cross-sections (inset) of a standard tree and a supertree, both with seven annual rings, shows the remarkable increase in wood production; the rings of the supertree are twice as wide as those of the standard variety – which means four times the volume of timber is produced. Quite simply, the supertree makes much more efficient use of the normal input of sunlight, rain and mineral nutrients in the soil

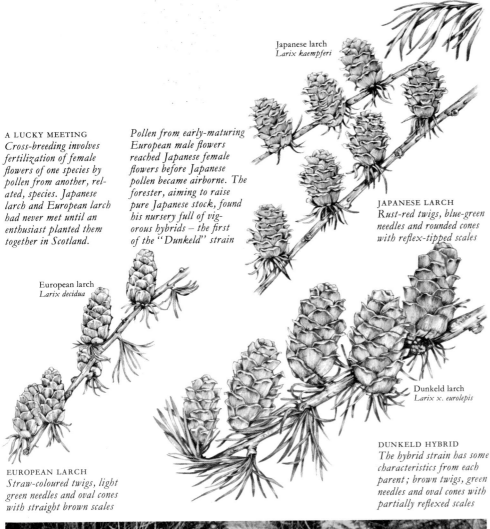

A LUCKY MEETING
Cross-breeding involves fertilization of female flowers of one species by pollen from another, related, species. Japanese larch and European larch had never met until an enthusiast planted them together in Scotland.

Pollen from early-maturing European male flowers reached Japanese female flowers before Japanese pollen became airborne. The forester, aiming to raise pure Japanese stock, found his nursery full of vigorous hybrids – the first of the "Dunkeld" strain

Japanese larch
Larix kaempferi

JAPANESE LARCH
Rust-red twigs, blue-green needles and rounded cones with reflex-tipped scales

European larch
Larix decidua

Dunkeld larch
Larix x. eurolepis

EUROPEAN LARCH
Straw-coloured twigs, light green needles and oval cones with straight brown scales

DUNKELD HYBRID
The hybrid strain has some characteristics from each parent; brown twigs, green needles and oval cones with partially reflexed scales

THE SUPERSEED HARVEST

In America, great advances in tree breeding have been made with southern pines, notably the loblolly pine, *Pinus taeda*. In a typical seed orchard, the trees are widely spaced and growth is promoted by fertilizing. All insect pests are checked by regular spraying, and squirrels, which can do enormous damage, by trapping

A tree shaker (above) ensures that seeds are harvested when ripe and before squirrels can attack them. The machine vibrates the tree, snapping the brittle stalks so that the cones shower to the ground. Many seeds, however, fall as the cones open naturally and these are collected with a giant vacuum cleaner (below). The valuable seeds are sifted from the debris ready for nursery use or, in Europe, for commercial marketing

CROSS-POLLINATION
Windblown tree pollen can travel freely through the air and when female flowers open they accept this male element from any chance source. Tree breeders must therefore protect female parent flowers, before they become receptive, by en-closing them within pollen-proof, but not air-tight, polythene covers. After planned fertilization, the covers are removed and the cones and seeds develop normally. Pollen is taken from desirable male parent trees and introduced into the protective covers with a modified "puffer" syringe (inset), which imitates the wind and carries the pale yellow pollen on to the female stigmas – the reproductive organs. Pollen can be cold-stored, but female flowers must be pollinated during the few days when they are fully receptive

Loggers and Buckers

By the year 1900, steam-age technology had overcome the problems of moving giant logs from America's northwestern forests and had supplanted the slow-moving oxen teams used in the early days. The steam locomotive, cheaply fuelled with waste wood, was the prime source of power. Railways, skilfully engineered to carry long heavy logs on flat cars round gradual curves, were built far up winding valleys, crossing rivers and gorges on soaring timber trestle bridges.

At intervals along the track, the logging foreman would choose a tall, sound and well-rooted tree as the "spar tree", to which the felled timbers were hauled in by wire ropes in the system known universally as "high-lead skidding" – forerunner of the techniques used in hill-country today.

An adventurous and highly skilled logger, called the high climber, ascended the chosen tree with the aid of climbing irons fixed to his boots and a strong sling round his body. First he topped the tree at a suitable height so that the crown could not cause the spar to sway dangerously in use. Topping was a hazardous task as the high climber had to dodge the falling stem as it heeled over, and hold on fast as the trunk vibrated. The next task was the fixing of the elaborate system of rigging and hauling lines; a typical "sky-line" rig needed a dozen fixed stays, fifteen running lines and fifteen pulleys. Power for the hauling lines came from a stationary wood-fuelled engine placed at a distance from the spar tree to avoid damage by travel-ling logs. The engine wound the cables over driving drums and through the pulley blocks, so drawing the cut timber in towards the spar tree – a process known as "yarding".

To load the logs on to the rail-cars, a "hay-rack boom" was constructed, pivoting on the spar tree about twenty-five feet above ground. Built from two massive logs, this device could raise a log from the ground and swing it into place over the car.

Felling, done entirely by hand, was a difficult and strenuous job handled by the loggers. The role of the buckers was to trim off any branchwood and to cross-cut, or "buck", the trunk into convenient lengths for the sawmills.

THE SPAR TREE
A selected Douglas fir, 150ft tall, has been topped and rigged out as a spar tree for high-lead skidding. Guy-lines add strength and lessen sway as the massive logs are hauled in, from either side, suspended from the aptly named skyline

HIGH-CLIMBER
Watched by the rigging crew the key high-climber begins his ascent of the spar tree. Behind him trails a line with which he hauls up his axe and cross-cut saw, the first hold-fasts, and finally the mass of rope and cable forming the main rigging

EXPERT WOODSMEN
The loggers stand on foot-boards notched into the trunk to enable them to cut at a point above the spreading base of the tree. Using heavy double-bladed axes, they have completed the first undercut at the front of the tree and are

now making the main cut; the bottle contains oil used to lubricate the saw. Wedges were also used to prevent the weight of the tree from jamming the saw. The tree finally falls – its direction controlled by the hinge of timber left between the two cuts

FOREST WORKHORSE
Until the advent of the petrol-driven engine, all power for the logging operation was supplied by steam engines fuelled simply and cheaply from waste branchwood and offcuts. Here, a steam crane loads logs on to rail-cars

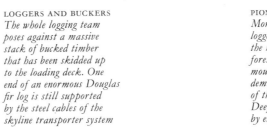

LOGGERS AND BUCKERS
The whole logging team poses against a massive stack of bucked timber that has been skidded up to the loading deck. One end of an enormous Douglas fir log is still supported by the steel cables of the skyline transporter system

PIONEER LOGGING
More often than not, the loggers' operations took the men deep into virgin forest-lands – rugged, mountainous country, which demanded all the skills of the railway engineer. Deep gorges were spanned by enormous trestle bridges

Mountain Forestry

The old-fashioned loggers were quite capable of tackling the forest's largest trees, but only at a relatively slow pace; the higher slopes of the mountains were beyond their reach, no matter how good the timber, because of the difficulty of providing transportation. Vast expanses of otherwise valuable forest were therefore written off as inaccessible, though accessible areas were often quite ruthlessly cleared of all standing timber.

Modern forest practice has changed all this. Concern for conservation and wise use of the available resource have helped in the development of timber harvesting techniques which not only maximize the forest's output of wood but also ensure continuity of growth and maintenance of wildlife habitats. The forester is now a farmer, replanting as he cuts, always looking to the future.

Bulldozers have made possible the building of roads high into the mountains, cutting deep into the hillside and piling up soil to create a precarious ledge. Road standards are high, for the loads are heavy and trucks must run without hindrance to ensure economic running costs. A modern twenty-wheeled log truck, itself weighing up to forty tons, may carry a dozen logs that together weigh a further eighty tons or more.

Up these precarious roads go the self-propelled, self-powered tower-yarders, each equipped with enough cable to enable them to haul in logs from ranges of up to half a mile. Their tubular steel telescopic towers can be extended to heights of 110 feet to give effective lift to the logs. They have almost replaced the old spar-tree as they are mobile, ready-rigged with self-tightening guy-lines, and can be erected and ready for operation within a few hours of reaching the new logging area.

Alternatively, a self-propelled yarder with a jib or small tower may be used to work a "skyline" system – an overhead line running far out into the forest, carrying a carriage that can be winched to and fro to haul timber over the most difficult terrain. Highly mobile wheeled or tracked log-loaders equipped with hydraulically powered jaws pick up logs weighing up to ten tons apiece and deftly stack them on the trucks ready for transportation down the mountain.

Once the operation is started, it proceeds with almost military precision. The yarding of the logs must keep pace with the felling, and loaders must always be on hand so that the trucks can maintain a continuous shuttle service between the logging area and the mill.

LOG LOADER
Small and extremely ma-
noeuvrable diesel-powered
loaders, equipped with
hydraulically powered jaws,
have now virtually replaced
the old "hay-rack" boom as
a means of loading logs
on to trucks and rail-cars
for onward transportation

CONFINED SPACE
To minimize the expense of
building access roads and
loading platforms, working
areas are kept to the safe
minimum. Here, a mobile
yarder hauls in timber as
a grapple-loader utilizes
its great manoeuvrability
to load a waiting truck

OLD-GROWTH TIMBER
High in the coastal hills of Washington State, left, a logging team operates in stands of fir and hemlock that have stood untouched for more than 300 years. To the logger, this would be medium-sized timber, not comparable to the huge coastal redwoods, which can often exceed 300 feet. Nevertheless, these trees will generally yield five, 35ft-long logs from each trunk, and the Douglas fir is particularly valued for its strong, knot-free wood. Here, timber is hauled up the valley side for loading

WEIGHT-LIFTER
In the log-yards far below the logging site the timber is lifted from the trucks by enormous hydraulic grabs to be sorted by species and size before being transferred to the sawmill. These weight-lifters are capable of unloading a truck in one "bite"

THE LONG HAUL
The building of roads, and their maintenance, is one of the logging company's biggest operating costs. In the upper photograph, a truck makes its way from the hill-country down to the coast: lower photograph shows one of the log-yards

Techniques of Logging

PACIFIC NORTH-WEST
Forest-land of the Pacific north-west still carries stands of Douglas fir, hemlock, western red cedar and allied conifers grown from natural seeding to maturity. Ages of 350 years or more, heights up to 200ft, and girths of 25ft are common

FELLING TECHNIQUE
Felling such giants without damaging their huge trunks creates a practical problem. Cuts are made from both sides, leaving an uncut hinge of wood across the centre. As rigid support is removed, the huge tree slowly heels over in a controlled fall

Modern methods of timber-harvesting vary with the terrain, the forest-management plan and the type of tree to be felled. The enormous high-powered machinery used in the American north-west and certain tropical forests is economically viable only where a large volume of wood can be clear-cut within a short period. Here the logs are so huge – around thirty feet long and six feet in diameter – that they can be handled only by heavy machinery, and must be removed unprocessed, with only their branches trimmed away. Intensive operations of this kind can be planned and concerted only by large companies with established sawmills and papermills at their back, ready to convert the timber for commercial outlets.

In Scandinavia, smaller trees growing on flatter land are selectively cut. Here giant machines fell, trim, debark and cross-cut the trees in the forest itself. This method was brought into use by a shortage of skilled labour, when woodsmen who had worked traditionally with axe, horse and sledge left their small farms and woodlots for the cities. Swedish lumber companies were therefore compelled to develop elaborate machines worked by far fewer trained operators, who could be paid high wages to compensate for the isolation of backwoods life.

So arose a new tribe of juggernauts with bewildering names, each ready to advance into the woods to do the work of a dozen men. *Fellers* or *harvesters* will cut through soft-wood stems up to two feet across, using shears instead of a saw. *Limbers*, *bunchers* and *processors* will strip off branches and bark, cross-cut stems to length, sort them automatically, and stack them neatly into piles for onward transport. Small timber, harvested in this way, is generally used for posts, pit-props and for pulpwood.

But for the many smaller, more scattered operations only simpler and less capital-intensive methods can pay. Felling here is a task for power-saws, and hauling a task for tractors or even horses. But on easy terrain the modern *forwarder* often proves cheapest. This is a load-carrying cross-country vehicle with large tyres, equipped with a crane with which to pick up its load deep in the forest.

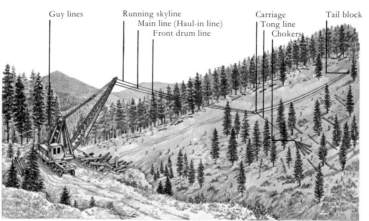

Guy lines | Running skyline | Main line (Haul-in line) | Front drum line | Carriage | Tong line | Chokers | Tail block

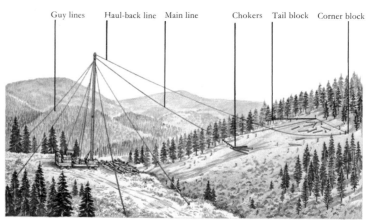

Guy lines | Haul-back line | Main line | Chokers | Tail block | Corner block

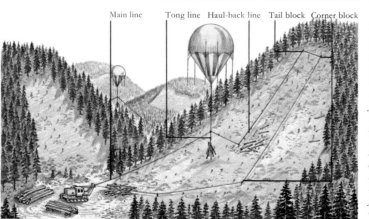

Main line | Tong line | Haul-back line | Tail block | Corner block

EASY ACCESS
Purpose-built tractors
called skidders are used on
level or gently sloping
ground where access is good.
They can negotiate narrow
tracks cut through standing
timber and haul out poles,
felled as thinnings, without damaging the crop

TREE HARVESTER
Mobile machines for felling
small trees prove cheaper
than hand-cutting on easy
ground. The harvester grips
the stem at its base, then
cuts through it with a pair
of hydraulic shears. The
entire tree is lifted and
passed back to a processor

TIMBER PROCESSOR
The processor strips the
branches from the tree far
faster than any woodsman.
The tree is topped, debarked
and bucked into standard
lengths by a cross-cut saw.
The lengths are graded for
pulpwood, or for use as
posts and mining pit-props

NATURE'S TRANSPORTER
In many parts of Scandinavia
and North America water
remains the cheapest means
of transporting logs. Rivers
carry them downstream as
single units, but in order
to cross still lakes and
wide estuaries, the logs are
assembled into rafts, held

together by floating booms
made of poles bolted end to
end with flexible linkages.
The rafts are towed and
manoeuvred by small tugs.
In the background of the
photograph above can be
seen the sorting channels
into which the logs are
graded for different uses

Forest Catastrophe

Natural fires caused by lightning, more rarely by volcanic eruption, were a part of the forest cycle long before Man appeared. But today, Man's activities, his camp fires, his forest machinery, and particularly his smoking habits, have increased the risks alarmingly. Every fire is a tragedy for the wildlife killed in the blaze that destroys its habitat and a major commercial disaster for Man. Constant vigilance by fire wardens using watch-towers and patrolling aircraft is essential if outbreaks are to be discovered promptly and tackled while still small.

The forest is a great storehouse of energy that can be released with catastrophic force by wildfire. Fuel is readily available in the form of standing timber, bark, branchwood, living and dead leaves held high above the ground and in the debris of fallen branches and leaf-litter on the forest floor. None of these burn readily when damp, or even in humid air, but a few days of drying wind can turn the whole forest into a tinder-box. One carelessly dropped match may ignite dry leaf-litter starting a spreading blaze that quickly ignites heavier fuels. Soon the fire "blows up", creating its own winds in the form of violent convection currents that draw in fresh air and send great columns of smoke billowing thousands of feet into the air.

Tackling such a holocaust is like a full military operation. Reinforcements, reliefs and food supplies must be planned for a campaign lasting days, or even weeks. Water can rarely be applied in sufficient quantity to extinguish a major blaze and the first aim is usually to contain the fire. Bulldozers tear fire-breaks, or traces, through the forest to halt the fire's advance, while aircraft "bomb" fiercely burning hot-spots with retardant chemicals, which cling to the foliage and branches – slowing the rate of burning long enough for teams of fire-fighters to arrive and tackle the blaze on the ground.

Often the battle is won only when heavy rain falls or the fire reaches a natural obstacle such as a river or mountain range. Some timber may be salvaged, but all too often little is left but ash and charcoal. No living thing survives; the forest and its life must be created anew from the sterile base.

THE OXBOW BURN
Starting in August 1966, the Oxbow Burn developed into one of the hottest and most destructive fires ever to hit the state of Oregon. Here, in the early stages, the fire traverses a hillside, taking hold and always threatening to "blow up"

THE "CATS" MOVE IN
Huge tracked bulldozers cut a swathe through the forest, removing timber and forest-floor litter in order to create a fire-break. Thick swirling smoke, and the fast-approaching firefront, make this extremely difficult and often dangerous work

RETARDANT BOMBING
Roaring in low over a hot-spot, a converted bomber releases the first of its four 500-gallon loads of fire-retardant. Though not able to quench the fire, the retardant clings to the foliage, holding back the fire until ground forces arrive

GLOW OF DESTRUCTION
As the fire increases in intensity, localized hot-spots develop where an abundance of dry timber lies in a ground configuration allowing powerful convection currents to form. These currents, generated by the fire itself, feed oxygen to the base of the fire, converting it to a white-hot inferno. For the fire-fighting teams to succeed, these areas must be tamed; if not, they rage out of control, sending burning debris high into the air and adding further impetus to the fire. The sight of such a hot-spot bodes ill for the forester; very little timber will be reclaimed from this area

SCARRED LANDSCAPE
The fire has raged up the flank of the ridge taking all before it; in the lea, patches of burned timber are intermingled with some that escaped destruction. These will help stabilize the soil and may also aid recovery by natural seeding

THE AFTERMATH
Trees are surprisingly resistant to fire and even after a severe burn a good deal of usable timber may be salvaged by the logging teams. However, at the site of a very hot burn, little is left but seared trunks and a carpet of white ash

43

Plywood Manufacture

STRIPPING THE BARK
*In this detailed view of
the debarking operation,
logs are fed in from the
right by chain conveyors.
The log being stripped is
rotated by ridged wheels
as the cutting head, on
its track, traverses the
log from end to end*

One of the most remarkable facts to emerge from studies of Man's utilization of timber resources is that, since the turn of the century, the total timber harvest has hardly increased even though our demand for wood-based products has increased many times over. What has changed, and changed dramatically, is the efficiency with which this resource is converted into a range of products.

Increased efficiency in wood utilization arises both from the great advances made in the technology of the sawmill and from the tremendous range of composite wood products developed in recent years. Plywood, fibreboard and blockboard all make use of wood fibre which might otherwise be wasted. Plywood, the most widely used of all the man-made boards, is now a major industry in its own right. America alone uses more than sixteen thousand million square feet every year – a meaningless figure until it is compared to a twelve-foot-wide pathway stretching from earth to beyond the moon.

The great advantage of plywood as an engineering material is that whereas nature makes wood in random fashion, with knots, splits and resin accumulations distributed throughout the fibrous material, plywood can be manufactured to precise engineering requirements with completely predictable and reliable mechanical properties. Knots are cut out and the holes patched, small splits may be filled with synthetic filler, and small pieces of veneer may be literally sewn together and used as core veneer without affecting the strength of the panel.

Waste material from the plywood peelers, the assembly lines and the trimming and sanding operations are all collected, as in the sawmill, for use in paper-making and in the manufacture of hardboard and particleboard.

A glance into the future suggests that in coming years even these man-made products may be regarded as outdated and wasteful. Current research is directed at reducing the whole tree – stem, branches, bark and roots – to a mass of fibre, and rebuilding this raw material into board and moulded products which, as well as having totally controlled properties, could be impregnated during manufacture with any desired additives.

PEELER LOG
Bucked log sections, clearly showing the marks of the debarking knives, are fed into the lathe-loader. The pale form of a log on the lathe, visible in the background, will be reduced to a six-inch core then sawn into lumber or chipped

RIBBON OF WOOD
As the cutting edge of the lathe is forced against the spinning log, the wood is effectively unwound in a continuous ribbon varying in thickness, depending on the type of board product, between one-quarter and one-sixteenth of an inch

VENEER STACKER
*As the sheet of veneer
emerges from the peeler it
is scanned automatically
and clipped into sections
of standard length (seen
at upper right). Veneer is
graded and stacked "green"
in readiness for transfer
to the drying ovens*

STITCHING VENEERS
*Small defects, splits and
variations in the moisture
content of the log cause
the veneer sheet to break,
but the short pieces that
this produces are never
wasted. They may be edge-
joined on the stitcher to
be used in core veneers*

PRESSING LINE
*Plywood panels stacked
in the foreground, above,
have been assembled with
the grain directions of the
alternate sheets at right
angles. The sheets are glued
and ready to be loaded into
the presses. Lastly they
are trimmed and sanded*

The Integrated Sawmill

The modern integrated sawmill is a unique blend of ancient skills and progressive technology brought together to ensure that optimum use is made of the forest resource. At every stage of the log's conversion to lumber the final decision rests on the experienced judgement of the sawyer, even though his work is now made infinitely more efficient by computerized measuring and monitoring systems and by mechanical handling machinery that can turn over a log weighing many tons with the ease of a man tossing a coin.

Nothing is allowed to go to waste: the bark, stripped from the log before it enters the mill, is pulverized and used in many ways. Some may be fed into the mill furnaces as low-cost fuel; it may be used in gardens as a soil mulch, and in the wood-products industries for the manufacture of hardboard and a variety of composite boards emerging from the experimental stage into full-scale industrial use. Sawdust is similarly collected from the bandsaws, circular saws and planers and sold as a raw material for particleboard manufacture and paper-making.

As the log is fed into the mill from the debarking centre it is measured by electronic scanners and cut off, or "bucked", into the most convenient lengths for processing. Large diameter logs then pass direct to the headrig for conversion (see page 48), while smaller logs may be routed to the smaller bandsaws or the chipper-canter, a recently developed machine which squares off timbers ready for conversion by the bandsaws.

Large dimension timber from the headrig passes to the re-saw stage, where bandsaws, often twin- or multiple-bladed, cut the wood to required dimensions, trimming off waste edges which are carried on conveyors to the chipping plant to be processed and sold as raw material for the pulping industry.

Processed lumber is transferred to drying kilns for seasoning so that it will be stable in use. Very large dimension timber, including beams, may be sold immediately after seasoning, but for most customers the lumber is finally passed through planers to be squared off and finished to standard dimensions. The lumber is then grade-stamped and automatically banded and wrapped for distribution.

DEBARKING PLANT
Each log in turn passes on to a system of rollers, which rotate it continuously as the cutting head traverses the length of the log, stripping off the bark. The operator can control the rate at which logs are fed to the mill

BUCKING CENTRE
An incoming log is viewed from inside the sawmill as it approaches from the debarking plant. The log is stopped automatically at a predetermined length and neatly bucked as the large circular saw swings down and across the log

HEADRIG SAWYER
The log is firmly clamped on to a carriage mounted on rails and free to traverse back and forth past the vertical blade of the band-saw. The sawyer's assistant uses a sharp hook to guide the severed slab of wood on to the rollers in the foreground as it falls on completion of the cut. As each log approaches the headrig station it is scanned by optical instruments, which measure its diameter at each end, its length, and the degree of taper. A computer then analyses these dimensions and displays, on a television monitor in the sawyer's control cab, a selection of cutting patterns. These, however, are used only as a guide and the final decision on how to cut the log rests with the sawyer, who must allow for internal features gradually exposed in cutting

CHIPPER-CANTER
One of the key developments in the streamlining of the lumber processing system is the chipper-canter – a newly developed machine which will accurately square off pieces of timber so that they can be fed direct to the resaw and cut to final dimensions

TWIN BAND RE-SAW
Large cants from the headrig are transferred to the twin band or multiple re-saw units to be reduced to smaller sizes. Roller and chain-link conveyors can be programmed or controlled manually to transport the lumber from stage to stage

PLANER
Once cut to size, the wood is seasoned in the drying kilns to ensure stability. The final machining process involves passing the wood through planers in which cutting heads remove any irregularities and trim the wood to accurate dimensions

The Head-rig

One of the most skilful and demanding jobs in the sawmill is that of the headrig sawyer on whose judgement and experience depends the successful conversion of the log into the optimum amount of sawn lumber.

Seated at his controls, often above the log, the sawyer must constantly revise his cutting plan as the unique internal structure of the log is exposed – taking care to avoid knots and splits, while bearing in mind the requirements of the mill's customers.

As the log is carried against the huge vertical bandsaw on its first cut, a slab is removed, rounded on the outer face and usually destined to be reduced to chips for the pulping process. Subsequent cuts remove a series of one-inch boards of clean, knot-free wood from the outer zone of the log, after which the sawyer will rotate the massive timber through 90 or 180 degrees and take a further series of cuts of high-quality timber. Nearer the centre of the log, the wood is more likely to contain knots and splits and from this heartwood region the sawyer takes heavier "dimension timber" and large beams.

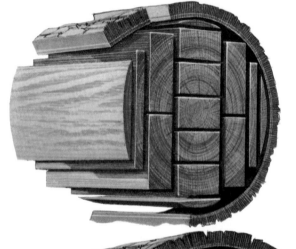

THE HEADRIG
In the upper photograph an outer slab is removed on the log's first traverse through the saw. The slab is pulled clear by an assistant. The lower photograph shows side three being worked for the clear, knot-free timber of sapwood and outer heartwood

HIGH-YIELD LOG
The sawyer opened the log on the upper side (as shown, left) and removed a number of high-grade boards, or selects. The log was then rotated through 180 degrees and the process repeated. This log consisted mainly of sound timber with very few knots and splits, so enabling the sawyer to cut a very high proportion of board lumber. The bark is used either as fuel in the mill or as an ingredient in some fibreboard products. Waste wood from the edges of trimmed boards is chipped for the pulp industry

SMALL TIMBER
Some sawmills specialize in processing small-sized logs – passing the whole log, in one movement, through a series of bandsaws and circular saws that reduces it to standard-width pieces two inches or four inches thick. These are then turned flat and re-sawn to 2x2, 2x4 and 2x6 lumber. The outside cuts, immediately after removal of the slabs, may yield wide, high-quality one-inch-thick boards

"ROUND THE LOG"
In "sawing round the log", each side is worked in turn. Here, the "lower" side (as illustrated) was cut first to yield, after the slab, six high-grade boards. The right-hand side (as viewed) was then cut to give two boards and a thicker piece of timber, which was resawn into smaller sizes. Side three yielded more board, while the last side was cut into larger dimension lumber. The heart of the log, now reduced to a 20in-square baulk, was converted into structural timbers in which knots do not significantly reduce the strength of the member

Treated Posts and Piles

Pressure-treated posts and piles may be seen in virtually every part of the world, carrying power and communication lines, supporting bridges and buildings, protecting coastlines and river-banks from erosion and providing moorings in the harbours and marinas ringing our shores. Once impregnated with creosote, the most widely used preservative, a softwood post should have a life of fifty years in all-weather use.

The movement of preservative through wood is dependent on the internal structure of the wood and may vary from one species to another and even between samples of the same species. Penetration is effected mainly along the fibre direction of the wood, while penetration across the grain, through the walls pits, is much slower. Rays and resin ducts are of minor importance in regulating the flow of fluids. Sapwood is penetrated relatively easily, but heartwood, in which chemical changes have blocked many of the internal passages, is resistant to passage of preservative fluids and in some species the heartwood is almost impervious.

PRESSURE TREATMENT
Modern pressure vessels may accommodate batches of poles up to 130 feet long. Batches of smaller timber are also processed and, if of heartwood (say, for use as railway cross-ties), the wood is incised to aid the penetration of the fluid.

A variety of preservatives may be used, but where the post is not required to be painted, creosote is most commonly used. Excess water is removed from the wood by circulating oil through the tank at more than 200 degrees F. In 40 to 50 hours 5,000 gallons of

water may be extracted from a load of green poles. Preservative solution is then run into the main tank and the pressure slowly increased until the required amount of fluid has been injected. The poles are steam-dried and then (inset) removed from the tank

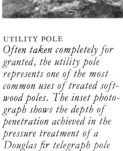

UTILITY POLE
Often taken completely for granted, the utility pole represents one of the most common uses of treated soft-wood poles. The inset photograph shows the depth of penetration achieved in the pressure treatment of a Douglas fir telegraph pole

RESISTING THE ELEMENTS
These groynes illustrate the great erosive force of sand and pebbles carried along a wave-swept beach. Without preservative treat-ment, erosion would be even more dramatic – with the ravages of marine borers an additional problem

The Hidden Resource

In a neat, easily transported package – the log – trees contain an abundant supply of fibre suitable for many industrial uses, the most important being the manufacture of paper and board. Without the bark, and depending on the species, the trunk of a tree consists of 65 to 85 per cent fibre, bound together with 15 to 35 per cent lignin. The purpose of the pulping process is to separate out the fibres preparatory to their reintegration in the final product.

Most pulping material comes from forests, normally as thinnings taken out to enable well-formed trees to reach full maturity, when they will be used for lumber. Pulpwood is transported by road or water, but most logs are floated down rivers, marshalled into large rafts and towed by tugs to the mill, where they are sorted.

Suitable logs are first fed to a de-barking machine. This, in its simplest form, is a forty-five-foot-long metal cylinder, slotted along its length, into which water is sprayed. As the drum slowly rotates, the logs tumble and crash together, and the bark is bruised

or broken off and washed away through the slots. Other types of machine use chains, knives and high-pressure water-jets to "clean" the logs. The bark, 7 to 9 per cent of the log's weight, is usually used as boiler fuel.

The clean logs are then pulped either mechanically or, after being cut into chips, chemically. In the mechanical process, grinders physically tear the fibres apart; in the chemical process the lignin which cements the fibres together is dissolved away. The semi-chemical process uses the two methods in combination.

Mechanical, or groundwood, pulp is weak: it contains lignin, and turns yellow with time. It is normally used for short-life products, such as newspapers and magazines. Chemical pulps are stronger and purer, but more expensive. The fibre yield of the log diminishes as the chemical treatment is intensified: yields range from 95 per cent of the clean log from mechanical pulping down to 45 per cent or lower for certain types of chemical pulp.

In the mechanical process the logs are

sawn to the required length and forced by hydraulic rams against a grinder, a circular stone with a burred rim spinning vertically in a chamber. Water is used to lubricate the grindstone and to wash away the torn and broken fibres.

Chemical processes, in which the lignin is softened or dissolved, may be acid, alkaline or neutral. So as to bring the chemicals into contact with the lignin as rapidly as possible, the logs are first cut into chips less than one inch wide and one-sixteenth of an inch thick by a chipper – a large spinning metal disc about nine feet in diameter to which eight or twelve sharp knives are bolted. A chute supplies the logs, on end, at an angle to the knives and the logs are reduced to chips as if they were pencils being sharpened. The chips are stored in piles before being fed to the chemical digesters. In the digester the prepared chemicals and the wood chips are pressurized by steam at a temperature of 170 to 190°C for ten to sixty minutes – depending on the species of the wood, the type of pulp required and the chemical process used. After cooking, the chemicals are drained off and reused. The chips are blown into a tank, where they may remain intact or explode, according to the severity of the process.

Finally, the pulp passes to washers which remove the cooking chemicals still present, and to further washers which remove sand and uncooked particles. The pulp may be used directly for the manufacture of paper, or first bleached. It may also be drained of water on a circular wire cylinder and formed into thick sheets, which are pressed, dried by hot air, and baled for shipment. Mechanical pulp, when it is not passed directly to the paper-mill, is drained on a similar wire and shipped as wet pressed rolls.

GATHERING PULPWOOD
Forest thinnings are the main source of pulpwood. Worked by one man, the Pika 52 picks up and de-limbs a felled tree, and cuts the trunk into programmed lengths. Prototypes also garner the limbs

PULPING
Mechanical and chemical pulping processes follow basically similar courses, with many stages common to both. Washing the pulp is especially important. Washed pulp may pass to a paper machine directly, by-passing the wire section

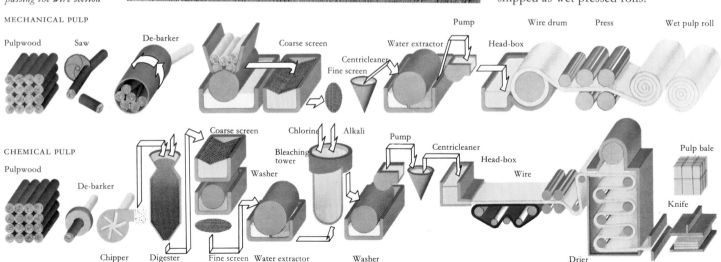

MECHANICAL PULP

Pulpwood — Saw — De-barker — Coarse screen — Centricleaner / Fine screen — Water extractor — Pump — Head-box — Wire drum — Press — Wet pulp roll

CHEMICAL PULP

Pulpwood — De-barker — Chipper — Digester — Fine screen — Coarse screen — Washer — Water extractor — Bleaching tower — Chlorine — Alkali — Washer — Pump — Centricleaner — Head-box — Wire — Drier — Knife — Pulp bale

PULP-MILL
Modern pulp-mills are vast, and are being built ever larger, often with a paper factory adjoining. At Umeå, on the Swedish east coast, pulpwood arriving by water awaits conversion into a mountain of wood chips, rising in the foreground

DE-BARKING
De-barking is the removal of the bark from the log, and is the first stage of the pulping process. Here sharp spikes on adjustable spinning arms tear off the bark, which falls below and is normally used as fuel for the mill boilers

WOOD CHIPS
Wood chips are about one inch long and one-sixteenth of an inch thick. They are blown out from the chipper into great piles which are evened out and managed by bulldozers. A series of chutes, through which they are blown, is visible here

WOOD PULP
Mechanical pulp, in which the constitution of the log is hardly altered, is brown in colour. Chemical pulps consist more or less purely of cellulose and even before bleaching are predominantly white, like a dense cotton wool

Paper and Board

A sheet of paper consists of a mat of randomly interwoven fibres formed into a sheet. It can be made from any organic fibre, but by far its most important source is wood pulp. In many mills, wet pulp passes direct to the paper-making plant, but in some the pulp is received at the paper mill in the form of pressed bales or rolls, in which case a slurry is first prepared by feeding the pulp into an open-topped cylindrical tank with a conical base, containing water. A vaned conical rotor in its base creates a vortex, which sets up cavitation forces. These separate the fibres and distribute them evenly through the water.

Before passing to the paper machine, the slurry is "beaten" and thoroughly cleaned. "Beating" is an important process determining the subsequent nature of the paper. The wall of a fibre consists of entwined fibrils, and during "beating" the fibrils comprising the outer layers of the fibre-walls are partially detached and unravelled, but not disconnected. This process, known as fibrilation, produces a stronger paper. When the paper sheet is formed, the un-ravelled fibrils increase the bonding between adjacent fibres. Papers of different properties are produced by varying the degree of fibrilation. Low-fibrilated paper, for example blotting paper, is bulky; high-fibrilated paper, for instance tracing paper, is dense.

The fibres are "beaten" in a series of refiners. The two types of refiner, conical and disc, both work on the same principle: the conical refiner is a solid cone which spins in a stationary conical shell – the gap between the cone and the shell being adjustable. Metal bars spaced radially down the inside length of the shell and the outside of the cone split, cut, bruise and impact the fibres passing between them, causing fibrilation.

Paper is formed when the slurry, made up of 99.9 per cent water and 0.1 per cent fibre, is fed on to a moving wire mesh. Most of the water drains away, leaving a tangled mass of fibre on the mesh. More water is removed by vacuum pumps drawing air through the mat, and by pressing the sheet through rollers. Finally the sheet is dried by passing it over huge steam-heated cylinders.

FROM PULP TO PRODUCT
The diagram shows the five main pulping processes, and the proportions in which each is commonly used to make different types of paper. It compares the hardwood and softwood intake of each process, and shows which is used in each wood-pulp product. For example, soda pulp, which is mainly hardwood and short-fibred, is used as a filler in printing paper. Sulphate and sulphite pulp

are both mainly softwood, and long-fibred; because they are strong, they are used for wrapping paper and tissue paper. Sulphite pulp has the additional quality of softening when bleached. Mechanical pulp is cheaper but also weaker than chemical pulps, and is used for short-life paper products. Brown semi-chemical pulp is used especially to make the corrugated infill and casing of cardboard boxes

FROM PULP TO PRODUCT

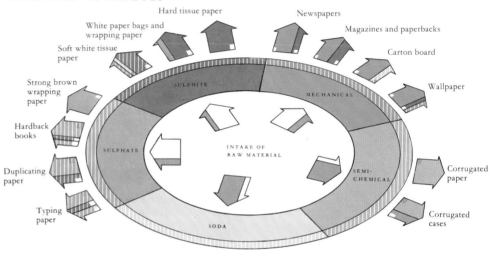

Hard tissue paper · Newspapers · Magazines and paperbacks · Carton board · Wallpaper · Corrugated paper · Corrugated cases · SODA · Typing paper · Duplicating paper · Hardback books · Strong brown wrapping paper · Soft white tissue paper · White paper bags and wrapping paper · SULPHITE · MECHANICAL · SEMI-CHEMICAL · SULPHATE · INTAKE OF RAW MATERIAL

Hardwood · Softwood · Bleached pulp

SOFTWOOD PULP
Chemical pulping preserves wood fibres intact, while dissolving away the lignin binding them together. The micrograph (x77) of softwood chemical pulp shows long, separate, undamaged fibres, which will provide a strong paper

HARDWOOD PULP
Because hardwood fibres are shorter than softwood fibres, hardwood paper is not as strong. But short-fibre paper has a closer structure and a smoother printing surface. The micrograph (x75), above right, shows aspen chemical pulp

PAPER-MAKING *(right)*
Pulp is mixed with water to make a slurry, which is beaten and cleaned before passing to the head-box of the paper machine. Paper is formed on the wire, then pressed, dried and wound through the calender on to a reel at the end

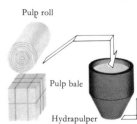

Pulp roll · Pulp bale · Hydrapulper

PAPER MACHINE
Pulp slurry is fed evenly from a slit in the bottom of the head-box on to a continuous band of wire mesh in constant motion, both forwards and from side to side. The rest of the machine – most of its length – consists of driers

THE FELT
The wet web of paper comes off the wire on to a felt. Before the web and felt pass together through the pressing cylinders, they travel over a perforated drum through which water is drawn off by suction. A water-jet trims the edges

DRYING CYLINDERS
Now containing 60 per cent water, the paper passes for drying over a series of steam-heated cylinders. The detail shows three cylinders, the paper, the felt holding it against the cylinders, and a blade which removes particles

THE CALENDER
The calender, consisting of a stack of shiny steel rollers, imparts a smooth finish to the surface of the paper. Very finely adjustable, it acts by a combination of pressure and friction. It is placed immediately after the driers

Sizing and colouring

Kaolin   Alum

Head-box

Conical refiner

Centricleaner

Wire

Press

Drying cylinders

Calender

Reel

The Chemistry of Wood

The woody substance of a tree contains an abundance of raw materials, exploited in virtually every area of technology. Besides the many traditional extractives, such as rubber from the rubber-tree, *Hevea brasiliensis*, and turpentine and rosin from coniferous trees, there are the direct and indirect products of the pulping industry, and the hydrolysates, or chemicals obtained by breaking down wood by hydrolysis.

The traditional method of tapping certain trees has in many cases been superseded by quicker methods of extraction or by the production of synthetics. An exception is gum arabic, a dried exudate of unhealthy acacia trees, which is used in foodstuffs, medicines, cosmetics, adhesives, paints, inks and textiles. Many volatile and essential oils, such as Canada balsam and cedarwood oil, which are used in perfumery and medicine, are still obtained in the traditional manner, though volatile oils are also largely synthesized.

"Naval stores", turpentine, rosin, pitch and tar, though still tapped in small quantities, are more commonly obtained by the steam distillation of the stumps of lumbered conifers. But by far the largest proportion of the world's supply comes from the "black liquors" resulting as a by-product of the sulphate pulping of softwoods. The liquors are distilled to recover sulphate turpentine, and the residue is concentrated to yield an insoluble "tall oil soap". Purified tall oil is a major source of rosin and a constituent of lubricants.

Turpentine itself is used in the manufacture of paint. It contains a high proportion of α-pinene, the raw material for the manufacture of chlorinated insecticides and of synthetic camphor.

Rosin was once used for caulking ships' hulls and to treat rope. It is nowadays more widely used as a paper size, and as a component of paints, varnishes, printing inks and adhesives. It is also used in the manufacture of plastics and synthetic rubber.

Tannin is used to cure leather, and nowadays also as a dispersing agent and for water treatment. It is obtained commercially by subjecting to boiling water the wood of the quebracho tree in Argentina and Paraguay, of the chestnut in France and Italy, and the bark of the black acacia in southern and eastern Asia.

For every ton of pulp obtained by the sulphite process, 500 to 1,000 lb of modified lignin are produced. These lignosulphonates act as dispersing agents, and are used in tanning, in rubber formulations, as a component of cements, and as a source of organic chemicals, such as vanillin, which is now a more common substitute for vanilla. Large quantities are used to impart the required viscosity to oil-well drilling muds. Neutral sulphite liquors from the semi-chemical process yield acetic and formic acids, and sulphate liquors are used as dispersing agents, and as components of foundry core-binders, ceramics, dyes and printing inks.

Wood can be broken down by acid hydrolysis to yield sugars, mainly glucose, which can be fermented to obtain ethyl alcohol. But for sugars and alcohol other sources are more economic, and, commercially, wood hydrolysates are more important in the production of special yeasts for animal and human foodstuffs, and in the fermentation of acetic acid, lactic acid, acetone and butanol. The more drastic hydrolysis of wood cellulose produces laevulinic acid; laevulinic acid derivatives are used as solvents and plasticizers. The hemicellulose in wood produces furfural, which is the starting material for the manufacture of moulding resins.

The most important wood pulp products are paper and board, but cellulose fibre has many further uses. In the viscose process cellulose fibre is "regenerated" to form rayon to make textiles and tyre-cords, and to form Cellophane. Modification of the cellulose molecules produces a range of materials, which may be either cellulose ethers or cellulose esters.

Cellulose esters are cellulose nitrate and cellulose acetate. Cellulose acetate is used to make wrapping and photographic film, and in the production of a wide variety of injection-moulded, sheet-fabricated and extruded plastics. Cellulose acetate fibre, acetate rayon, is superior to viscose rayon as it absorbs much less moisture and, when moist, is stronger; it is also less susceptible to wrinkling. Treated with camphor, cellulose nitrate was the first successful synthetic plastic, Celluloid, still widely used for moulded articles such as piano keys and table tennis balls, and as a surface coating. Highly nitrated cellulose is used to make guncotton and cordite. Both cellulose acetate and cellulose nitrate are used in solution as adhesives and lacquers.

Cellulose ethers are used as thickening and emulsifying agents in foodstuffs and paints and as sizes and adhesives for the paper and textile industries. The most widely used is carboxymethyl cellulose, which has applications as a stabilizer for oil-well drilling muds, as an adhesive and pigment binder, as a strengthener in unfired ceramics, and in laundering to facilitate dirt removal.

MAPLE SYRUP
Whenever a thaw follows a frost, a dormant American maple will exude sap from a wound. Nowadays collected from a stand in a system of plastic tubing, the sap is evaporated to yield a syrup used in confectionery and for flavouring tobacco

NATURAL RUBBER
Natural rubber production, though less than that of synthetic rubber, amounts to millions of tons annually. A milky latex exuded from the cut bark of the tree is coagulated and nowadays vulcanized to form strong, elastic, non-sticky rubber

THE SOURCE

The cellulose used for the manufacture of textiles and film comes from spruce, eucalyptus and sweetgum trees, which are pulped by the sulphite process. The high-purity pulp is then bleached, pressed into sheets and shipped in bales

ALKALI CELLULOSE

In the factory the pulp sheets are steeped in caustic soda and pressed. The alkali cellulose which results is then ground into crumbs and stored to allow oxidation by the oxygen of the air. This is known as "pre-aging"

VISCOSE

When carbon disulphide is added, cellulose xanthate results. This is dissolved in caustic soda to form viscose. The viscose is "ripened" to the correct viscosity for spinning, and filtered to remove any undissolved particles

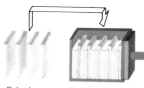
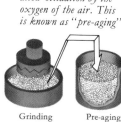

Pulp sheets   Steeping and pressing   Grinding   Pre-aging   Xanthation   Dissolving   Ripening   Filtration

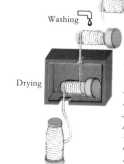

SPINNING

The viscose is extruded through the fine holes of a spinning jet into a bath containing sulphuric acid and salts. The original cellulose is regenerated in continuous filaments, which are drawn together and wound on to a reel

Extrusion

Washing

Drying

COMPLETION

On the reel the filaments form a single yarn. As it passes through a system of further reels, the yarn is first thoroughly washed, then dried in an enclosed container. It is wound on to a bobbin for shipment

GRINDER

Large rotating knives break up sheets of alkali cellulose into crumbs

VISCOSE

Newly formed viscose is poured into a vat before ripening and filtration

CELLOPHANE

In the manufacture of Cellophane sheet, fluid viscose is extruded through a fine slot and is thoroughly washed by a series of water jets as it passes over rollers to the final take-up reel

SPINNING RAYON

After being filtered and deaerated, the viscose is forced through thousands of tiny holes in a jet to form filaments. These are drawn up through an acid bath on to "godets", then combine with fibres from other jets into a "tow"

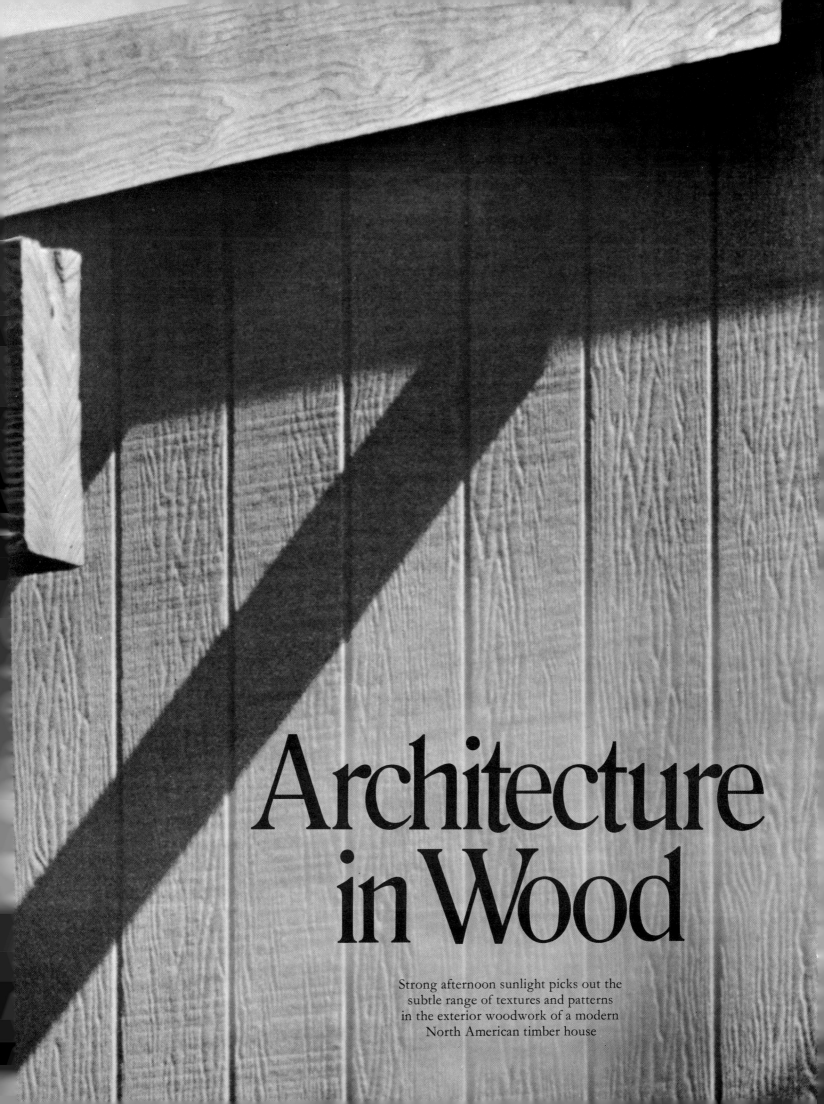

Architecture
in Wood

Strong afternoon sunlight picks out the
subtle range of textures and patterns
in the exterior woodwork of a modern
North American timber house

The Basic Shelter

ABORIGINE SHELTER
*Always on the move, the
hunter–gatherer Aborigines
of Australia build only the
simplest of temporary
shelters; often little more
than a windbreak; some-
times a simple hut of bark
and twigs supported on a
framework of branches*

Even the most advanced buildings can usually be traced back to their generic ancestors in both concept and building technique, but in some cases the initial concept fulfilled its requirements so effectively that the design has remained virtually unchanged through many hundreds of years.

Where light organic materials were available for building they were invariably used in conjunction with a timber framework of some sort. The limited requirements of an Aboriginal lean-to shelter, for example, were easily met by a simple framework of rough branches covered with leaves, grass or bark. The few poles providing support to such a simple shelter are the ancestors of the more complex post-and-truss frames which developed in other parts of the world where societies placed ever-increasing demands on their buildings.

The methods of joining together the parts of a simple house-frame are conditioned primarily by the presence or absence of sophisticated tools. In many parts of the world, few joints are used, other than simple notches, and frame members are lashed together with roots, leaf fibres or climbing plants. Lashed joints are extremely strong and, due to their natural flexibility, ideal for use in areas prone to high winds. In the use of plant material for covering building frames, many devices have been used which eventually developed into more sophisticated forms such as reed matting, woven mats of palm fibres, and wattle, which later became the usual supporting matrix of the mud or clay wall. Reed huts have been developed in many parts of the world and their use continues today in Sudan and southern Iraq.

The great variety of buildings built with simple braced frames, lashed together with natural plant fibre, speaks for their soundness of design and structural reliability, and they represent, perhaps, the most economic use of readily available building materials in the history of building. Some of these structures, like the yurts of Central Asia, have reached a level of development which can hardly be improved by modern technology, and this alone has ensured their survival well into the twentieth century.

HIGHLAND HUT
*In many primitive societies,
where whatever material is
at hand is used for building,
one of the commonest forms
of shelter is the round hut.
Here, large pieces of readily
accessible wood are being
made into a women's house
in the New Guinea highlands*

PREHISTORIC TOOLS
The tools still used by the Aborigines for building their shelters differ very little from those used by Stone Age man. Illustrated are a stone knife, two forms of stone-bladed axe, and a knife made from a fragment of a broken bottle

VERSATILE LATTICE
The yurt's most practical feature is its latticed frame, the sections of which fold flat for ease of transportation. The number of sections used may be varied, allowing the size of the yurt to change according to the needs of the day

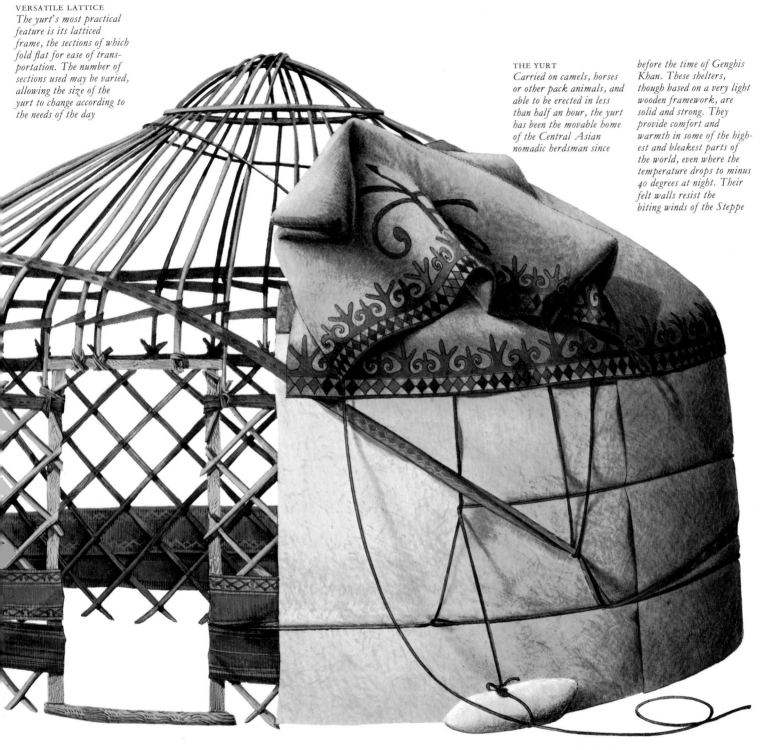

THE YURT
Carried on camels, horses or other pack animals, and able to be erected in less than half an hour, the yurt has been the movable home of the Central Asian nomadic herdsman since before the time of Genghis Khan. These shelters, though based on a very light wooden framework, are solid and strong. They provide comfort and warmth in some of the highest and bleakest parts of the world, even where the temperature drops to minus 40 degrees at night. Their felt walls resist the biting winds of the Steppe

ROOF AND DOOR
A frame is lashed into the lattice wall to carry the stout wooden door, the last item to be added. The roof-covering is supported on a framework of wooden ribs, lashed to the wall-lattice and slotted into a strong, wheel-like, central support

INSULATED COVERING
The nomads' herds provide the material for the outer covering of the yurt. Thick felt mats are stretched over the frame and lashed in place or weighted with stones. In severe weather, as many as eight layers of felt may be needed for insulation

Houses of Pine

There is archaeological evidence that log building was practised in Neolithic times, and Roman writers, including Tacitus, seem to have been acquainted with structures in Germany that were made of square-hewn logs. According to Herodotus, the Scythians also used log construction in their royal burial chambers in the first millenium BC. The wealth of coniferous forests in northern and eastern Europe made timber the principal building material in these regions, and it is known that around 700 BC, at Biscupin in Poland, there was a settlement of log houses.

In Norway, log building was known in the fourth century AD, and in Scandinavia generally houses built of either vertical or horizontal logs were in use by AD 1000. The horizontally placed logs, which crossed each other in a variety of joints at the corners of the buildings, eventually became more widely accepted as they offered greater stability to the structure.

One of the principal drawbacks of log building, however, is the difficulty of making the horizontal logs weather-proof. A variety of methods, such as stopping the gaps between the logs with coloured woollen cloths, or, in poorer households, earth and moss, achieved only moderate success. Eventually, builders resorted to external boarding, and occasionally shingling was applied.

Another problem is that log construction offers no protection for the end surfaces of the logs, which are exposed at the corners of the buildings. It is there that decay first attacks the wood and, to protect the main wall, a good deal of waste timber must be left projecting beyond the overlapping joint.

With the improvement of sawing techniques, and the establishment of saw mills operated by water power at the beginning of the sixteenth century, it was possible to produce planks more readily and to provide them with mortises or tenons along their length. Thus, weather-proof joints could be made and rectangular interlocking logs replaced the round timbers which previously could be joined together only at the corners. With the use of such logs the stability of the structure increased considerably and five- and six-storey buildings became feasible.

MULTI-STOREYED HOUSES
In Switzerland, and notably in the Valais, log houses are traditionally five or six storeys high with masonry bases to safeguard against fire. Despite frost, sunshine, rain and snow, many houses in this region have stood for more than four centuries

AMERICAN LOG CABIN
Log-building techniques were carried to North America by Swedish, Finnish and German immigrants during the late 17th and early 18th centuries. Simple to erect, and based on locally available materials, the cabins made ideal homes for the pioneers

JAPANESE SACRED BARN
Early Japanese storehouses, like this example standing beside the Sangatsudo temple at Nara, were built with horizontal notched logs. Probably dating from the 7th century, this barn may have been built by Korean craftsmen

INSULATING ROOF
The roofs of the log houses were made from planks covered first with over-lapping strips of birch bark and then with turf. This dense thatch kept the house cool in summer and held the winter snows, creating a thick insulating covering

TRADITIONAL TOOLS
For many centuries, log houses were built with only the simplest of tools, the most important of which was the axe. A variety of axes was used for specialist jobs such as felling, shaping and scoring. The unusual saw was used in stair building

THE OLDEST LOG HOUSE
Dating from about AD 1250, Raulandstue is Norway's oldest surviving log-built house. Originally from Numedal, the building now stands in the Oslo Folk Museum. The fir-trees selected for the building were "topped" and then left standing for two years before felling

NATURAL MOVEMENT
Simple log houses had doors but no windows: the opening for the door was cut after the walls and roof had been built and left to settle for about a year. During this time the logs would shrink and settle into place, stabilizing the building

FOUNDATIONS OF STONE
Raulandstue, like many log houses in Norway, was built on stone rubble foundations. Other buildings, notably outhouses that were used for storing grain and fodder, were often raised on stone or wood pillars to prevent the entry of rats

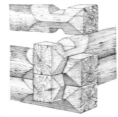

CRADLE JOINT
The stability of a log house depends on the skill with which the corner joints are constructed. The most common type of joint was made by carving a hollow in the upper surface of each log, a few inches from the end, to act as a cradle for the next log

NOTCHED JOINT
Log houses are found in many parts of the world, particularly in Norway, Canada and Russia, and their joints are remarkably similar. One widely used type involves the cutting of notches in the top and bottom of each log to give a secure interlock

HALF-DOVETAIL JOINT
The introduction of additional tools, like the mallet and chisel, made more sophisticated joints possible. The half-dovetail joint was one of the most secure joints — the logs fitting together so closely that very little in-filling was needed

Building on Stilts

The development of architecture passes from the primitive to the vernacular stage and then, given sufficient technology, beyond. The countries of the industrialized world have passed into this advanced stage, but many societies have not yet developed the use of powered machines, metals, plastics and glass. Yet the vernacular architecture existing, and still employed in Asia, Africa, Polynesia and South America is among the most impressive in the world.

Wood is still used occasionally in the living form – a practice which survives from the Stone Age. Tree houses are used in New Guinea, Malaysia and in parts of India, though such structures are more commonly used as temporary, rather than permanent, shelter. Though dependent on a restricted range of tools, like the polished stone axe and the shell adze of Polynesia, vernacular buildings show considerable sophistication and complexity of design.

The meeting-houses of some New Guinea villages are sixty feet in height, more than 130 feet long, and constructed entirely of wood. Huge bamboo poles are set deep in the ground and bent over in the shape of Gothic arches to carry the thickly thatched roof, creating some of the boldest structures built with minimal equipment and technology.

Prefabrication is another technique in widespread use. In Samoa and in parts of Africa and India, roofs are constructed, with or without their thatch, and are then lifted into place. In Cameroon, the wooden frames of the enormous Bamileke houses are constructed flat on the ground and are then erected by special house-building teams.

The practice of standing buildings on stilts, on land, or on heavy piles sunk into the beds of lakes, belongs to a tradition stretching from the Valais and the Pyrenees to Indonesia, the Philippines and Peru. Examples of stilt villages exist today in China, Africa, particularly in Dahomey, in South America and in many parts of SE Asia. Once the floor was conceived as a wooden platform, it was a natural development to raise it above ground – protecting the house from animal pests and lifting it clear of flood waters.

BATAK HOUSE
The boat plays such an important role in island cultures that its shape is often reproduced, symbolically, in the construction of houses. The boat-shaped Batak house in Indonesia is built on piles as a defence against vermin and seasonal floods

SEPIK MEN'S HOUSE
*In the Sepik River region
of New Guinea, boat-
shaped houses similar to
those of Indonesia are built.
This Kanganaman men's
house emphasizes the
traditional shape, though
the pointed gables betray
some Oriental influence*

HOUSES ON STILTS
*The fishermen of Ganvié, in
Dahomey, build their
houses on Lake Nokwe
using exactly the same
techniques in building as
they use in making their
nets and baskets. The stilt
houses are thatched with
straw and palm fronds*

LASTING TRADITION
*The Torajas of Sulawesi
believe their ancestors came
from South China or
Cambodia, so their houses
are sited facing north and
are built in the shape of
their boats. The tall,
narrow houses are built on
piles, with walls of plaited*
*palm fibres, bamboo or
palm-wood floors, and
thatched gables. Beneath
the thatch, the roof is of
split bamboo rods. The
frame and scaffolding
visible in the partially
completed building are
also made of bamboo or
palmwood*

63

The European Oak Frame

Timber-framing was developed in Scandinavia with the stave structures, but it was overtaken by log construction in the fourteenth century. In Britain and Europe, frame construction was first used in Roman times, and many old Italian towns once had an abundance of timber-frame houses closely huddled together and dangerously exposed to the hazards of fire.

The basic principles of timber-framing were probably known in Europe in the Bronze Age. Though laborious to produce, square-hewn lengths of timber were well within the scope of the limited range of tools available, and because a frame needs fewer such members than a log or plank structure, framing eventually became the pre-eminent method of building in wood. Universally, the frames were made with oak timbers.

The understanding of basic forms of bracing and the principle of triangulation, achieved either by joining together a vertical and a horizontal timber with a diagonal strut or by crosswise strutting in the form of St Andrew's crosses, or saltires, also dates from earliest times.

In English practice the subdivision of buildings into framed bays, joined together by beams and braces, was developed in a most logical form, and both in England and in Europe the use of cantilevered upper stories or jetties was well appreciated because it allowed a reduction in the size of the load-bearing timbers and made frame building easier.

To increase the span of buildings a special frame called a cruck was invented by medieval English carpenters. The cruck frame, consisting of two curved vertical members – often the matching halves of a single cleft tree trunk – tied together at the apex by a ridge and stabilized by a tie-beam at first-floor level, was widely used in England.

Later on in the Middle Ages, the increasing skill of the carpenters enabled them to construct six- and seven-storeyed timber-frame buildings. Indeed, some of the largest and most imaginative buildings of the Middle Ages and the Renaissance had timber frames, and many of them have lasted every bit as well as others built in stone or brick.

THE SKELETON FRAME
The bare beams of this medieval house, dismantled from its original position and here reconstructed, show the basic box frame, its roof separate from the walls. Only when the whole frame is erected is the wattle and daub applied

MEDIEVAL BOX FRAME
The gatehouse at Stokesay
Castle in Shropshire has
curved and decorated braces
on the upper storey. The
strong corner posts (often
squared-off tree-trunks)
required to support the
weight of the upper storey
were decorated with carving

THE FRAME
Early tent-like frames,
left, with two branches tied
to a ridge pole, led to the
simple cruck frame, its
split timbers rising from
stone plinths. Collar and
tie beams strengthened the
cruck, developing into the
post and truss frame

THE DRAGON BEAM
To achieve the jettied construction of overhanging
storeys, joists were extended beyond the lower wall,
and were held by the
weight of the upper wall.
A dragon beam at the corner
carried the overhang round
two sides of the building

THE MASTER-BUILDERS
The use of projecting beams
was developed to extremes
in the timber-framed buildings of medieval Germany.
The town hall at Esslingen,
built in 1430, illustrates
the number of jetties which
can safely overhang the
ground-plan of the building

The Art of Infill

Much of the appeal of European timber-frame architecture of the late Middle Ages and the Renaissance lies in the wall elements of the buildings. These were without exception constructed purely as infillings while the frame carried all structural loads.

The most ancient materials used for infilling were sand and clay applied to a close lattice-work of interwoven branches or twigs. The woven lattice was firmly attached to the main members of the frame and the infill material plastered on to both the inner and outer surfaces. Where the framing members were closely spaced, as is often the case in English frames, such lattices could easily support the weight of the infill, but with wider spaced frames, more substantial lattices of woven laths were used. Shrinkage and twisting of the frame, common where green oak was used, also tended to crack the plaster and the natural drying of the panels could lead to gaps appearing at the edges.

The shortcomings of such infilling methods eventually led to their replacement by stone and brickwork, both of which could be supported on the width of the frame crossmembers. In Sweden, and later in Holland, infillings of thin masonry were adopted, but a far-reaching change in technique came with the introduction of brickwork, which not only provided a lasting and weatherproof surface, but could also be used to decorative effect. Paycocke's House in Coggeshall, Essex, has brick infill laid to a herringbone pattern, while the stately Portan House, near Hamburg, displays on its gables a considerable variety of brown brick panels with white joints and highly inventive patterns. Sometimes, brickwork infill was rendered and whitewashed.

With the increased availability of glass in the seventeenth century, the advantages of timber framing could be further exploited. Windows became a prominent feature of town houses, easily accommodated within the traditional frame, while the infill panels were used to even greater decorative effect below and to either side. During this period, timber framing reached its highest development and design features were evolved that recur in twentieth-century architecture.

MEDIEVAL DOORWAY
Made of stout vertical boards backed by horizontal planks, the door of the timber-frame house often had elaborately carved side posts, as can be seen in this East Anglian guildhall. Sometimes a smaller door was set into the main door

WATTLE AND DAUB
The walls of this 15th-century farmhouse, right, have wattle and daub infillings. Inside, the daub has mainly fallen away, revealing the woven wattle, left. Outside, the daub was covered with a thin layer of plaster, below right

BRICK NOGGING
Brick infilling, or nogging, was more watertight than wattle and daub and also more decorative: the bricks could be laid in ingenious patterns. Both the diagonal and herringbone designs were used for the market hall, above and right

LATH AND PLASTER
Many timber-frame buildings have lath and plaster infillings. Two layers of lime plaster were applied to the wooden laths, as can be seen in the detailed picture above. In some areas a final coat of plaster covered the timbers as well

67

Balloon Frame and Platform

The first European settlers found in the New World an abundance of coniferous and deciduous trees, which enabled them to continue their tradition of timber-frame building. In the United States some eighty seventeenth-century timber-frame buildings have survived and the oldest example, the Jonathan Fairbanks House at Dedham, Massachusetts, dates back to about 1636. In design and construction it closely followed contemporary English practice, including the post-and-beam frame with bracing, and the jettied or cantilevered upper storey.

The mother countries continued to exercise a stylistic influence on the settlers' buildings without introducing changes in the methods of frame construction. The American carpenters had no difficulty in producing, in timber, the equivalent of a medieval, Queen Anne or classical building made of stone. One indigenous contribution to the internal planning of houses was a much-used layout in which the staircase encircled the massive central chimney-piece.

With the introduction of mechanical sawmills, great quantities of planks and boards became available. Then, in the 1830s, the Americans mastered the mass production of nails. In the wake of these technological advances there came the development of two new timber-framing techniques. For the first method, called the Balloon frame, the posts were two storeys high with the vertical members running unbroken from the ground to the roof. For the other type of frame the posts were only one storey high and each storey was built separately, the lower serving as a platform for the upper one. Both framing systems required bracing. Traditional diagonal bracing at the corners of the building was replaced by diagonal boarding across the entire outside of the frame. To this boarding another layer of horizontal or vertical boards was then applied with a variety of interlocking edges.

For the outside cladding, the carpenters often used painted softwood, or redwood and cedarwood, which remained untreated at first. The development of preservatives, however, paved the way for attractive finishes which required little maintenance.

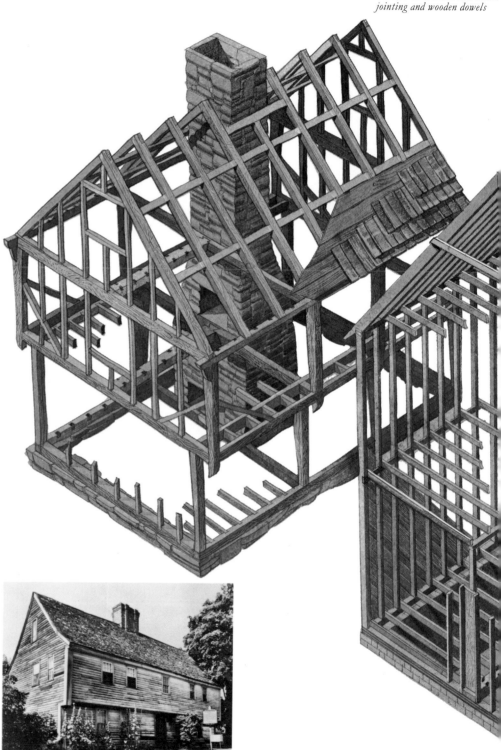

THE BOARDMAN HOUSE
A typical New England timber-frame house, built in 1651, the Boardman House demonstrates the single central chimney, the steep, gabled roof, the pronounced overhang and the unpainted clapboard cladding of early American construction

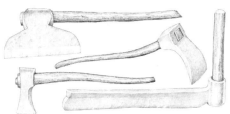

THE AMERICAN TECHNIQUES

THE BALLOON FRAME

The frame consisted of joists fixed to vertical studs running the whole height of the building (usually two storeys), from the ground sill or plate to the head plate supporting the roof rafters

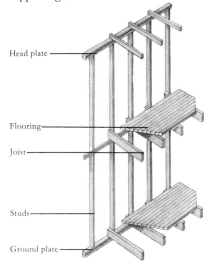

Head plate

Flooring

Joist

Studs

Ground plate

THE PLATFORM FRAME

Though the platform frame retained elements of the balloon frame, its vertical members were only one storey high, each finished floor providing a working platform for constructing the next storey

THE BALLOON FRAME HOUSE
With the increased use of saw-milled planks, and nails, a lighter frame was evolved in the 1830s, able to be erected by one man using only hammer and saw. The simple frame of closely positioned studs and joists is strengthened by diagonal boarding

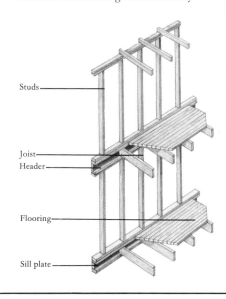

Studs

Joist
Header

Flooring

Sill plate

Clapboard and Shingle

Wood is the most plentiful and the cheapest of all building materials in the United States, and to this day, although many European countries rejected long ago the ancient timber traditions in favour of heavier materials, the majority of Americans still live in wooden houses.

In the late eighteenth and early nineteenth centuries, during their "Georgian" period, American architects and carpenters reached standards of design and execution which compared favourably with the best that Europe could offer in brick and stone.

These timber-frame houses were invariably fini ed on the outside with horizontal weatherboarding. Before the advent of cheap interlocking boards the protection of the end grain of overlapping boards had always presented difficulties, and thus in many buildings, end-grain decay could not be halted. This is evident in many buildings of the "Georgian" period, and even in the uncommonly well-built houses of the religious community of the Shakers, which was founded in the late eighteenth century.

In about the middle of the nineteenth century, when the battle of styles was raging fiercely in Europe, America was overtaken by the general confusion and timber building was forced to adjust itself to the changing modes of the day. By then, however, indigenous framing techniques had established themselves and permitted extensive freedom of design. This was often misused and the twists and turns of this unsettled period produced many deplorable, though durable, results. During this time, techniques from the past were resuscitated, sometimes for the wrong reasons. Thus architects like Henry Hobson Richardson revived wall shingles in spite of their extreme vulnerability to end-grain decay, and in preference to the technically superior horizontal boarding which, by then, was available with interlocking edges.

Yet there were other architects, like Bernard Maybeck and the Greene brothers, who could still handle timber building with a practical and sensitive understanding of the material, and these men paved the way for the great changes that were to come in American timber building in the twentieth century.

THE MODERN IDIOM
Shingling and weatherboarding have been creatively used in this recently built house in America. The colour and texture of the wood enhance the unusual sweep of the roof and the strong line of the vertical wall cladding

CLAPBOARDING

Unlike their counterparts in Europe, American timber-frame houses after the 17th century were clad with horizontal, unpainted boards. Oak and cedar boards were cleft from the log radially (right) to give wedge-shaped boards; pine was more usually sawn. Nailed or pegged to the frame timbers with their thicker edges exposed, the overlapping boards (far right) are called (a) feather-edge or (b) rebated feather-edge boarding. Flat, interlocked boards are called shiplap (c) or V-jointed (d)

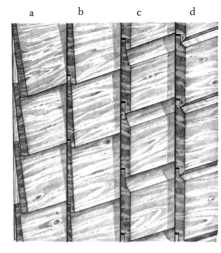

GEORGIAN ELEGANCE

Built by John Vassal in 1759, and used by the poet Longfellow, this framed house in Massachusetts has horizontal clapboarding. Unlike earlier weather-boards, these are painted to enhance the symmetry of the formal classical façade

STOUGHTON HOUSE

Shingling became a style in its own right in America in the 1880s, in New England and California. The Stoughton house was built in 1882 by H. H. Richardson, and illustrates the uniform covering of cypress shingles typical of the style

SHINGLE CLADDING

Like slates and tiles, the wooden shingles and shakes shed water from the surface of the building on the same principle as the overlapping feathers of a bird. Shingles may be sawn but shakes are always cleft using a froe and wooden mallet (right).

Today, shingles are usually made from cedar, though oak, cypress and redwood have been used extensively in the past. The wood must be well seasoned and roofs steeply pitched, at least to an angle of 45 degrees, so that water is shed quickly to avoid warping the wood

Buildings of Asia

The domestic architecture of Japan, like that of many other countries of eastern Asia, owes much to the influence of mainland China – a powerful influence that spread rapidly in the wake of Buddhism. Features which the dwellings of these countries have in common are the post-and-pile construction with panelled infill, the bracketed, non-triangulated roof, and the wooden floor raised on low columns, which are themselves supported on stone foundations. The cypress, the cedar and the tsuga, a type of spruce resembling hemlock, are common building materials throughout the area.

However, while the buildings of the mainland tend towards solidarity and heaviness, and are often richly decorated, Japanese domestic architecture followed a course of development of its own. The climate of the Japanese islands, with a high rainfall, hot and humid summers and cold, dry winters, was an important factor. A Japanese dwelling must have openings to allow free circulation of the air and to combat the effects of damp.

Buddhism reached Japan in the sixth century, and its influence created a trend towards extreme simplicity. Buildings, however, relied on methods of construction such as the total rejection of bracing, which considerably reduced the stability of the structure and put additional stresses on its complex joints.

The "shinden-zukuri" style, associated mainly with the houses of the aristocracy, appeared well before the eleventh century. Its plan was based on a central hall, the shinden, around which other rooms and passages were grouped in strict symmetry. No building survives from this period, but parts of the nineteenth-century Imperial Palace in Kyoto were built to conform to this style.

During the eleventh century, when the élite Samurai, the military class, had consolidated its power, the "shoin-zukuri" style emerged, free and asymmetrical in plan and with widely overhanging roofs and verandas. Translucent paper was used as a covering material for sliding windows and doors, while movable shutters of wood appeared a little later. This style eventually came to dominate Japanese domestic architecture.

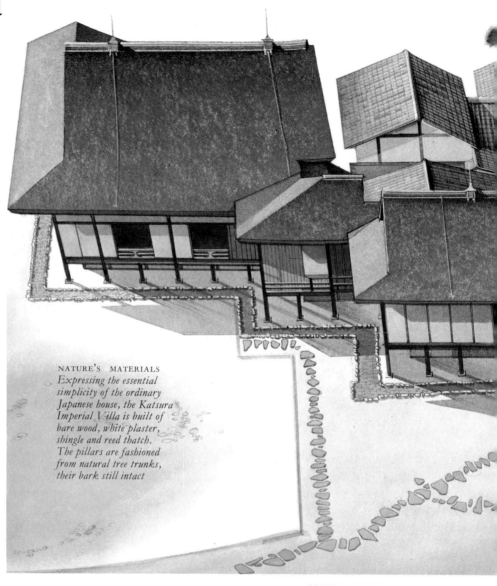

NATURE'S MATERIALS
Expressing the essential simplicity of the ordinary Japanese house, the Katsura Imperial Villa is built of bare wood, white plaster, shingle and reed thatch. The pillars are fashioned from natural tree trunks, their bark still intact

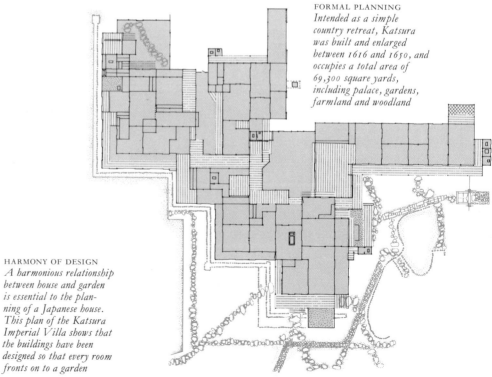

FORMAL PLANNING
Intended as a simple country retreat, Katsura was built and enlarged between 1616 and 1650, and occupies a total area of 69,300 square yards, including palace, gardens, farmland and woodland

HARMONY OF DESIGN
A harmonious relationship between house and garden is essential to the planning of a Japanese house. This plan of the Katsura Imperial Villa shows that the buildings have been designed so that every room fronts on to a garden

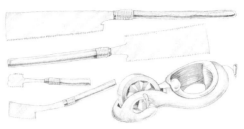

ORIENTAL TOOLS
*Like their skills, which
were handed down from
father to son, the tools
devised by Japanese
carpenters are still in use
today. Illustrated here
are various saws and an
inker-line for drawing
directly upon timber*

SKILLS OF THE CARPENTER

During the 14th century the carpenter's
square became popular in Japan. Its skilful
use made possible the precise construction
of complicated joints for use in house
building. Shown below are (a) the arigake
corner joint; (b) the kanawa-tsugi joint,
used in columns, and (c) the kama-tsugi
joint, used like (d) the ari-tsugi joint, for
eaves, beams, purlins and ridge beams

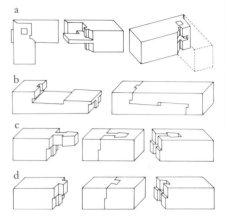

The solid walls of a Japanese house consist
of bamboo lattice-work fixed between uprights and plastered on both sides with mud
plaster and an admixture of finely chopped
straw, which binds the material together

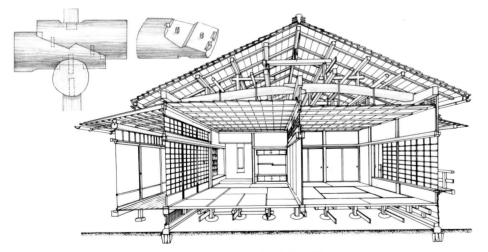

DOMESTIC BUILDING
*The ordinary Japanese
house, as this elevation
shows, is a framework
building with no corner
ties or diagonal bracings.
In framing, four- to five-
inch-square posts of sugi-
wood, a damp-resistant
hemlock, are set in stone*

*foundations. Cypress is
mainly used for exterior
surfaces and red pine for
roofing beams. Ceilings of
battens with sugi-wood
panels are hung on straps
from the rafters, and
thatching is laid in thick
layers on bamboo battens
and clip-finished*

Roof Beams and Arches

In its most elementary form, the concept of shelter is synonymous with that of the roof. As far back as the Neolithic period, Man had attained great skill in solving the structural problems encountered in providing a basic form of shelter and, today, the huge bee-hive houses of some South American tribes, and the long-houses of New Guinea, show the scale of roof-building possible with only the simplest of tools and techniques. The structures themselves appear deceptively simple, yet their builders have mastered many sophisticated engineering problems.

Timber was the first material used in roof-building and throughout the history of architecture has remained one of the most essential and versatile materials. In the construction of the large, steeply pitched roofs of central, western and northern Europe, the carpenter's craft developed apace, while the low-pitched and flat roofs of the Mediterranean countries set their builders less stringent tasks.

In Italy, as in Ancient Greece, the purlin-supported roof with its low pitch met all requirements and remained in use until after the Renaissance. Elsewhere in Europe, roof-building was dominated by the trussed rafter method – a fundamentally different concept in which the rafters are held together by collars and tie beams stressed in tension. Built of oak, the performance of these roof structures was remarkable, culminating in the multi-storeyed roofs of the sixteenth- and seventeenth-century European town houses and the huge, steeply pitched roofs of the Westphalian farmhouses.

The peak of the development of roof structures was reached in the Middle Ages with the invention, in East Anglia, of the hammer-beam roof – a massive structure which could be used to span widths of nearly seventy feet, and a major feature of many of the Great Halls of England.

In Georgian times, a period not known for taxing its carpenters, the Shaker community in America produced an outstandingly inventive circular barn roof. The structure was supported by a central ring of eight timber columns, tied together with cross-members to form a central ventilation shaft.

HAMMER-BEAM ROOF
Short horizontal beams, or hammers, supported by wall-posts and braces, project from the face of the wall to support the main arch. The total span is thereby effectively reduced, allowing spans of as much as 70 feet to be bridged without intermediate supports and without the necessity for very long timbers. Among the many famous roofs constructed by this method are those at Westminster Hall, built in 1395, and at Eltham Palace (illustrated above), built in 1405

FRENCH DOVECOTE
During the medieval period, doves were commonly used to supplement the dull winter diet. This large dovecote, at the Château de Vayres in France, has a conical roof supported by masonry walls and a central timber column with bracing crossmembers

THE ENGLISH BARN
This typical 17th-century barn roof at Jordans, in Buckinghamshire, was built with oak timbers from the Mayflower in 1624. The advantages of using old ships' timbers were that the wood was well seasoned and of good quality

DECORATED TIMBERS
The people of the Sepik region of New Guinea are renowned for their painted and carved decoration, both of which are used extensively in their buildings. This meeting-house is built in the style of one of the traditional spirit-houses

GUYANESE HOUSE
Locally grown bamboos and hardwoods have been used to build the roof of this richly decorated Guyanese house. The basis of the roof is a simple timber frame, like the European frame, over which is laid a dense thatch of plaited bamboo fronds

Composite Roof Timbers

PARABOLIC ARCHES
Laminated parabolic arches allow large open areas to be spanned without the need for intermediate supports. Laminated Douglas fir beams, clad with cedar planking, have been used to span the passenger arrival concourse at the Southampton docks

Until the advent of modern adhesives, very considerable spans were bridged by traditional methods of timber construction, most of which were developments of the arch and the lattice truss. Rapid advances were made in the field of wide-span building at the beginning of the twentieth century, and a number of impressive buildings were erected with timber structures made from such high-grade woods as Douglas fir, hemlock, Oregon pine and various European pines, that could rival their counterparts in steel. For example, the huge Westfalenhalle sports centre in Dortmund, Germany, which with arches and lattice trusses bridged a span of two hundred and fifty feet, was, at the time of its construction, one of the largest timber structures in the world.

However, while in steel construction the plated girder appeared with its stiffened web, no equivalent was available in wood engineering until the development of lamination techniques. The concept of deep wooden beams to bridge vast spans depends entirely upon the use of adhesives permitting the fabrication of large sections from small pieces of lumber. With such methods, timber beams became available that offered not only superior properties with regard to resistance to bending, but also to deflection under their own weight. The parabolic, laminated timber arch is even more successful than the flat laminated beam, because the stresses are taken by the arch itself, making the construction more stable.

With laminated timber construction the whole load taken by the beam is resisted by the adhesive and the only problematic element in this process is the possibility of glue fatigue. However, the reasons for some early failures of laminated timber beams have now been reliably established: the main fault lay with the architectural design rather than with the adhesive or with the wood.

Notwithstanding such problems, the potential of laminated construction in timber-building is very great. This is shown by recent developments of such promising elements of structure as the I-beam, which in some ways resembles the plated steel girder, and the box-beam with its hollow core.

BARREL-VAULT ROOF
A series of stressed plywood skins has been used in the construction of the barrel-vault roof of this school assembly hall in Berkhamsted. The outer laminates are of beech, giving the roof a warm, pleasant finish and an easily maintained surface

COMPOSITE TIMBERS

The roof detail below illustrates the combination of large-scale laminated beams with both flat-sawn boarding and ply panelling. The main beams are secured at their intersections by steel bracing members set flush with the wood surface

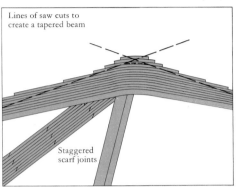

GLULAM SYSTEM

Using laminations, above, structural timbers may be made to any shape, while the grade of timber used may be varied according to loading and aesthetic requirements. Softwoods are generally used, particularly Douglas fir, southern pine, hemlock and redwood, but hickory, oak and other hardwoods are also used. Laminates are glued and then pressed together and heated to set the adhesive. Scarf joints, either plain or hooked, are staggered to avoid creating lines of weakness in the composite timber product

THE GRIDSHELL

Among the most exciting new developments in the field of timber roof-span construction is the grid-shell structure developed in Europe by Frei Otto. In essence, the gridshell is conceived as a hanging net, which is then "frozen" in position and overturned. The lattice illustrated is made of Siberian yellow pine, but western hemlock might be preferred for its superior flexing properties. The mesh is light, strong (as each member is in compression only) and is readily transportable. The grid-shell shown above is one of many small experimental models, but the technique is readily adaptable to large-scale structures. One of the largest built to date is the exhibition hall in Mannheim, built in 1975 by Frei Otto and spanning more than 200ft

REDWOOD PILLARS

Composite beams and pillars are frequently used in large municipal buildings both for their aesthetic appeal and their structural versatility. The roof of this swimming pool is supported on laminated pillars of European redwood

The Appeal of Wood

The many species of wood that can be used for building combine great variety of tone and colour with an organic texture which no man-made material can equal. Veneer and plywood, developed in this century, have greatly extended the range of timber building techniques in both form and scale. Whereas formerly the timber wall relied on the log, the sawn plank and the cleft shingle, wide panels, either flat or curved, offer new scope to the builder while presenting grain patterns endowed with an infinite variety of linear decoration. Success in waterproofing has, furthermore, made plywood a suitable material for exterior use provided that the endgrain can be adequately protected against penetration of moisture.

To show the natural beauty of the wood has become one of the principal aims of the architect working in timber, and in this he is greatly helped by vastly improved methods of wood preservation. Until recently, exposed woodwork required extensive and frequent maintenance; the only sure protection was achieved by painting the wood with oil-based paints, an effective technique but one which hid the wood from view. New techniques of pressure-impregnation allow the wood to be used in a more natural state while giving long-term protection to the surface. Similar techniques may also be used to impregnate structural and cladding timbers with fire-retardant chemicals, so removing one of the oldest and most serious problems inherent in timber buildings.

Though it is customary to expect timber cladding to be smoothly planed, some modern architects, notably the American Frank Lloyd Wright, have made extensive use of unplaned timber – attracted by its rustic qualities and by the contrast to be gained from setting it against smoothly finished joinery on door and window frames.

Despite their high costs, log structures are still built in many regions of the northern hemisphere, but the interlocking board, with its variety of sizes and finishes, is today the most commonly used cladding material, and one which has effectively resisted the competition offered by plywood sheeting and brickwork skins laid over timber frames.

SOFTWOOD DETAILS
The softwoods of North America and Europe offer the architect a wide range of cladding material. Douglas fir, hemlock, spruce, pine, larch and cedar may all be prepared in the form of flat-sawn boards or with machine-finished textures; *in natural colours or after staining; as standard-sized sheets, tongue-and-grooved panels, or shingles. Their versatility in exterior use may be greatly enhanced by pressure treatment with preservatives, to combat fungal and insect attack, and with fire-retardants*

REDWOOD DECKING
A common design feature of timber-built houses in western North America is the exterior decking – a timber platform, generally at first-floor level, which provides an open leisure area. Here, the decking is of richly toned redwood

WARMTH AND COLOUR
Building research studies show that timber cladding, particularly that with an internal air-cavity, has insulation properties far superior to those of either concrete or brick walls – a feature utilized in the chalets of Alpine Europe

The Versatility of Wood

HOUSE FRAMING
*The illustrations below
demonstrate the ease with
which even a large domestic
building may be erected in
timber. The main frame of
standard-dimension lumber
is supported on concrete
or stone footings or on
treated timber foundations*

Wood requires the lowest energy input of all the materials used in the construction industry, and with proper husbandry and intelligent use of timber technology it need never be in short supply. Though the development of new techniques has taken place mainly in the United States, northern Europe, too, is becoming aware of the advantages of lightweight construction in timber.

In countries like Switzerland, with an indigenous tradition, the grafting of new methods of timber building tends to be a slow process; in Scandinavia, where timber is an essential national asset, modern timber technology has developed of necessity. In the United States, where the majority of the population lives in timber houses, there are naturally very powerful incentives to exploration and improvement. As a result, the average American home can offer, for the same outlay, considerably more than its European counterpart.

It is true that a lightweight structure does not possess the same thermal storage capacity as one built in heavy materials, and it loses warmth more rapidly. But a lightweight structure requires much less input of heat to restore and maintain a level of comfort, and lightweight structures are therefore more economical to warm.

In most industrial countries the wide variety of timber products and man-made materials for the interior finish of buildings can, with planning, be used in their marketed sizes. Interlocking boards, sheet materials, plain and decorative plywoods, wallboards, fibreboards and laminates all offer distinct advantages over traditional finishes such as plaster and paint. Most need little maintenance and many possess excellent thermal insulation properties.

Building in timber has many striking advantages, including low labour costs, short building periods, adaptability coupled with low investment in plant, ease of standardization, and the possibility of meeting a wide range of requirements within a very small organization. The most spectacular results have been achieved in the United States, but timber building has great potential for further development in Europe.

EXTERIOR CLADDING
*The frame of the building
is first clad with a skin
of heavy-gauge plywood to
form a structurally strong
unit and to provide good
thermal insulation. Outer
cladding of more decorative
boarding, either stained or
textured, is then overlain*

FLEXIBLE SYSTEMS
The partially completed house below illustrates the versatility of modular systems. The main frame is complete; the roof cladding protects the interior, and the wiring circuits can be installed within the wall thickness before cladding

COMPOSITE BOARDS
Composite wall units for the wooden house may be prefabricated in factories and then transported to the site for immediate use. The sections may be simply of frame and cladding or may incorporate layers of additional insulation

MAN-MADE BUILDING MATERIALS
The development of composite wood products has greatly increased Man's efficiency in utilizing the forest resource. Core veneers in plywood may be of inferior grade to the face veneers without loss of structural strength, while waste lumber and the fibre waste of the sawmill may be used as core material in a wide range of panels

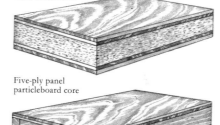

Five-ply panel
with veneer core

Five-ply panel
particleboard core

Five-ply panel
with lumber core

Five-ply panel
with edge-banding

Multiple-veneer plywood panels may be edge-joined either by tongue-and-groove joints or by the more complex interlocking finger-joints. In either case, strength may be increased by pinning through the joint

MODULAR FLOORING
Standardization of sizes for most building materials has led to the development of very efficient assembly techniques. Here, adhesive is being applied to floor joists carefully spaced to take standard-sized sheets of heavy-grade plywood

TONGUE-AND-GROOVE
The sheets of plywood used for the flooring may have tongue-and-groove joints along their margins. This flooring may suffice alone if subsequent loadings are light, but will probably be used as a sub-floor beneath conventional floor-boards

PROTECTIVE BATTEN
Great care must always be exercised in the handling of boards with machine-cut tongue-and-groove edging as damage to either will ruin the sheet. Here a protective batten takes the blows of the hammer as a plywood sheet is driven into place

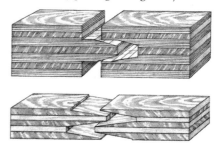

MOBILE HOMES
The lightness of a wood-framed building makes it possible to manufacture a complete house in factory conditions and then transport it, as a unit, to the site, where it can be linked in to the main electricity, water and drainage systems

Sacred

Buildings

The projecting beam-ends of hammer-beam roofs provided the medieval wood-carver with a perfect place to display his craftsmanship

Western Church Architecture

Many countries of the Western World possess a rich heritage of timber-built churches dating back to the eleventh century and earlier and some of these superb structures survive today, notably in Scandinavia, central Europe and in France, where some two dozen remain.

In the past, Europe was largely covered by immense mixed forests; conifers dominating in the north and in the Alps, while in central Europe, France and particularly Britain, forests of tall, strong oaks supplied large-dimension timbers for constructional use. It may be assumed that by the Bronze Age, a variety of mortise-and-tenon joints were known, and the carpenters of the Christian era used most of the tools employed by their successors.

In very early European churches built of timber, the walls were made of split logs, one of the finest examples being the church at Greensted in Essex, England. In even older buildings these trunks were set directly into the ground, but the introduction of ground beams or sills, eventually on stone beds, led to a method of building of remarkable strength that survived for many centuries. Standing on a grid of sills and held together by knee braces, St Andrew's crosses (or saltires) and clamping beams, the vertical members that developed from tree trunks into vertical posts were able to support complex many-tiered roofs, which frequently made use, as in stone buildings, of the lean-to and derived structural strength and stability from its bracing effect. For a period of more than 300 years this system of construction endured – aptly named, in Norway, the "stav" or stave method after the vertical member which forms its main structural component. Unfortunately the Black Death prevented knowledge of the stave-building technique from being passed on.

The church building of the early American settlers in New England could rely on the established European framing techniques, which were carried to the richly wooded new colonies. The stave construction, however, may be regarded as the predecessor of timber framing, for it anticipated many of the problems of structural stability in framing that faced the builders of the following ages.

SOLID TIMBER WALLS
Possibly derived from palisading, early church walls were made of split tree-trunks or logs, placed vertically and held together by wooden tongues. Greensted church in Essex (right and above) is one of the oldest examples still standing

TWISTED SPIRE
Erected in the late 14th century, the spire of Chesterfield church developed a marked twist, which gives the church a unique character. The frame of the spire is of oak, a timber susceptible to warping when used "green", or unseasoned

THE OLDEST CHURCH
The parish church at Greensted in Essex was built, according to ring-dating tests, in AD 845. Recent examinations suggest that many of the massive logs may have come from the same tree, a huge oak, itself more than 600 years old

STAVE CHURCH

The church at Borgund is one of 25 stave churches remaining from a former total of 700. Built in 1150 of coniferous timbers, it has six shingled roofs, each gable decorated with crosses or dragons' heads, old Viking symbols

THE GROUND SILL

Though early forms of stave building relied on posts set directly into the ground, later constructions included ground sills on which the roof-bearing inner posts could stand. The wall posts stand at the edge of the massive ground sill

TIMBER-FRAMED CHURCH

Lacking the heritage of European monumental buildings, early American churches were constructed much like the contemporary framed houses, with large windows and weather-board cladding. This church is typical of New England

The Domes of Khizi

Eastern Europe is the home of log-building – its extensive forests offering a great variety of constructional timber including fir, pine, larch and some of the finest spruce in the world. Of hardwoods there is excellent maple, beech, lime, elm and oak, the latter being used almost exclusively for the Church of Saint Sophia in Novgorod, built in AD 989.

The tools used by carpenters working on these early churches were extremely limited, but the craftsmen learned to handle them with consummate skill and many old churches are decorated with fine carving executed with the axe. In tools and techniques, eastern Europe remained conservative in the extreme and the mass-production of nails, for instance, was not mastered until the early part of the twentieth century.

The dominant feature of many eastern European churches is the division of the building into three parts; the central space formed the main body of the church, to the east lay a smaller annexe housing the altar and to the west a covered porch. The buildings were almost invariably constructed of round logs and, in order to gain the maximum interior space using logs of limited length, an octagonal ground plan was adopted. Structural stability was ensured by the polygonal plan and the horizontal rigidity of the log construction, which required no additional bracing.

Often the octagons were set one above the other, surmounted by onion-shaped domes, while the numerous gable ends, also having the characteristic onion shape, add to the decorative effect. Few of the wooden churches surviving from the seventeenth century have much additional adornment; their decorative effect derives almost entirely from the structural form and from the texture of the wood itself.

One of the most remarkable features of these churches are the roofs, clad with shingles fixed in place with wooden pegs. The Church of the Transfiguration at Khizi, in Karelia, built in 1714, has twenty-two such domes, clad with aspen shingles. Built to replace an earlier church destroyed by fire, this monument represents the highest attainment in Russian wooden architecture.

KARELIAN MASTERPIECE
The four flights of cupolas of the Church of the Trans-figuration lead the eye up-wards to the central cupola, whose supporting drum rests directly on the octagon formed by the walls of the main building. Though more ornate than the traditional tent churches which preceded it, the whole structure is designed to preserve the traditional pyramidal form

Using elongated lozenge-shaped shingles, the Russian carpenters completely mastered the difficulties of cladding the complex curved surfaces of their domes. The shingles, in this case of aspen, were fixed in position with hardwood pegs

EAVES OF PINE
The interlocked corners of log-built structures are always prone to weathering and decay. Here they are given some protection from rain and melt-water by the overhanging eaves of cleft softwood planking

CARVED DETAIL
Contrasting with the simple log structure of the main building, the boards along the edges of the roof, the rain-water spouts and the poles supporting the porch canopy are all decorated with carved perforations and scalloped edges

Temples of the Orient

The plan shapes of religious buildings in eastern Asia lack the complexity of their European counterparts. They are generally simple rectangular halls with few internal subdivisions, or many-tiered tower structures often standing in walled and colonnaded precincts. Chinese temple buildings are laid out in strictly axial symmetry, but in Japan asymmetrical arrangements seem to have been preferred throughout the ages.

With the spread of Buddhism, the Chinese influence came to predominate in eastern Asia and the art and techniques of Chinese carpentry were adopted in all the neighbouring countries. From the varied and excellent building timbers of the area's extensive forests a wealth of impressive structures was created, of truly monumental scale though also often of great simplicity. Cypress trees like the Japanese hinoki, an aristocrat among conifers, and the cedars of eastern Asia, which allowed eighty-foot columns four feet in diameter to be set up as roof supports, endowed these temples with a grandeur not found elsewhere. They permitted also the construction of multi-tiered pagodas with centre columns of similar proportions.

Red pine was commonly used for beams, and for decorative carved details the area provided a host of attractive woods, including maple, which in China grows to a height of fifty feet, mulberry, Japanese saw-tooth oak and Chinese sandalwood.

The basic structure of most eastern Asian temples is that of post-and-beam set on a base of stonework, sometimes with additional tiers of columns raised on top. The tier construction of their complex roofs entirely eschews triangulation; were it not for the great weight of the roof-covering of pantiles or thick layers of thatch, in addition to the mass of the supporting timber frame, such reliance on compressive strength without bracing could not suffice. Triangulation is not unknown in the East; it is used during construction for temporary support, but is removed on completion. Its rejection is based on tradition, for the Chinese have always used the short beam or bracket, supported by close columns transmitting all loads vertically to the foundations.

HORYUJI TEMPLE
Standing within a wooden enclosure pierced by two ceremonial gates, the seventh-century pagoda and the kondo, the temple hall, of the Horyuji Buddhist temple at Nara in Japan, are probably the world's oldest ceremonial buildings constructed of wood

SHRINES OF ISE
Thought to preserve the style of the ancient aristocratic houses of Japan, the Shinto shrines at Ise were built in AD 478. Built of pale hinoki wood they stand on stilts, while wooden cylinders along the ridge protect the thatch from damage by high winds

ORIENTAL ROOF CONSTRUCTION
The archaic Chinese roof consisted of beams placed one above the other and decreasing in length towards the ridge. They were separated by struts above which purlins supported the rafters, ensuring a vertical thrust downwards upon the pillars. The building could be widened by increasing the number of pillars. Later, the upward- and outward-expanding bracket cluster increased the area of support afforded by each pillar

Archaic design with narrow span

Wider version with additional pillars

Bracket clusters reduce the number of pillars

Tou-kung, or brackets, support eaves, rafters and purlins

NAN-MU COLONNADE
Thirty-two complete trunks of nan-mu, a fragrant Chinese sandalwood, support the projecting roof of the Hall of Sacrifices of Emperor Yung Lo, near Peking. Each column, three feet in diameter, rises more than 33ft from floor to roof

A WEALTH OF DECORATION
Wooden ceremonial buildings in the Far East, like this building of the late Yi dynasty in Korea, were richly carved and painted. Korean buildings were always painted while the Chinese adopted the custom of laquering the surface of exposed woodwork

Painted and Gilded Wood

Religious buildings in the West have contained examples of the wood-carver's art from as long ago as the days of the Ancient Greek and Roman civilizations. Few carvings survive, however, from much earlier than the beginning of the Middle Ages, when Europe in particular was in a frenzy of church-building, and wood-carvings decorated screens, choir stalls, rafters, beams and almost every sort of church furniture.

In England and northern France, the church wood-carvers worked mainly in oak. South of Burgundy and in Italy walnut was easily available and frequently used, although fig-tree and cork wood were sometimes selected for their lightness. The German and Scandinavian wood-carvers took advantage of their vast, native pine forests – slabs of pine were used as a framework for intricate patterns of scrollwork and mythical beasts.

After the wood-carving had been completed in the medieval churches, it was nearly always painted with brilliantly glowing, pure colours. Blue and gold were favourite colours for roofs; elsewhere green and red, black, white and yellow not only decorated individual areas of the church, but also, when enough money was available, contributed to a dazzling, all-over colour scheme.

Unfortunately, such gaiety did not appeal to the English Puritans, who covered much of the woodwork of the Middle Ages with dark, dull paint. Perhaps because of their inaccessibility, however, many painted roofs with gilded bosses escaped such harsh treatment and the original colouring still survives to delight the eye.

In Asia and the Far East, there is little large religious furniture other than the statues in Buddhist and Hindu temples, though the minutely elaborate doors of some Hindu temples were made from sandalwood. In Burma, teak, shisham (harder but inferior to teak) and deodar, an excellent softwood, were used for statuary. The columns of Chinese temples are made from nanmu-wood, a tree grown specially for its tall, straight trunks, and in Japan superb examples of 17th-century woodwork can be seen in the decorations of the mausoleums at Tokyo, Shiba and Nikko.

FLIGHTS OF ANGELS
The timber-clad tie-beam roof of Blythburgh Church in Suffolk was constructed by medieval craftsmen using wooden pegs and mortise-and-tenon jointing techniques. Only in this century has it proved necessary to strengthen the main crossmembers.

The original red, green and black paint remains visible and the rafters are decorated with delicate floral motifs and the sacred monogram. Cromwell's troops were responsible for extensive damage to the church and many of the carved angels were badly damaged

ORIENTAL EFFIGY
Religious figures in the Far East were also commonly painted. This carved effigy of the Bodhisattva, who was believed to take care of those in need, is from 11th-century Japan. Here, the figure has taken on the guise of a mendicant priest

ROOD SCREEN
Its wood intricately carved, this screen in the church of La Faouet, in France, dates from 1480. It stands in the Chapel of St Fiacre, its bright colours hinting at the glorious painted woodwork characteristic of churches of the Middle Ages

THE VINTNER
Though misericords were rarely painted, this one, depicting a vintner with his barrels, is crudely but clearly carved and painted in subtle colours. It came originally from the church of Saint Etienne des Tonneliers in Rouen, France

FINNISH PULPIT
Scandinavian medieval church interiors were often highly coloured. Wooden ceilings were painted; interior walls covered in frescoes. This pulpit in Hattula, Finland, was carved with a number of simple figures and painted in the 16th century

"WINE-GLASS" PULPIT
One of the few remaining painted medieval pulpits in England is in the church of Burnham Norton in Norfolk. It survived the ravages of Cromwell's troops and also escaped the Georgian fashion for staining wood in imitation of figured mahogany

The Master Craftsmen

During the early part of the Middle Ages, the wood-carver held a lowly place in the social scale and was considered inferior to the stone-mason. Indeed, the wood-carver himself generally considered his work, and his material, a poor second to that of the master mason and probably hoped that once painted or gilded his work would be indistinguishable from stone.

By the beginning of the fifteenth century, however, the situation had been reversed and the wood-carving crafts had attained a new importance. The reason for this was that as the wood-carver followed behind the mason, utilizing many of his techniques and designs, it became apparent that certain decorative styles, such as tracery, could be executed far more successfully in wood than had ever been possible in stone.

A second development which encouraged the wood-carver to view his work with a renewed respect was the growing popularity of carved foliage as a decorative motif in contrast to the stiff and unnatural stylized designs that had predominated in the early Middle Ages. These flowing, graceful lines were used to great effect by wood-carvers and were used extensively on misericords, pews and screens. The choirstalls of the great cathedrals of Europe provided the wood-carvers with a superb opportunity to display their skills. The canopied choirstalls of Chester and Winchester cathedrals are masterpieces of finely detailed tracery while, in contrast, the wood-workers of Italy and elsewhere in southern Europe developed the techniques of inlay, notably of intarsia and marquetry.

The austere, enclosed, high-backed pews introduced by the Victorians reflected a marked change in social attitudes. Lacking any decoration and made simply of pitch-pine planks, they form a great contrast with the heavy, low oak benches of earlier times which were often exuberantly carved with mythical beasts, foliage, figures and ornate crosses. Many of these medieval benches, built using oak planks more than four inches thick, still show the marks of adze and chisel and often carry the discreet sign of the craftsman who made them.

WALNUT CAPITALS
Though used today almost exclusively for veneers, walnut can be obtained in large dimensions and in the past has been used in the solid. This capital, taken from an Italian column of the 12th century, depicts the prophet Jeremiah

HIDDEN GEMS OF THE CARVER'S ART

Among the most interesting pieces of wood-work found in churches and cathedrals in France and England are the misericords. These are carved ledges on the undersides of the tip-up seats of choirstalls, which provided a support upon which exhausted monks and priests could rest during the interminably long services of the Middle Ages. They hold an important place in the history of ecclesiastical wood-carving, as they record valuable information about medieval life, costumes, work, tools and so on. Because the underside of a seat was generally hidden from view, the craftsman was free to indulge in themes which might appeal to him personally. As well as Old Testament stories, myths, popular tales and romances were frequently chosen and illustrated with great humour

Amiens Cathedral in France is known as the Bible of Amiens because of the many biblical tales recorded in its wood-carvings. Here Joseph's cup is discovered in Benjamin's sack

The misericords of Worcester Cathedral date from 1390 and include several subjects from the Old Testament. Here, between two bunches of foliage, David grapples with a lion

THE MADONNA
In place of the mosaics of Byzantine art, the Russians used wood to represent religious subjects at a popular level. They felt they were reconstituting nature itself in the glory of God. This carved, painted fragment (left) comes from Vologda

CHOIRSTALLS OF SIENA
The choirstalls of the cathedral of Siena demonstrate the art which was born in that city. From this famous centre the masters of intarsia went out to decorate churches and cathedrals throughout Italy in the 15th and 16th centuries

PAINTINGS IN WOOD
The intarsiatori reached such perfection in their craft that it was often difficult to distinguish the wood inlay from perspective painting. This detail of a rabbit comes from one of the choirstalls in Siena Cathedral

Wood Mosaics of Islam

It is paradoxical that within Islam's western sphere of Syria, North Africa and Spain, Egypt, so lacking in native wood, should produce some of its finest woodwork. Partly just because wood was scarce and precious, Egyptian wood-carving from the tenth to the fifteenth century has the minute and patient delicacy of chased ivory.

It was a peculiarly Egyptian practice to use on their furniture a multitude of very small panels to cover a large area, producing a complex miniaturized surface of stunning cumulative effect. For this there was a practical as well as an aesthetic reason: the hot sun, alternating with the cool night, would warp and split a large board and distort its joints.

The necessary use of small pieces of wood to make up a large whole appears at its most extreme in the *mushrabiahs*, or lattice-work grilles, which served as windows on many Arab houses. These lattices consisted of hundreds or thousands of turned balls and dowels, which were not rigidly nailed or glued together, but, to allow for distortion, were socketed loosely into one another. The greater the number of balls and dowels within a small area, the more effective their diffusion of the glare of the sun, and up to two thousand separate pieces of wood might make up a single square yard of the lattice screen.

The Egyptian craftsmen were also proficient in techniques other than that of panelling and *mushrabiah*. Their lofty wooden ceilings, for instance, might be so delicately fretted, coffered and encrusted with stalactitic forms that the dazzled eye could hardly appreciate their variety.

The Egyptians seem to have used most commonly oak and pine from Turkey, but for decorative inlay, and sometimes for more massive use, they imported teak from India and ebony and other decorative woods from the African interior. But in their golden age, the Egyptians relied more simply on the effect of their miniscule reliefs covering every vacant area – and demonstrating remarkable patience, since every panel was an individually conceived work executed with the most elementary tools.

IBN TULUN
Now happily restored, the minbar, or pulpit, in the Ibn Tulun mosque in Cairo dates from 1296. Probably made of Turkish oak, it embodies every variety of Egyptian carving – the stalactitic headpieces, the carved
inscription over the door, the arabesque reliefs, the mushrabiah railings and the supreme virtuosity of the panelling of the sides. Each polygonal panel has a different shape and is held in an asymmetrical pattern without the use anywhere of glue or nails

MUSHRABIAH

Mushrabiah, meaning literally "balcony", where it was very commonly used, is a form of lattice-work consisting of a vast number of tiny turned oval balls connected by radiating links. The elements, which were arranged in every sort of pattern, were cut, in Egypt, usually from 20-ft lengths of pitchpine from Turkey. The craftsman sat cross-legged before a lathe set close to the ground, and used both hands and feet to manipulate the work

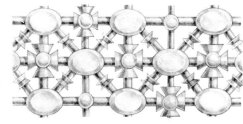

PANELLING DETAIL

Extraordinarily detailed arabesques adorn the numerous different panels, which fit into their moulded surrounds by hidden tongues. The edges are made up of a number of pieces of wood, and are inlaid with light woods

Living
with Wood

The style and elegance of a previous age
repose intact in this early nineteenth
century Chinese chair. Its bracket
ornament and the gloss of its heavy
varnish set a tone of tasteful simplicity
in the house of its present owner

The Ancient World

Fortunately for posterity, the Ancient Egyptians used to bury furniture and other personal possessions with their dead in order to provide comfort in the afterworld. Thus a number of splendid pieces have survived to tell us how the Egyptian craftsman worked and the materials he used.

Because Egypt was poorly endowed with natural timber, it was necessary to import much of the wood used in the construction of fine furniture. A record exists stating how, during the 3rd dynasty (c. 2686–2613 BC), as many as 40 ships laden with timber came to Egypt. These ships probably sailed from Syria, where so many of the woods used by the Egyptians grew, and their cargoes would have included cedar, cypress, ash and box.

Ebony was probably the most highly prized wood and was often used as tribute payment from other lands; Herodotus claims that it was offered as a tribute from Ethiopia. Certainly it was used for making a wide variety of articles such as chests, whips, statues and harps, and in about 1390 BC King Amanophis III sent four beds, a head rest, ten stools and six chairs, all of ebony, to the King of Babylon.

Of the timbers indigenous to Egypt, tamarisk is one of the most familiar, but among the others are sidder, acacia and carob (the locust-bean), and of course the date-palm, not used much in furniture because of its tendency to split. Ebony was often used to veneer a less prized wood, and in conjunction with ivory to form rich and intricate patterns. Ornamental techniques were highly developed and imaginative. The art of veneering had been known at least since 3000 BC, and marquetry and inlay decoration were used to great effect.

So little Ancient Greek and Roman wooden furniture has survived that most present knowledge has been gleaned from the study of vase paintings and relief sculpture. Plenty of timber was available to the Greek and Roman craftsmen, one of the most valuable being citron, which was imported from Mauretania and North Africa. Other more common timbers included maple, yew, holly, oak, willow, lime, zygia and beech. The latter was thought particularly suitable for beds,

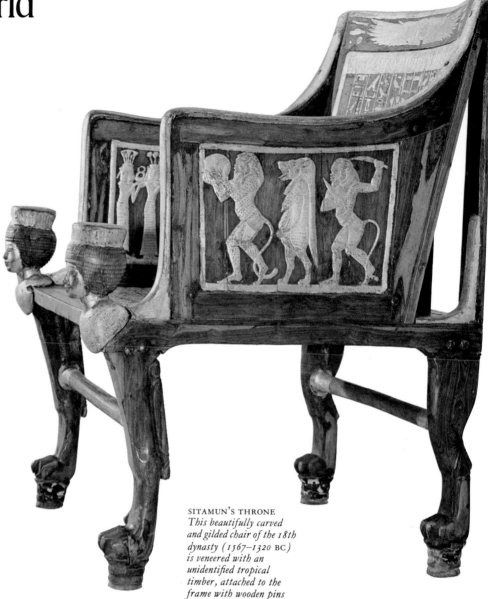

SITAMUN'S THRONE
This beautifully carved and gilded chair of the 18th dynasty (1567–1320 BC) is veneered with an unidentified tropical timber, attached to the frame with wooden pins

chairs, tables and chests, for, as Pliny comments, it was "easily workable, but brittle and soft". To achieve a polished surface, such woods were rubbed with skate skin.

Like the Egyptians, the Romans and Greeks used mortise-and-tenon joints and wooden dowels, and glue was in common use. Lathe-turned chair and table legs were popular, and, as in Egypt, veneering and inlay were used to decorate finer pieces.

The forms of Greek furniture were influenced by those first developed in Egypt, but animal legs on seat furniture gradually became less common and were replaced by rectangular and turned supports. An elegant style of chair, known as the Klismos chair, with curving legs and backboard, had by the 5th century AD become very popular, and was to inspire furniture designers some 1,300 years later. The Romans evolved a heavier form of Klismos chair, and also new types of seat such as panelled benches and wickerwork chairs. The Egyptians and Greeks used chests and boxes to store their belongings, and it was the Romans who first introduced the chest-of-drawers. Although the barbaric invasions of Rome did much to destroy classical traditions, some nevertheless survived to inspire countless furniture designers and craftsmen in later centuries.

EGYPTIAN MUMMY-CASE
Conforming to their belief in the after-life, Egyptians used to embalm the bodies of the dead to preserve them for the life ahead. Mummies were placed in wooden coffins, each shaped and decorated to represent the dead. This 6th-century BC *softwood coffin of the "Milk-bearer of the House of Amun" is decorated only with an elaborate collar and a band of hieroglyphs containing the conventional prayer to Osiris and the gods of the underworld. The rest of the case is left undecorated*

EARLY EGYPTIAN JOINTING

The common use of the mortise-and-tenon joint by Egyptian craftsmen is shown in the way this bedleg is fastened to the timber frame

Another method of securing joints was to pass leather thongs through holes in the leg and rails. This would have been used to secure the mattress

A ROYAL BEDLEG
This fragment of a royal bed of the 18th dynasty is made from fine hardwood decorated with gold cobras. It clearly shows how wooden dowels were used together with the mortise-and-tenon joint

ROMAN BED
Unearthed from Pompeii, this richly decorated ebony bed is inlaid with ivory, paste and semi-precious stones. Beds and couches were used for sleeping and reclining at meal-times

KLISMOS CHAIR
This detail from an Attic red-figure vase demonstrates the elegance of the Klismos chair, a popular type with curved back and legs, developed by the Greeks in the 5th century BC

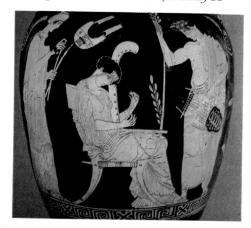

TWO STOOLS
Stools were by far the most common form of seat in ancient times and varied greatly in design. Among the most popular was the lattice type (left), from Thebes. The lattice bracing is both structural and decorative and would *have originally been painted white on a gesso ground. The other example is a folding stool with legs terminating in finely carved ducks' heads inlaid with ivory. The seat would have been made of leather or of woven cords or rushes*

Elements of Furniture

Many small-scale societies can obtain wood with ease; some can even fell large trees without difficulty. From such trees, and from fallen branches, they make the simplest of all furniture: logs dragged into place to form seats and branches stuck vertically into the ground on which they hang containers, clothing or weapons. There are cases when the distinction between natural wood and human artefact is even harder to draw. Many West African societies, for example, plant large-leaved trees (usually of the Ficus family) to provide shade for their elders' meetings. The trunks and protruding roots are gradually trimmed and worn down to provide comfortable seats and back-rests, and the living tree becomes the village's main item of furniture. In hunting and gathering societies, like the Australian Aborigines or the Bambuti pygmies of central Africa, a few branches may be pulled down to form a simple bed.

The natural characteristics of trees to throw out branches at an angle to the main stem is utilized in many ways. Among the Acholi and other Sudanese groups, forked branches form the uprights for sleeping-hut platforms: a simple bed is made by fastening cross-pieces in the forks. Triple-forked branches, with the main branch from which they sprang, were used to make back-rests for chiefs among the Ashanti of Ghana. In New Guinea, head-rests were made by chopping off a length of branch from which lesser branches grew: three or four of these on roughly the same side were trimmed to form legs, while the main stem was smoothed to form the body of the head-rest.

Single blocks of wood are often worked with an adze into simple three- or four-legged stools. Among the southern Akan groups in West Africa, a simple rectangular billet of esese (*Funtumania* sp.) is split from a log, smoothed down and a small protrusion left at one end as a handle. Great ingenuity is displayed by the cedar storage boxes of the Indians of the northwest coast of America. Logs were split with wedges into planks and then the base and four sides of the box cut in a single piece from the plank. By scoring and bending, the sides and base were brought to their correct position and fastened together.

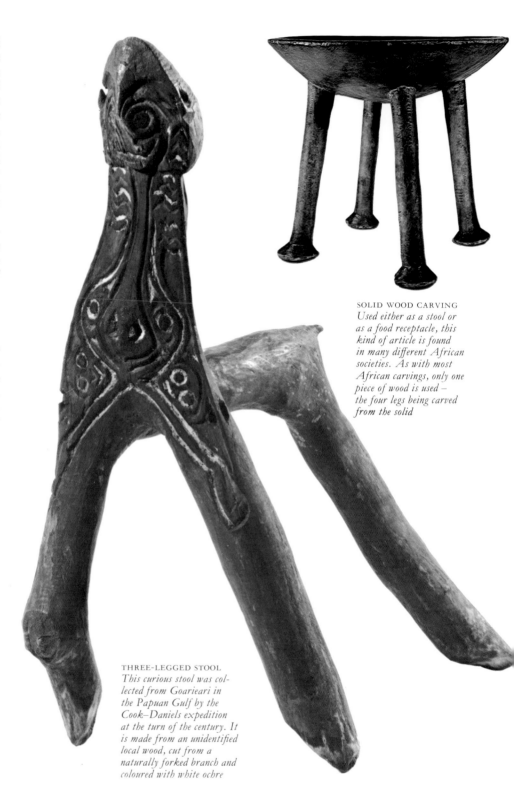

SOLID WOOD CARVING
Used either as a stool or as a food receptacle, this kind of article is found in many different African societies. As with most African carvings, only one piece of wood is used – the four legs being carved from the solid

THREE-LEGGED STOOL
This curious stool was collected from Goarieari in the Papuan Gulf by the Cook–Daniels expedition at the turn of the century. It is made from an unidentified local wood, cut from a naturally forked branch and coloured with white ochre

NATURAL BACK-REST
A piece of furniture can emerge from almost any naturally grown piece of wood. This many-branched item stands firmly on three legs while its main stem and two outspread arms form a comfortable back-rest. It comes from the Wele district of Mangbetu in Zaire and is decorated with copper wire and iron nails

HEAD-RESTS OF WOOD
The use of wooden head-rests is widespread in many parts of the world where they are used both for comfort and as a means of protecting elaborate hairstyles. In form, material and decoration they are infinitely variable; they may be carved, painted, inlaid or may simply use the natural form of the wood itself. The three examples below are, from top to bottom, a sleeping pillow from Aneityum in the New Hebrides; a Tongan head-rest made of a single piece of hardwood, possibly bent; and a second Tongan head-rest made of bamboo with hardwood supports

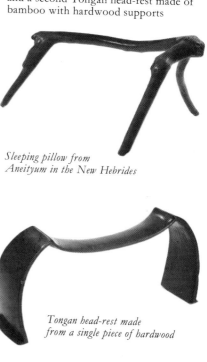

Sleeping pillow from Aneityum in the New Hebrides

Tongan head-rest made from a single piece of hardwood

Tongan head-rest of bamboo with hardwood supports

RED OAK TABLE
This unusual table, standing in an inn in the English county of Northamptonshire, is made from two pieces of the same red oak tree. Part of the gnarled trunk and the main root-mass forms the central support while a cross-section of the stem forms the table top. The tree was probably cut down in the ancient forest of Whittlebury more than 100 years ago. To protect the wood from the studded boots of the canal workers using the inn, the lower part of the stem is protected by hundreds of nail heads

The Mood of the Middle Ages

To modern eyes, an early medieval interior, even one dwelt in by a nobleman and his entourage, would appear extraordinarily sparse. The great hall was the most important room, and because of the many and varied activities which took place there, it was important to keep the central space as free as possible. Furniture was therefore limited in quantity, and generally placed round the sides of the room. At a time when living conditions were far from settled, it was important that furniture should be easily transported. Trestle tables could for instance be dismantled and stacked away after meals, even though the wood used, solid oak, would have been very heavy. Chairs were often made to fold, while benches and stools could be moved around at will.

With timber beams, straw matting on the floors, and wooden furniture, fire must have been a constant hazard, especially as, in the early Middle Ages, it was frequently the practice to light an open log fire in the centre of the room and let the smoke escape through vents in the timber roof.

Though medieval man lived thus surrounded by wood, he in fact valued more highly the tapestries, embroideries and carpets which often hid walls and covered furniture. This panelled room illustrates a late medieval room, bereft of some of its elaborate textiles.

THE BOX CHAIR
One of the first forms of armchair, the box chair developed from the chest, and is of panelled construction. A hinge on the back of the seat itself, or a hinged door below, gave access to storage space

LIVERY CUPBOARD
Made to contain food and drink, these cupboards were carved with pierced Gothic tracery

ADDED COLOUR
Though wood was the most common material used for construction and furniture, textiles had greater value. Tapestries gave colour and warmth, cushions were placed on beds and seats, and floors were covered with rush matting

OAK TABLE
The most common table to be found in the great hall was a massive trestle construction made of oak planks, firmly held in place by hardwood pegs and wedges

Plank and Panelled Furniture

Very little early medieval furniture has survived, and indeed, by more modern standards, very little was made. From contemporary paintings, illuminations and the few extant examples, it can be seen that it was generally clumsy and crudely constructed of thickly hewn timbers, usually oak, while its decoration was largely inspired by architectural motifs, such as tracery, columns and arcades. The chip-carved patterns so often found on medieval pieces usually imitated those first invented by the stonemason.

Though the Gothic style in architecture evolved national characteristics, it is difficult to attribute furniture with any confidence to one country or another until the end of the Middle Ages. Generally, however, it can be said that in northern Europe, where hardwoods, especially oak, were used in the making of furniture, deeply cut sculptural ornament was more favoured than in southern Europe, where the softer woods of Alpine forests were less suitable for high-relief carving. For this reason, in Italy and southern Germany, painted decoration was more sophisticated than in the north.

While the decoration of furniture might have come from stonemasons' patterns, joints were the furniture-makers' own craft. And this, after all, was the Age of the Joiner. Early crude carpentry would have been hidden by bright-coloured paint, but gradually more sophisticated joints were employed, culminating in the panelling construction as used on chests, chairs and walls.

The prime consideration in the construction of medieval furniture was probably practicality. The chest, by far the most common piece of furniture during the Middle Ages, certainly possessed this quality, as it was not only used for storing all kinds of goods, but could also function as a seat, table or even a bed.

Stools and benches were more often used than chairs, which were usually reserved for those of importance, such as the Master of the House. The bed was considered the most important single article of furniture, and those which did not have their wooden framework hidden by textile hangings were often elaborately carved or inlaid.

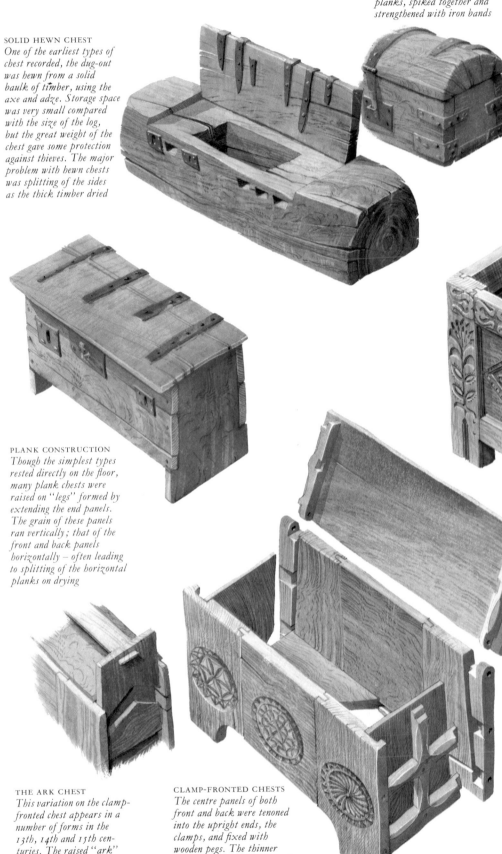

SOLID HEWN CHEST
One of the earliest types of chest recorded, the dug-out was hewn from a solid baulk of timber, using the axe and adze. Storage space was very small compared with the size of the log, but the great weight of the chest gave some protection against thieves. The major problem with hewn chests was splitting of the sides as the thick timber dried

PLANK CONSTRUCTION
Though the simplest types rested directly on the floor, many plank chests were raised on "legs" formed by extending the end panels. The grain of these panels ran vertically; that of the front and back panels horizontally – often leading to splitting of the horizontal planks on drying

THE ARK CHEST
This variation on the clamp-fronted chest appears in a number of forms in the 13th, 14th and 15th centuries. The raised "ark" lid is made from three planks, held in place by the shaped end-pieces, and was hinged on wooden pegs

CLAMP-FRONTED CHESTS
The centre panels of both front and back were tenoned into the upright ends, the clamps, and fixed with wooden pegs. The thinner side panels were also tenoned and were additionally braced by the stout cross-pieces. The lid pivoted on dowels

THE DEVELOPMENT OF FURNITURE
Throughout the Middle Ages, and well into the 18th century, quality furniture was restricted to the wealthy; the common folk had little more than benches, a trestle table and chests for storage. The earliest furniture was all plank-built, but the development of panelling techniques gave rise to a variety of versatile box constructions

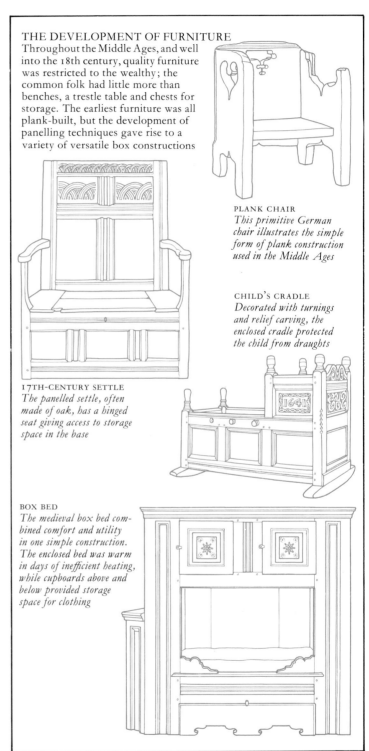

PLANK CHAIR
This primitive German chair illustrates the simple form of plank construction used in the Middle Ages

CHILD'S CRADLE
Decorated with turnings and relief carving, the enclosed cradle protected the child from draughts

17TH-CENTURY SETTLE
The panelled settle, often made of oak, has a hinged seat giving access to storage space in the base

BOX BED
The medieval box bed combined comfort and utility in one simple construction. The enclosed bed was warm in days of inefficient heating, while cupboards above and below provided storage space for clothing

THE PANELLED CHEST
Curiously, although frame-and-panel techniques were known much earlier, their use in chest-building did not become widespread until the middle of the 16th century. The inset panels were generally quite narrow, to avoid jointing them in width, and these were set in grooves without gluing or pinning so that the wood could expand and contract without splitting. The degree of decoration and the standard of craftsmanship vary enormously, but very few chests were made without some form of chip-carved ornamentation on the main framing members

Walls, Floors and Ceilings

As the majority of us live or work within a framework of floors, walls and ceilings, their construction and decoration are both matters of some importance. In spite of the competition from wallpapers, plaster, linoleum, carpets or other fabrics, wood must surely remain one of the most pleasing, natural and versatile materials traditionally used for this purpose.

At one end of the scale is simplicity and functionalism, the tepees or log cabins of North America and simple primitive huts made from branches and twigs being obvious examples. Crude plank floors have been made since ancient times, with the gaps between the boards sometimes filled with oakum and bitumen in the manner of a ship's deck, and thick unadorned beams have proved durable supports for ceilings in almost every country in the world where wood is available and easily extracted and converted into workable timber.

Timber thus used in its natural, or near natural, state is attractive for its texture and colour and indeed for its very simplicity. When, however, the craftsman or artist intervenes, walls, floors and ceilings, while retaining their basic function, can and often do become works of art in their own right. One has only, for instance, to look at the remarkable intarsia-panelled rooms which can be seen in some sixteenth-century Italian palaces or monasteries, the richly carved ceiling of an Oriental palace, or an inlaid or parquetry floor in a French eighteenth-century hotel to see evidence of the lasting appeal of wood.

Wall panelling has undergone many stylistic changes from the linenfold pattern so popular in England and the Low Countries in the fifteenth and sixteenth centuries to the eighteenth-century panelling with its undulating curves and overflowing boiserie decoration. In the twentieth century, the Scandinavians have had a great influence on the rest of Europe and North America due to their imaginative and effective use of wood in interior decoration, using particularly pine and other indigenous softwoods for panels throughout their houses on walls, ceilings and floors.

GHENT BALLROOM
The glorious floor in the Hôtel d'Hane-Steenhuyse took from 1776 to 1781 to build. Six woods were used – mahogany, rosewood, ebony, walnut, sycamore and palisander. It is polished every month to preserve the woods

GEOMETRICAL INLAY
Some of the most elaborate ceiling panelling to be found in Europe is Arabic in influence. This fusion of Muslim and Christian art glows from the 15th-century ceiling of the Monastery of San Juan de Los Reyes in Toledo

OAK PANELLING
Oak was used extensively for wall panelling in Elizabethan times, and some of the oldest in England lines the walls of the Long Gallery at Hever Castle. The wood's curious grain resembles tortoise-shell

Artist and Craftsman

From the Middle Ages, when carved decoration on northern European furniture was largely inspired by Gothic and Romanesque architecture, the art of the carver has played an important role in the development of interior decoration and the embellishment of furniture. Shallow chip-carving was used on plank-built furniture and linenfold carving appeared not only in panelled rooms but also on cupboards, chests and box chairs. By the sixteenth century, when the ideas of the Italian Renaissance had filtered through to other European countries, Italianate heads, acanthus leaves, cupids and other classical ornament became mingled with Gothic tracery and imaginatively carved beasts.

In later centuries, carved decoration sometimes became so much the dominant feature on furniture that it is tempting to describe some seventeenth-century Italian or French pieces as sculpture rather than furniture. In England, Grinling Gibbons and his followers, strongly influenced by the Flemish school, were responsible for some of the most lively and finely executed carvings of the period. Soft woods were the most suitable for the complex and sometimes delicate compositions that typically included fruit, flowers and musical instruments, and Gibbons himself seems to have favoured limewood. The most fashionable tables, chairs, stands and looking-glasses of this period generally have richly carved legs, stretchers and frames.

The Rococo style, at its height in the middle years of the eighteenth century, was a style superbly suited to the art of the carver. Furniture increasingly became an integral part of the interior scheme of decoration, so that the carving on a chair might reflect that on the panelling behind.

Carvers working in the nineteenth century were able to use a wider range of woods than their predecessors. They produced some remarkable results, such as the "naturalistic" furniture carved with highly crowded allegorical scenes.

In this century, the penchant for clean lines and pure form, and the fact that much furniture is mass produced, has brought about a marked decline in the art of carving.

HIGH-RELIEF CARVING
Not only furniture was embellished with rich carvings during the 18th century. Made of hard woods able to be carved deeply, this 18th-century mahogany and oak staircase shows the influence of Grinling Gibbons

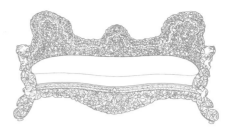

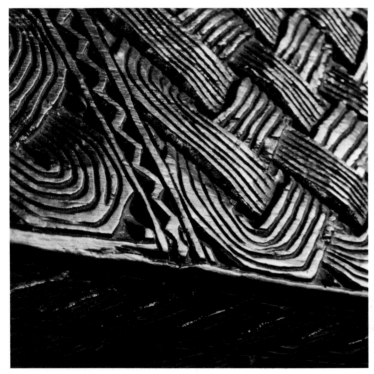

AFRICAN CARVING
The African wood-carver is often one of the important members of society, and for centuries has produced fine works of art. This 19th-century chest from Benin is roughly but effectively carved in imitation of woven basketwork

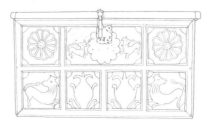

SOFTWOOD CARVING
The carved decoration of soft woods, due to their loose grain, is necessarily shallow. The pine furniture of New Mexico, like this chest, was decorated with simple scratch and gouge carving carried out with a V-shaped chisel

Wood-turning

The ancient craft of wood-turning was almost certainly developed in Egypt, where decorative turning on the legs of chairs and stools reached a degree of sophistication and high craftsmanship not surpassed until the early seventeenth century.

In the simplest form of lathe, the wood is held between two pointed centres and is revolved by means of a cord looped round the wood and attached to a bow worked either by hand or by foot. The wood can only be cut on one stroke, the return movement serving only to spin the wood back in readiness for the next power stroke.

The pole lathe, widely adopted by folk-furniture makers in many parts of the world, is an extension of the bow concept; the wood is again spun by a cord, but the lower end is attached to a treadle below the lathe and the upper end to a long springy pole.

The main tools used by the wood-turner are the gouge and chisel, and various forms of scraping tools. The chisels, bevelled on both faces, and the gouges, are strictly cutting tools and are always used in turning soft wood when the waste wood is removed in long continuous shavings. Very hard, and dense, woods require a different technique and are usually scraped, with the edge of the tool making contact near the centre-line of the spinning wood.

In Europe, the late sixteenth and early seventeenth centuries saw a renewed interest in wood-turning, not only for the legs of tables and chairs, but also as applied decoration in the form of split turning. The wood to be turned was first cut along its length and then glued back together with a layer of paper between the two halves. The wood was then turned and the pieces separated to give matching ornaments used to great effect on the fronts of cabinets.

Bulbous table legs and ornamental vases would have involved tremendous waste of valuable wood had they been turned from the solid and, for these pieces, a composite block was first built up to the required diameter – great care being taken to ensure that the grain directions of the individual elements were aligned to avoid the risk of the wood splitting during turning.

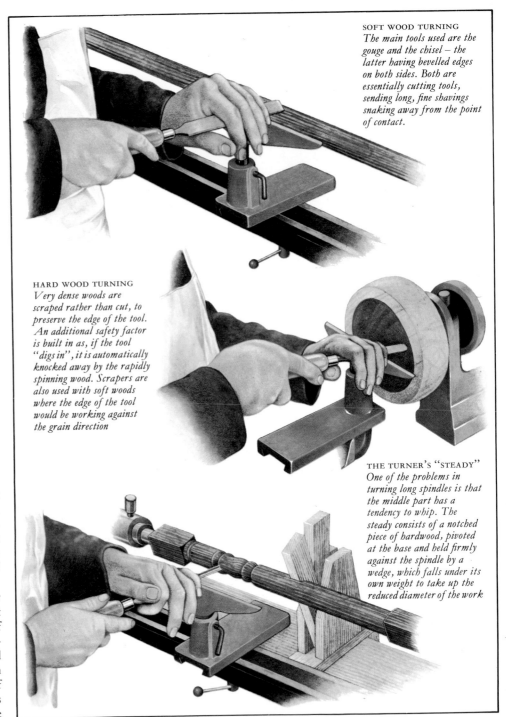

SOFT WOOD TURNING
The main tools used are the gouge and the chisel – the latter having bevelled edges on both sides. Both are essentially cutting tools, sending long, fine shavings snaking away from the point of contact.

HARD WOOD TURNING
Very dense woods are scraped rather than cut, to preserve the edge of the tool. An additional safety factor is built in as, if the tool "digs in", it is automatically knocked away by the rapidly spinning wood. Scrapers are also used with soft woods where the edge of the tool would be working against the grain direction

THE TURNER'S "STEADY"
One of the problems in turning long spindles is that the middle part has a tendency to whip. The steady consists of a notched piece of hardwood, pivoted at the base and held firmly against the spindle by a wedge, which falls under its own weight to take up the reduced diameter of the work

ERROR OF JUDGEMENT
The chair leg illustrated, left, shows an error common to the beginner; the wood has been cut away too deeply at the "necks", weakening the leg

DECORATIVE TURNING
Though wood-turning is known from pre-Roman times, the techniques were not fully developed until the 17th century. Chair legs of the mid-17th century were often partly turned but invariably retained large sections of squared timber to take the tenons of the heavy cross-rails. Later developments in style and technique brought in much lighter legs with more sophisticated curves and the squared sections were greatly reduced. Decoration took the form of turned ridges and grooves

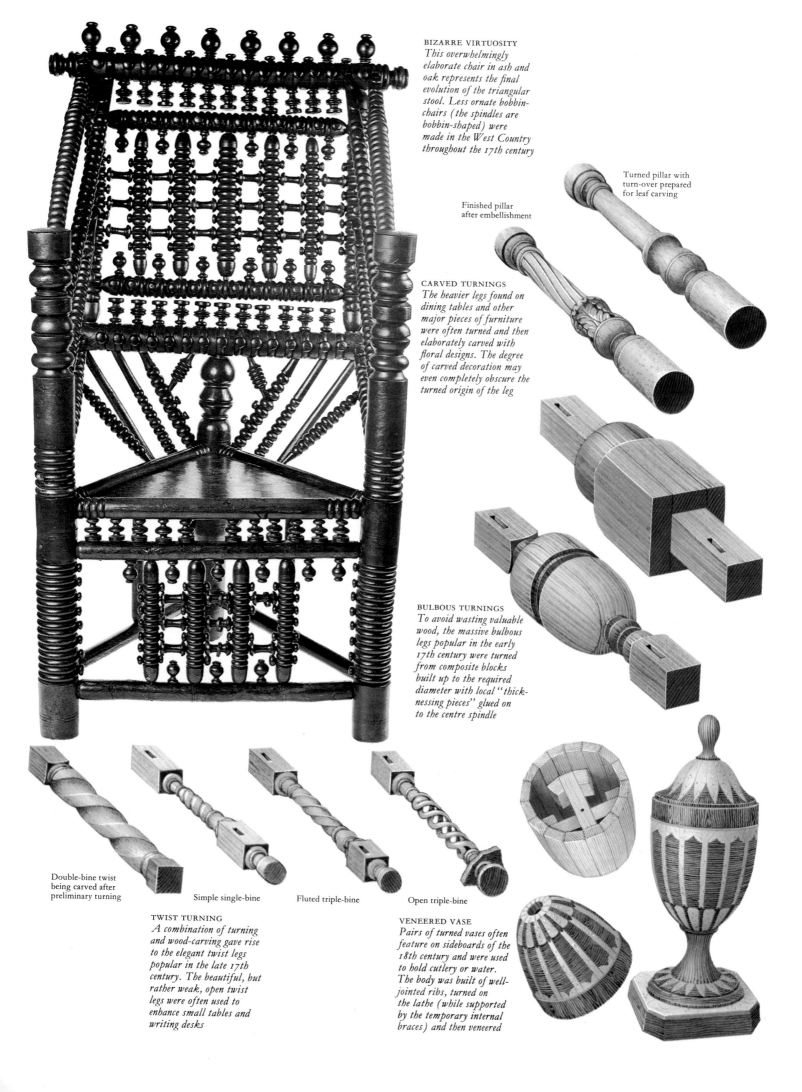

BIZARRE VIRTUOSITY
This overwhelmingly elaborate chair in ash and oak represents the final evolution of the triangular stool. Less ornate bobbin-chairs (the spindles are bobbin-shaped) were made in the West Country throughout the 17th century

Turned pillar with turn-over prepared for leaf carving

Finished pillar after embellishment

CARVED TURNINGS
The heavier legs found on dining tables and other major pieces of furniture were often turned and then elaborately carved with floral designs. The degree of carved decoration may even completely obscure the turned origin of the leg

BULBOUS TURNINGS
To avoid wasting valuable wood, the massive bulbous legs popular in the early 17th century were turned from composite blocks built up to the required diameter with local "thick-nessing pieces" glued on to the centre spindle

Double-bine twist being carved after preliminary turning

Simple single-bine

Fluted triple-bine

Open triple-bine

TWIST TURNING
A combination of turning and wood-carving gave rise to the elegant twist legs popular in the late 17th century. The beautiful, but rather weak, open twist legs were often used to enhance small tables and writing desks

VENEERED VASE
Pairs of turned vases often feature on sideboards of the 18th century and were used to hold cutlery or water. The body was built of well-jointed ribs, turned on the lathe (while supported by the temporary internal braces) and then veneered

The Chair Bodger

BEECHWOOD BLANKS
The beechwood logs are sawn into suitable lengths and cleft into wedges. The bark is removed and the knotty inner core-wood, which would split on the lathe, is cut away. The wood is finally trimmed to a five-sided billet with the axe

Until the Elizabethan period, only those of high social rank – the court, lords and senior clergy – would have possessed chairs; those of lesser rank had to be content with stout oak benches and stools.

Gradually, during the sixteenth century, the stool was developed and a back-rest was added. These new pieces of furniture, tactfully known as backstools because the nobility still considered chairs to be the prerogative of their class, were the forerunners of the Windsor chair – the first chair to become readily available to the cottager. The Windsor chair differs from other folk furniture in its basic construction: instead of a heavy wooden frame of squared timber, the legs and back supports are turned and are set into holes bored in the elm seat.

Inevitably, as the demand for chairs became more widespread during the eighteenth century, a folk industry grew up in those country towns surrounded by woodland. In England, the town of High Wycombe, in the beechwoods of Buckinghamshire, became such a centre – and to this day retains an important furniture-manufacturing industry. From the very beginning, Windsor chairs were made by a form of mass production. Specialist craftsmen made the individual parts, which were then assembled in cottage workshops and, later, in factories. Many of the craftsmen worked in their own homes, but the "bodgers", the men who made the turned legs and stretchers for the chairs, overcame the problem of transporting their raw materials by erecting temporary workshops in the forests where they lived and worked.

The bodgers usually worked in pairs. A "stand" or "fall" of timber would be bought by the senior man and felled; the logs would then be converted to plank for the seats, an arduous task using the heavy, seven-foot pitsaw, or sawn into short logs and cleaved into billets ready for turning.

With the introduction of fast and efficient power lathes, the days of the chair bodger were numbered. Initially his superior-quality craftsmanship ensured a ready market, but as factory techniques improved so his share of the market declined, and today few representatives survive of this once-thriving industry.

TRADITIONAL WOVEN FURNITURE
The manufacture of woven furniture is closely allied to the ancient craft of basket-weaving. The most commonly used material is the osier, or willow rod, and the chair may have a frame of stout willow rods or be entirely of woven construction. Willow furniture is extremely light and, due to the natural flexibility of the material, comfortable and durable

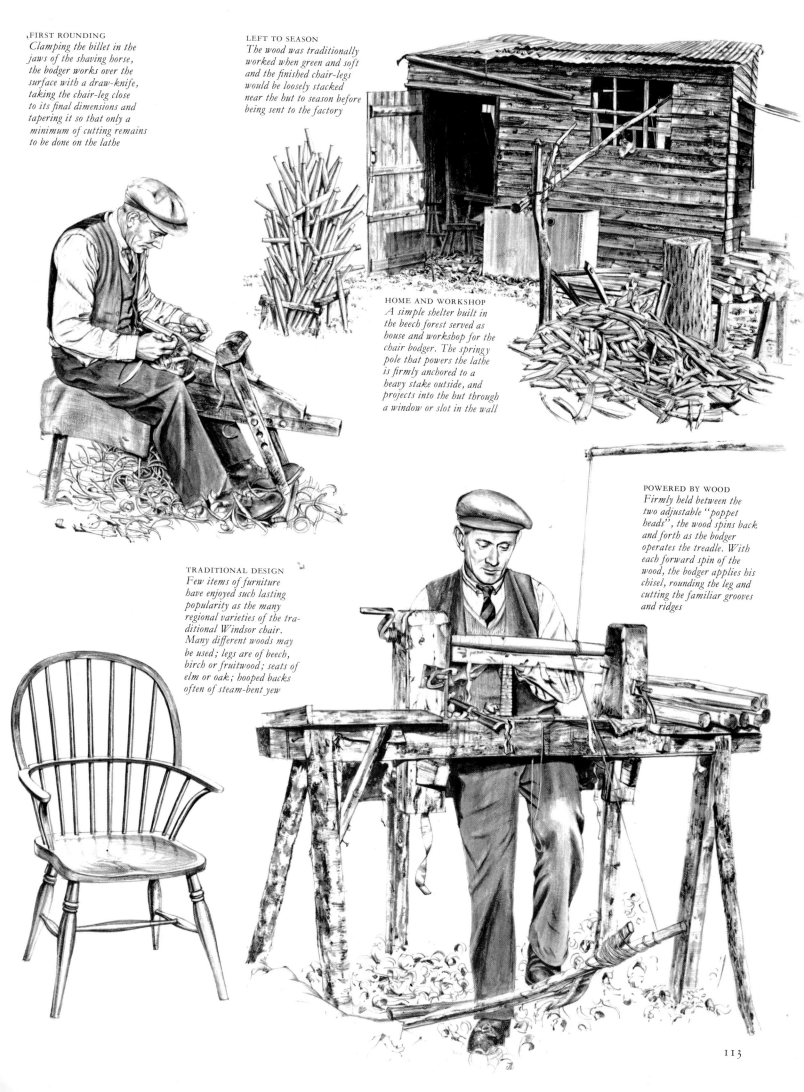

FIRST ROUNDING
Clamping the billet in the jaws of the shaving horse, the bodger works over the surface with a draw-knife, taking the chair-leg close to its final dimensions and tapering it so that only a minimum of cutting remains to be done on the lathe

LEFT TO SEASON
The wood was traditionally worked when green and soft and the finished chair-legs would be loosely stacked near the hut to season before being sent to the factory

HOME AND WORKSHOP
A simple shelter built in the beech forest served as house and workshop for the chair bodger. The springy pole that powers the lathe is firmly anchored to a heavy stake outside, and projects into the hut through a window or slot in the wall

POWERED BY WOOD
Firmly held between the two adjustable "poppet heads", the wood spins back and forth as the bodger operates the treadle. With each forward spin of the wood, the bodger applies his chisel, rounding the leg and cutting the familiar grooves and ridges

TRADITIONAL DESIGN
Few items of furniture have enjoyed such lasting popularity as the many regional varieties of the traditional Windsor chair. Many different woods may be used; legs are of beech, birch or fruitwood; seats of elm or oak; hooped backs often of steam-bent yew

Formal Harmony

The traditional Japanese house, made of "wood, paper and bamboo", is by Western standards sparsely furnished. The Japanese have little interest in beds and chairs, preferring to sleep and sit on the floor, and as a result any furniture which is produced tends to be low and small in scale.

The walls usually consist of nothing but a wooden framework covered with painted paper panels, which can be slid from one position to another, thus creating rooms of different shapes and sizes, as no room has any one specific function. The Tokonoma, or picture recess, is a feature typical of the Japanese house in which is placed a picture, decorative wooden ornament, bonsai tree or flower arrangement, the display being frequently changed.

There is no shortage of timber in Japan, so that wood is one of the most important materials used for building houses. Unpainted wood is often employed, as the Japanese like to feel close to nature. Sugi-wood, a soft, reddish-brown cedar, is usually used for ceiling panelling. Furniture is often made from mulberry wood or Paulownia, the latter being a pleasant silvery colour and very resistant to the damp climate. Among other woods seen in Japanese houses are ebony and ironwood and Indo-Chinese woods such as sandalwood and red sandalwood which are highly prized.

Japanese rooms may be uncluttered but they are not austere, for the Japanese are masters at using natural light, colour and materials with subtlety and sensitivity.

NATURAL WOOD
Narrow wooden battens are hung on slats from the rafters to form the traditional ceiling. At right angles to the battens are laid panels, often of sugi-wood. Woven strips of wood or rushes with bamboo supports may also be used

FITTED CUPBOARDS
The many built-in cupboards eliminate the necessity for movable furniture. Cupboards usually have sliding wooden doors, while a more modern arrangement is a whole wall fitted with drawers, which are often made from kiri-wood

MATS AND BOARDS
Floors are raised above the ground, leaving a space underneath. For warmth, thick rice-straw mats are laid down, their standard size determining the room size. Where floors remain uncovered, hinoki-wood boards are used

FIXED WALLS
Immovable walls are constructed with bamboo lattice-work strengthened by pierced uprights. The bamboo is covered on both sides with mud and straw plaster, making a wall about 2in thick. The surface is left unpainted

WOODEN BEAMS
The threshold and frame of the Tokonoma are usually emphasized by using a wood different from the rest of the room. The natural trunk of the sugi tree is often used, either stripped of its bark and polished, or left untreated

THE TOKONOMA
The picture recess was developed from the tradition of Buddhist pictures hung over low tables. Gradually a special recess was built, and secular hangings replaced religious ones. The floor of the raised platform is either matting or of plain hardwood

PAPER WINDOWS
To a room with little or no decoration, the diffused light obtained from the use of translucent paper instead of glass provides a soft and peaceful atmosphere. As they are not air-tight, permanent natural ventilation keeps the room fresh and cool

Furniture of the East

Although the mythological joiners P'an Ku and Lu Pan managed to achieve divine status, cabinet-making was not considered an important art in China. The Chinese were, however, skilled wood-workers from very early times, and by the Han dynasty (206 BC– AD 221) the production of household furniture was well established.

China was reasonably rich in indigenous woods, but timber was also imported, much of it from the Philippines and the East Indies. Wood was expensive nevertheless, and because labour was cheap and plentiful old pieces of discarded or broken-down furniture were often "reclaimed" and re-employed on other pieces. The most highly prized woods used for furniture in China were the hardwoods hua-li mu, huang-hau-li and tzu-t'an, which are broadly speaking varieties of rosewood (*Pterocarpus indicus*). Tzu-t'an translated literally means purple sandalwood, and was a wood much favoured by Kublai Khan (1216–94), who had it imported for the building of his palace halls.

Another wood valued by the Chinese has the delightful name of chicken-wing wood, which looks somewhat like satinwood, though it matures to a much deeper brown. The grain is rather coarse but, to the cabinet-maker, it has the advantage of being tough and durable. Among other woods used by the Chinese for furniture are nan-mu, classified as cedar, yü-mu (elm), much used for strictly functional and practical pieces, hsiang-sha-mu (fragrant pine) and chang-mu, or camphor-wood, which, as in the West, was a favourite wood for the making of high-quality chests.

Joinery is the outstanding aspect of Chinese cabinet-making. Nails were never used, dowels rarely, and glue was abhorred. From 400 BC an elaborate mortise-and-tenon joint, combined with a mitre, and occasionally a dovetail joint, was used. The hardwoods used were susceptible to expansion and contraction, therefore "floating" panels were essential to avoid distortion of the furniture.

Because Chinese traditions and social customs differ fundamentally from those in the West, they have evolved many types of furniture that have no counterpart in Europe or America. Tables of all shapes and sizes were made to serve a wide variety of purposes; chairs also differed greatly in design, some having a box-like frame, others a deeply curved back, while others were made with cane-woven seats with palm-fibre under-webbing. Ornamentation included elaborate fretwork design, lacquer, paint or inlay with precious materials. The Chinese made an enormous number of different kinds of chests and cupboards, and wherever possible seem to have preferred to store their belongings away rather than leave them out.

The Japanese, unique in major civilizations, have never developed furniture to any extent. Preferring always to sit on the floor, even the few low tables or boxes they do have are stored away in fitted cupboards when not in use. The low table, with folding legs, was used for eating, and because it was never covered, the quality of the wood from which it was made was of aesthetic importance. The wood used for chests in which the kimono was carefully folded, or manuscripts rolled, was that of the pawlonia tree, which swells in humidity, thus preventing damp from reaching the precious contents.

ROSEWOOD ARMCHAIR
Of all the timbers used by Chinese cabinet makers, rosewood was the most highly regarded. The seats of this type of chair were usually made of woven soft cane or palm fibres, and the separate movable frames were of hardwood, secured by wooden pins

SIMULATED BAMBOO
Though bamboo has generally been used for more expendable furniture in China, its shapes and textures often appear in carved decoration on more expensive hardwood furniture. This set of occasional tables, with carved legs and borders, dates from the 1800s

K'ANG TABLE
Dating from the Han dynasty (206 BC to AD 221), and originally reading or writing tables, these items of furniture were placed on the k'ang – a raised platform used for sleeping or simply reclining at leisure

KIMONO RACK
The traditional kimono rack has been used for centuries in Japan. This elegant item of furniture is gilt-laquered and the joints and ends are ornamented with chased metal-work. To ensure stability, a very heavy wood must always be used for the base

CHINESE CUPBOARD
Clothing, working materials, crockery and other household goods were all stored out of sight in cupboards and chests. This beautifully figured rosewood cupboard, with brass hinges and handles, was made in the 17th century

INTERLOCKING CONSTRUCTION

Countless centuries of experience have given the craftsmen of China a unique mastery of the ubiquitous bamboo. These chairs, made entirely of bamboo, are constructed without recourse to nails or glues; the main framing members are notched and tapered to fit together like the pieces of a puzzle and despite continuous use hold firm indefinitely. The seats are commonly of slender bamboo slats or of interwoven rushes or reeds

Veneering

Before the 19th century, veneers would have been cut with a hand saw. This illustration, taken from L'art du menuisier ebeniste, dated 1774, shows two craftsmen cutting veneer with a multiple-bladed frame saw. The log, specially chosen for the beauty of its grain, would be held upright in a strong screw vice fixed to a workbench. As the cut progressed, the log would be raised in the vice and the ends of the veneers tied together to prevent them splitting

Though Pliny dismissed it scornfully as "bestowing upon the more common woods a bark of a higher price", the art of veneering has been a highly skilled and popular way of utilizing the decorative qualities of wood from pre-Egyptian times right up to the present day.

Veneering consists of gluing a thin layer of an attractively figured wood on to a groundwork of plain but structurally stable wood. The technique allows more economical use of the expensive decorative woods and also leaves the craftsman free to create composite patterns by quartering or by setting burr or curl veneers into sheets of contrasting wood.

The earliest veneers were cut by hand using a large multiple-blade saw. By the beginning of the nineteenth century, power saws were available and the veneers were cut on a saw bench with large diameter circular saws. Modern veneers are much thinner than their predecessors and are generally sliced from the log on the flat, quarter, or half-round. Veneers may also be rotary cut into a continuous sheet by rotating the whole log against the knife, but rotary-cut veneers show a rather wild and unnatural grain and are generally used only for plywood.

Furniture made before the turn of the century was usually veneered on a groundwork of yellow pine, plain mahogany or oak. In some cases, strips of wood about two inches wide were joined side by side with the heart-wood, facing alternately up and down to counteract the natural tendency to warp. Today, veneer is usually laid on one of the man-made boards, plywood, laminboard or chipboard, as these are readily available in large sizes and are stable in use.

Veneered boards are now mass-produced in multi-plate presses in which thermosetting adhesives are cured in minutes by the application of heat and pressure. The older types of hand-operated screw presses are, however, still used in small workshops and for some high-quality work. Though the age of technology has brought about many changes, the ancient veneering crafts are still to be seen in specialist workshops where antique furniture is restored.

To achieve the convex and concave shapes on articles such as this 18th-century burr walnut commode from the Netherlands, it was necessary to dampen the veneer sheet and press it to the required shape with the aid of hot sandbags

For shaped work, matching cauls are used to ensure an even spread of glue between veneer and base

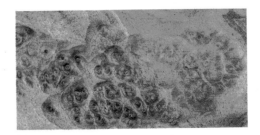

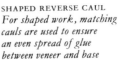

These most highly prized veneers are cut from the end-grain of irregular outgrowths found on the trunks of many trees. They are commonly cut from elm, oak, ash, walnut (above) and other hardwoods. Length of sample – 8in

Veneer from the junction of a branch and the main trunk gives the attractive curl figure shown by the mahogany sample above. Beautiful figured veneers are also cut from the main root members of some trees. Length of sample – 22in

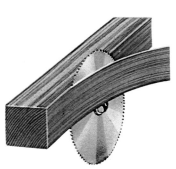

19th-century veneers were cut with circular saws – a highly skilled, but wasteful, operation Thickness: 1.1 to 1.6mm

Flat-cut sliced veneer Thickness: 0.7 to 0.9mm

Half-round sliced veneer Thickness: 0.7 to 0.9mm

Rotary-cut veneer Thickness: 0.3 to 9.5mm

CAUL-LAID VENEER

A flat wooden plate, faced with zinc, is clamped tight on to the glued veneer on its groundwork. The caul is preheated to keep the glue liquid and allow it to spread evenly as pressure is applied to the clamps

SINGLE VENEER

When veneer is laid on one side of the groundwork only, it tends to shrink and pull the board into a concave shape. It is therefore laid on the heartwood side to counter the wood's natural tendency to warp on drying

COUNTER VENEER

The best work is always counter veneered – the first layer having its grain running across that of the groundwork; the final veneer lying parallel to it. Both sides of the composite groundwork are counter veneered to produce the absolute stability needed

HAMMER VENEERING

In traditional hammer work both veneer and ground are glued. The work is then dampened, heated with an iron and gently beaten down with zig-zag movements of the hammer, starting at the centre of the work

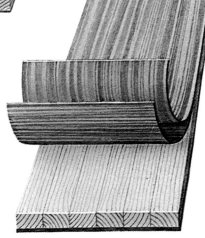

SLICED VENEER

The half-log is softened in water prior to cutting. Slices taken across the crown of the log expose the marbled heartwood, sometimes with contrasting pale sapwood at the margins. Sample is of Rio rosewood. Length of sample – 18in

QUARTER-CUT VENEER

After softening by soaking or steaming, the quarter-log "flitch" is mounted on a movable carriage which brings the flitch across a fixed knife-blade at a shallow shearing angle. The sample above is sapele. Length of sample – 15in

ROTARY-CUT VENEER

Softened, and "rounded-up" on a debarking lathe, the whole log is brought against the knife and the veneer cut in a continuous sheet. "Bird's-eye" maple (above) is one of the few woods rotary-cut for decorative use. Length of sample – 15in

The Techniques of Inlay

MARQUETRY

This delicate form of veneer inlay was first introduced into Britain from Europe at the restoration of Charles II. Since the design is cut out with a very fine-bladed saw, the technique allows scope for the creation of intricate and elaborate patterns. Traditional marquetry was cut on an appliance known as a donkey; the frame of the saw was moved horizontally and the device incorporated a simple foot-operated clamp with which the veneer was held steady. The individual pieces are glued, placed on a groundwork, and then clamped firmly in a caul until the glue sets.

PRICKING OUT
The outline of the master-copy of the design is first pricked out with fine holes. Bitumen powder can then be dusted through on to additional paper sheets and heated to give any number of identical copies

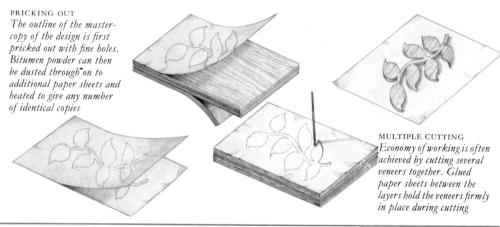

MULTIPLE CUTTING
Economy of working is often achieved by cutting several veneers together. Glued paper sheets between the layers hold the veneers firmly in place during cutting

PARQUETRY

The characteristic feature of parquetry is its regular, repeating pattern, built up either of thin veneer for use in furniture or from thicker blocks for use in decorative flooring. The shapes are invariably geometric and are usually designed so that the veneers can be cut into lengths, assembled side by side, and then cross cut so that, by sliding one strip against the next, a variegated "chequer-board" pattern can be created. Paper sheets, glued over the surface of the blocks, hold the individual elements of the design in place while they are pressed in the caul.

PREPARING THE BLOCKS
Two contrasting woods are chosen and cut in standard-width strips. The strips are then laid side by side, alternately, and held in place with glued paper

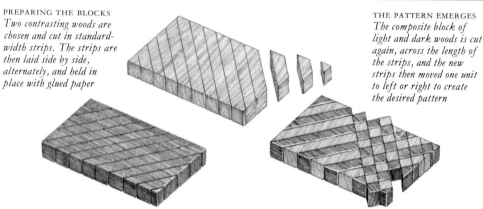

THE PATTERN EMERGES
The composite block of light and dark woods is cut again, across the length of the strips, and the new strips then moved one unit to left or right to create the desired pattern

INLAY AND BANDING

The art of inlay consists of recessing a number of contrasting woods into a ground-work of solid wood. Since the recesses must be cut out with gouges, the degree of detail is limited. The design may be first cut out of veneer, using a fine fret-saw, and the outline then marked on the baseboard; alternatively, the rebates may be cut and the design transferred to the veneer using a paper rubbing. Bindings and strings are a specialized form of composite inlay work often used to ornament the edges of large pieces of furniture like tables, desks and cabinets.

RECESSED GROUNDWORK
The design is either drawn on the wood or on to paper glued over the surface. Recesses, to take the inset woods, are first cut out with chisel and gouge and are finished with a router

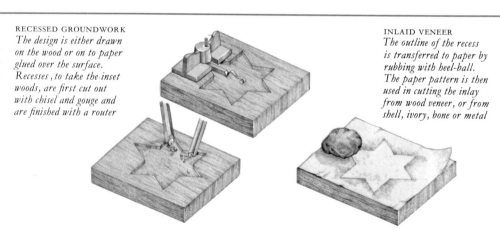

INLAID VENEER
The outline of the recess is transferred to paper by rubbing with heel-ball. The paper pattern is then used in cutting the inlay from wood veneer, or from shell, ivory, bone or metal

TUNBRIDGE-WARE

This popular form of wood-mosaic decoration was developed during the nineteenth century in the English town of Tunbridge Wells. The mosaic is built up of small, square-section rods of differently coloured woods to create the desired pattern or picture. The rods are glued and then tightly bound until set into a solid block from which veneers as fine as one-sixteenth of an inch may be cut. Tunbridge-ware was very popular with the Victorians as a decoration on tables, desks, work-boxes, snuff-boxes and other trinkets.

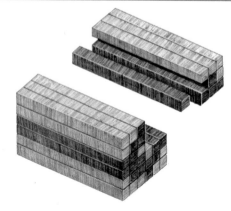

PICTORIAL MOSAIC
The individual elements in the mosaic consist of the ends of dozens of slender wooden rods; the colour may be natural or have been altered by staining or by heating the rods. The grain of the rods usually lies parallel to the face of the mosaic to facilitate clean cutting of the veneers

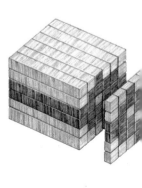

THE DONKEY
The traditional marquetry-cutter's donkey incorporates a simple form of foot-operated clamp with which the work is held steady. The fine, horizontal saw-blade has some lateral movement, but curves are negotiated by moving the veneer

FINAL ASSEMBLY
The various elements of the design, perhaps consisting of a number of pieces cut from contrasting woods, are then assembled on a paper sheet and carefully marked out. This second pattern is used, as the first, to cut the background veneer

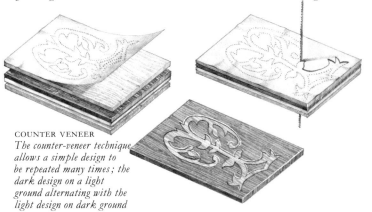

COUNTER VENEER
The counter-veneer technique allows a simple design to be repeated many times; the dark design on a light ground alternating with the light design on dark ground

HAND-SET DESIGNS
Many parquetry designs cannot be created by the strip method and must be built up by hand from separate blocks which have been machine-cut to standard sizes. Floor coverings are frequently built up by this method

HERRING-BONE BANDS
Two layers of wood slices, with their grain directions forming an acute angle, are placed between veneers to form this popular band

BANDS AND STRINGS
A wide variety of banding and string decoration may be made from blocks of wood "sandwiched" between veneer sheets. Strips about two millimetres thick are cut from the edge of the composite block using a fine-bladed circular saw

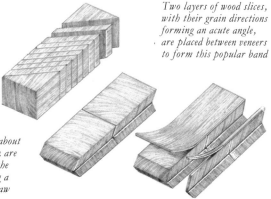

TRADITIONAL SAW-BENCH
This curious hand-driven circular saw is, with the exception of the blade, made entirely of wood. It was used for cutting the fine rods for use in Tunbridge-ware and probably also for making bandings

Decorative Inlay

The art of inlaying developed enormously in the fourteenth century, particularly in Italy, where this type of decoration, known as intarsia, was used extensively on such articles as boxes and coffers. At first, contrasts were achieved simply by using light woods like boxwood or spindle, together with dark woods like ebony and walnut, but by the end of the fifteenth century, shading effects were obtained by staining or by blackening the wood with a hot iron. During the second half of the fifteenth century, the *intarsiatori* produced some remarkable panels depicting architectural ruins or purely geometrical patterns – a result of their new understanding of the rules of perspective. Landscapes, still-lifes, animals and *trompe-l'oeil* effects were also produced at this time.

Elaborate marquetry became a distinctive feature of sixteenth-century German furniture and such pieces as cabinets and dressers were decorated with complex compositions of beasts, birds, strap-work, classical columns and numerous other fanciful details. In Holland and England there was a fashion in the latter half of the seventeenth century for the brightly coloured floral marquetry that is seen on so many cabinets, long-case clocks, tables and looking-glass frames made in this period. John Evelyn mentions in his book *Sylva*, published in 1664, that acacia and rosewood were used for yellows and reds, "with several others brought from both Indies". By the early eighteenth century this use of brilliant colour had given way to more subtle contrasts.

The French *ébénistes* (cabinet-makers – a name which lingered from the seventeenth-century vogue for ebony) were superb *marqueteurs*, and their furniture is typically decorated with marquetry combining geometric cubes or a trellis pattern with floral or pictorial designs. Among the finest examples of marquetry ever produced are those by the German cabinet-maker David Roentgen (1743–1807), who developed a fluid and painterly style. By this time, and indeed throughout the following century, a wide variety of woods were used in marquetry decoration – rosewood, palisander, mahogany, olive and boxwood being but a few.

THE NONESUCH CHEST
*By the end of the 16th
century, inlay was prac-
tised all over Europe. The
so-called Nonesuch chests
were inlaid with formal
designs supposed to represent
Henry VIII's Palace of
Nonesuch, demolished in
the 17th century*

MODERN MARQUETRY
Marquetry adapts itself to all styles of art. This panel showing a girl in fancy dress is art deco, probably Portuguese, inlaid with more than ten different woods. Today, marquetry is an extremely popular hobby

SCENES IN MOSAIC
A Victorian jewel cabinet is decorated with two typical Tunbridge-ware designs. The scene on the lid depicts the ruins of Beynham Abbey, while the doors are embellished with a floral pattern imitating the popular Berlin wool embroidery

FLORAL MARQUETRY
A detail from a 1670 English box showing the bright colours fashionable then. Walnut is inlaid with woods which were scorched to give shaded effects, or stained to give green leaves. Green could also be obtained from certain diseased woods

OPTICAL ILLUSIONS
On this mid-18th-century Dutch chest of drawers, the cube-like effect of the parquetry, a popular feature, is achieved by using different woods, here walnut and box-wood, their grains running in opposite directions

123

An Age of Discovery

The discovery of new lands during the great period of European expansion often led to a productive mingling of indigenous artistic tradition with well-established European ideas of comfort and beauty in furniture. New timbers were discovered and shipped home, creating important timber trades, often the bases of successful colonies. New woods influenced furniture produced in the home country, while immigrant craftsmen imported and adapted their traditions and styles, so creating a general interaction of design and custom.

In Europe itself, as early as the eighth century, the Moors, renowned masters of woodwork, had invaded Spain. They imposed their styles on architecture and furniture, in particular the mudejar style of wood panelling. These geometrical patterns, often of contrasting woods, sometimes inlaid, were later further transferred to furniture and interiors in the Spanish colonies in Central and South America. But the most important aspect of the story of wood in South America in undoubtedly that of mahogany. Growing in the West Indies and in Central America down to northern Colombia and Venezuela, this remarkable wood was first noted in 1595 by a carpenter travelling with Sir Walter Raleigh. At first it was used generally for ship-building and house-building, appreciated for the fact that it did not warp, and its size enabled vast widths to be cut. Not until the late eighteenth century, however, did it have such a profound effect on the furniture of elegant Europe, when it was realized that its close grain allowed the carving of, for example, narrow tapered legs.

Another wood popular with furniture-makers in South America was cedar; Cobo, a Spanish naturalist working there in the early 1800s, wrote: "Almost all the curious and lasting things made in this land are of cedar, such as . . . cabinets, chests and a thousand other things." Precious woods were brought home and used in marquetry, while exotic woods like jacaranda and pausanto replaced oak and walnut in the construction of Portuguese chairs. The sheer exuberance of the carving so characteristic of Spanish-Peruvian Baroque furniture demonstrates a refreshing mingling of styles, Inca and Indian elements like the sun and snake motif being woven into the European Baroque idiom.

While Spain and Portugal were involved in South America, more Portuguese adventurers were revealing the delights of the East. They settled first along the west coast of India, later to be followed by the Dutch and English. Like most eastern societies, the Indians used little furniture themselves, though there is much architectural evidence to prove their skill as craftsmen in wood. Under European influence, however, the Indian wood-carvers adapted their expertise to produce furniture. They were sent to Lisbon to learn the trade, a policy which resulted in one of the more lasting mixtures of styles, the Indo-Portuguese. By the end of the sixteenth century, towns like Goa and Malabar had become flourishing cities full of furniture workshops, using sal, teak, ebony and other indigenous woods.

The earliest North American colonial furniture dates from the late seventeenth century, and is so close to English early seventeenth-century examples that it is described as Jacobean. As with all colonial furniture, there is an element of time lag between the original style in the home country and its becoming popular in the new colony. Americans followed particularly the English fashions, though Dutch and German examples, especially in Pennsylvania, are still in use. Though in England high-quality furniture was invariably made in London, in America recognizable regional styles soon became apparent, and were to a considerable extent influenced by the local timbers available to the cabinet-maker. In the south, magnolia and tulipwood were used, while maple, its curly grain attracting more urbane tastes, was favoured in the north.

In Quebec in Canada, French styles, especially the more provincial ones, were adapted to the readily available pine and birch to produce some of the most sophisticated rustic furniture of the colonies.

"BEAUTY RESTS ON UTILITY" — SHAKER MAXIM

Immigrants to America included the Shaker community in 1774. Their furniture, like their way of life, was modelled on ideals of simplicity and functionalism. A pine pegboard for coats and furniture ran round the room. The famous slat-back chair, produced commercially, was made of maple, pine or beech, and the oval hatbox (right) was of maple slats, coiled round and secured by small hardwood pegs

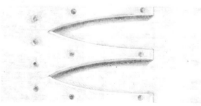

CABINET OF THE GODS
This gilded and painted soft-wood bureau from Peru was made by Indian craftsmen under Spanish tutelage in the late 1600s. Inca motifs decorate its European shape – the feet are the claws of the sacred Condor

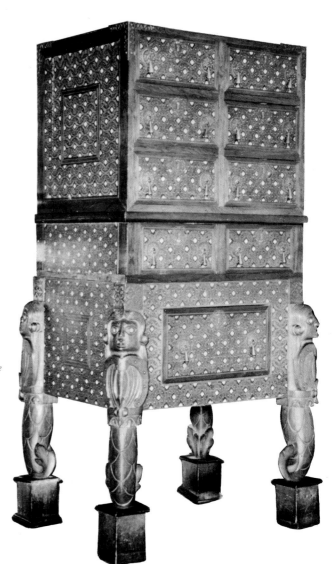

INDO-PORTUGUESE
A perfect example of the blend of Indian and Portuguese styles, this mahogany chest is inlaid with rosewood, ebony and ivory. The design is typical of the output of the furniture workshops in Goa, which excelled under the Portuguese in the 1600s

SUNFLOWER CHEST
One of the first truly "American" pieces of furniture, this type of oak and pine lidded chest with drawers was made in the late 17th century in Connecticut. Its name refers to the low-relief carved flowers on the central panel

PORTUGUESE INFLUENCE
Carved in India in the 19th century, this rosewood cabinet combines pierced and relief work. Made for westernized Parsees, this type of furniture is said to resemble articles from Brazil – another early Portuguese colony

Staining and Polishing

OAK: POLYURETHANE
*Finishes serve to bring out
the beauty of the grain and
figure of the wood, and to
prevent the ingress of dirt.
Matt or gloss polyurethanes
produce a hard and durable
surface, and are especially
water-resistant; they tend
however to darken with age*

It must have become obvious quite early in Man's use of wood that some form of finish was desirable, partly to seal the grain against the atmosphere and so limit the tendency to warp and shrink; to protect the surface from becoming marked or dirty, especially on items constantly being handled; and to enhance the beauty of the grain.

Early oak furniture appears to have been finished in various ways. When left in its natural state oak fades to a rather sombre colour, but some dark items were probably stained or darkened by the application of liquid ammonia, or by exposure to ammonia fumes. In both cases the surface was afterwards given several coats of a drying oil such as linseed, either in its natural state or darkened by the immersion of bruised alkanet root in it.

Walnut was also given a preliminary oiling, generally followed by varnishing, and several successive coats were applied, each of which was rubbed down with a fine abrasive powder applied with a felt rubber. Afterwards, constant rubbing with wax polish built up a fine protective sheen. Mahogany had much the same treatment, but again the oil was sometimes darkened with alkanet root. Early mahogany was of the dark Cuban variety and was probably treated with clear linseed oil, but later eighteenth-century mahogany was more commonly of the lighter Honduras species and, although oil was used, it was sometimes clear – accounting for the soft honey colour of many items. Sheraton, in his *Cabinet Dictionary*, states that mahogany was finished by rubbing brickdust and oil over the surface with cork, but his writing makes no mention of varnish.

Any really hard wood takes on a natural sheen when burnished by rubbing a hard, smooth tool over it. Carvings frequently show this feature as the bevel of the carving tool gives a natural burnish as it follows the cutting edge. It is quite possible that woodworkers noted this early on and burnished other parts to produce a natural sheen without recourse to oils and polishes.

French polishing, based on the use of shellac, was first used in the early nineteenth century and soon became one of the most popular and universal finishes. It is a highly skilled process in which the polish is applied with a pad of cotton wool encased in a cloth cover, the polish being gradually built up by the movement of the rubber over the semi-hard polish beneath.

Black polish is also used in the "ebonizing" process. In this, the wood is first stained black, the grain is filled in with a black filler, and the surface finished with black polish. Originally the black used to darken the polish was made by suspending a piece of tin-plate above a gas burner, producing a sooty deposit which was scraped off and mixed with the polish. Today, black aniline dye is used instead. A major disadvantage of all French polishes, however, is that they are easily marked by heat, water or spirit.

Although most period walnut and mahogany furniture was finished in natural colour, except for the use of oil, much Victorian and later furniture was darkened with stains or chemicals. Of the former, Vandyke crystals gave a basic stain imparting a rather cold brown shade, dark or light according to strength. Ammonia was usually added before use as it helped to drive the stain into the grain, so deepening the shade. Of the chemicals, bichromate of potash was used chiefly for mahogany, which it turned a rather warm brown shade, deepening almost to black depending upon the strength of the chemical and the variety of the mahogany.

The economic pressures of fast production-line manufacture, and public demand for hard-wearing, trouble-free finishes, has led to the development of a number of very durable synthetic finishes. Resistant to scratching and heat, and to staining by damp, cellulose, polyurethane and polyester finishes have now largely superseded French polishing. Most of these finishes can be quickly applied either by brushing or spraying and require only a fraction of the time and effort involved in hand-finishing work. The traditional methods are still used, however, in the manufacture of high-quality furniture and nowhere are these skills more in demand than in the restorer's workshop, where antique furniture, damaged by damp, heat or ill-use, is carefully restored to its former splendour.

BEECH: TEAK OIL
*Oils tend to darken the wood
more than do other finishes,
some of which would hardly
alter the colour of a wood
such as beech. More durable
modern oils with synthetic
resins dry more quickly than
linseed oil, with less risk
of picking up surface dust*

OGANY: PLASTIC
...use mahogany, like
...ut, darkens on the
...ication of a finish of
...kind, it is impossible to
...n the lighter red of the
...al wood. Clear plastic
...hes adhere well and are
...ble, resistant to heat
...vater, and do not age

RESTORER'S WORKSHOP
Beneath the circular
rubbing of the restorer's
cloth, the once discoloured
table surface is brought
back to a splendid gloss.
Brass inlay sets off the
figured rosewood, which
gleams under a French
polish and a wax finish

SHELLAC POLISHES
The many varieties of French
polishes are all made from
shellac, an incrustation on
the twigs of certain trees,
commonly fig-tree, mainly in
India. Shellac is the dried
secretion of Laccifer lacca,
a parasitic insect. It is
scraped off, washed, winnowed

and strained whilst molten
to form buttons and flakes,
and dissolved in industrial
alcohol to produce polish.
French polish, which is made
from the flakes, is usually
much darker than the purer
button polish. Transparent
and white shellacs are made
by bleaching and de-waxing

NUT: DANISH OIL
...lnut, like mahogany,
...kens when finished, but
...s its rich colouring.
...ed, like teak oil, on
...hetic resins, Danish oil
...additional ingredients,
...h impart a soft lustre.
...are easily applied by
...ing with a rag

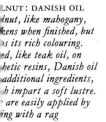

TELL-TALE SIGNS
The circular top of this
19th-century mahogany
table has been damaged on
at least two occasions.
Wood fillers tend to darken
with age and even after
careful finishing it is
almost impossible to com-
pletely hide the repair

STAINING NEW WOOD
The damaged box inlay on
this walnut-veneered 18th-
century Italian bureau has
been replaced, and the new
wood is stained to match
the old. The restorer mixes
and experiments with his
colours until he achieves the
exact patina needed

LIVING WITH WOOD
Tone and Texture

In keeping with the trends of modern architecture, fashions in house decoration have, in the years following World War II, exulted in the use of new materials. Recently, however, there has been a marked revival of older and more traditional values. The striking effect of glass, steel and plastic furniture has been largely lost – its impact reduced by familiarity – and the use of these materials has been challenged by a renewed creativity and interest in wooden furniture, wood panelling and in the aesthetic and practical applications of new wood products.

The wave of open-plan design which brought light and space to the dark cubes in which an earlier generation had been at home, has now broken. A nostalgic movement harks back to homeliness and to the warmth and easy familiarity of natural materials.

Manufacturing developments have brought wood-panelling within the reach of every home budget; large areas, once panelled with numerous small elements, may now be covered with veneered plywood sheets. Technical innovations have vastly improved the natural acoustic, insulating and fire-retardant properties of wood and new finishing techniques have rendered it more durable while enhancing its grain and colour.

Once regarded purely as structural elements in the home, the timber members of floors, ceilings and staircases are now more and more frequently left uncovered – their form and colour adding a new dimension to the open areas of the house.

DINING-ROOM
Cane chairs are simple, and easily adaptable to modern shapes: the bent wood used for the frame offers wide scope to a designer. The blinds, made of laths of a light-coloured wood, match the canes, and allow the easy passage of light

WICKER CHAIR
Comfortably creaking wicker-work armchairs lend the room an air of easy informality, and add textural variety to its dominant brown tones. A low table beside the chair has a square oak top set into legs of turned beech. The floor is of clear-waxed pine

WALL VENEER
Many kinds of wood are cut into veneers, one of the most attractive being quartered walnut. Cheaper core veneers reduce the overall cost. Matching veneers from the same flitch can provide variations on a theme for use on large surfaces

FIRE-RESISTANT CEILING
Wood surfaces char as they burn – restricting the rate of burning so that they only burst into flame in great heat. This natural safeguard is supplemented by treatment with fire-retardant chemicals

RATTAN TABLE
A low table of split rattan provides a hardy surface, and produces, with the wickerwork chair, an easy contrast to the softwood tones of the walls, floor and ceiling. A third theme is provided by occasional hardwood panels and the hardwood tables

DRIFTWOOD SCULPTURE
The tangled white form of a piece of driftwood is an evocative reminder of the outdoors, and reflects by contrast the pleasurable warmth of an insulated room. It is a gentle reminder also of wood's buoyant versatility

Tradition and Technology

Furniture of the nineteenth century was inspired mainly by that of earlier periods; thus Renaissance, Baroque, Oriental and Neo-Classical styles were constantly revived and reinterpreted. In England, cabinet-makers continued to favour mahogany and rose-wood, while ebony was used with great effect as an inlay in lighter woods. Early Victorian furniture was generally plain, but after the Great Exhibition in 1851 designers became more adventurous, using more carved decoration and a greater variety of woods.

Techniques of cutting veneers had greatly improved by the early nineteenth century, and French designers tended to concentrate on the decorative effects to be achieved with the natural grain, colour and figure of light woods, "bird's-eye" maple being one of the most popular of these "bois clairs".

However, despite the continuing influence of historical styles, the nineteenth century saw the development of many important new styles and techniques, including the process, developed by Belter in New York, of laminating panels or strips of wood together in such a way that they could then be steam-bent into the required shape. Inevitably the once-prized qualities of craftsmanship deteriorated as a result of rapidly increasing mechanization, though the traditional methods survived, particularly in Scandinavia, where industrialization was less advanced, and in parts of Europe, where the followers of William Morris and other reformers made a sustained effort not only to restore high standards of craftsmanship but also to re-establish the status of the craftsman in the eyes of society.

Although mass-produced furniture is generally associated with the nineteenth and twentieth centuries, rudimentary assembly lines have existed since the late seventeenth century, when craftsmen began to specialize; one making chair legs, another concentrating on carved decoration and so on. In England, vast numbers of Windsor chairs were assembled in the factories of High Wycombe, using legs and stretchers made by chair bodgers in the beech forests of Buckinghamshire.

Modern furniture manufacture involves not only very sophisticated production-line technology but also a wide variety of composite wood products. Solid wood may be used for high-quality furniture – the parts machined automatically but assembled by hand and then finished by specialist craftsmen: in making less expensive items, costs are reduced by using blockboard or fibreboard panels, veneered with hardwood or covered with a synthetic material, often with wood-grain finish. In many factories, particularly those manufacturing kitchen cabinets and wall units, where the individual items are made to standard dimensions, the parts move continuously from station to station on the production line, carried on conveyor systems and processed almost entirely by machines.

Laminated woods have assumed an important role in furniture design and manufacture: they allow the wood to be used efficiently, as only the facing laminate needs to be of the highest quality, and they are versatile in that they can be moulded into complex curved shapes in heated high-pressure forms, giving the designer a freedom of form not easily attained using solid wood.

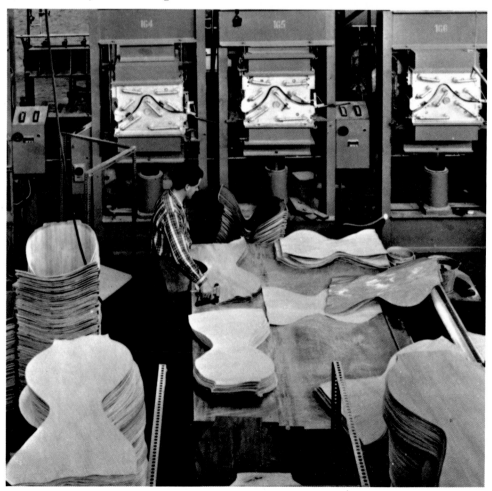

THE FACTORY METHOD
A wide variety of woods may be used in the ply-wood moulding process and in this Danish factory nine laminates are combined in making a chair. The two face laminates are of figured oak, beech, teak or mahogany, while the inner laminates are all of beech, two having their fibres running across the chair, the others running longitudinally. The laminates are coated with a waterproof glue, resistant to changes in temperature and humidity, and are then pressed in heated moulds

YEW FOUR-POSTER

At John Makepeace's country workshop, Andrew Whateley carves one of the four feet of what will be a yew four-poster bed. Working with a small group of creative craftsmen, John Makepeace designs each piece of furniture to be made from one of the massive whole logs kept in stock. Each log is studied carefully, its grain, colour and even its imperfections assessed before a piece of furniture, or even a group, is designed to make full use of the wood's own particular qualities. This bed was designed to be made from a yew log of exceptional width and length. Optimum use of the timber is assured by building up the massive feet with laminations; they are then carved to their final shape (inset) so that their solid form captures the gnarled and irregular shape of the living tree

THE DESIGN

The four posts of the bed curve slightly inwards to create the impression that they are held in tension by the silk tester. This tension lends unity to the whole while the massive feet counterbalance the slender grace of the columns

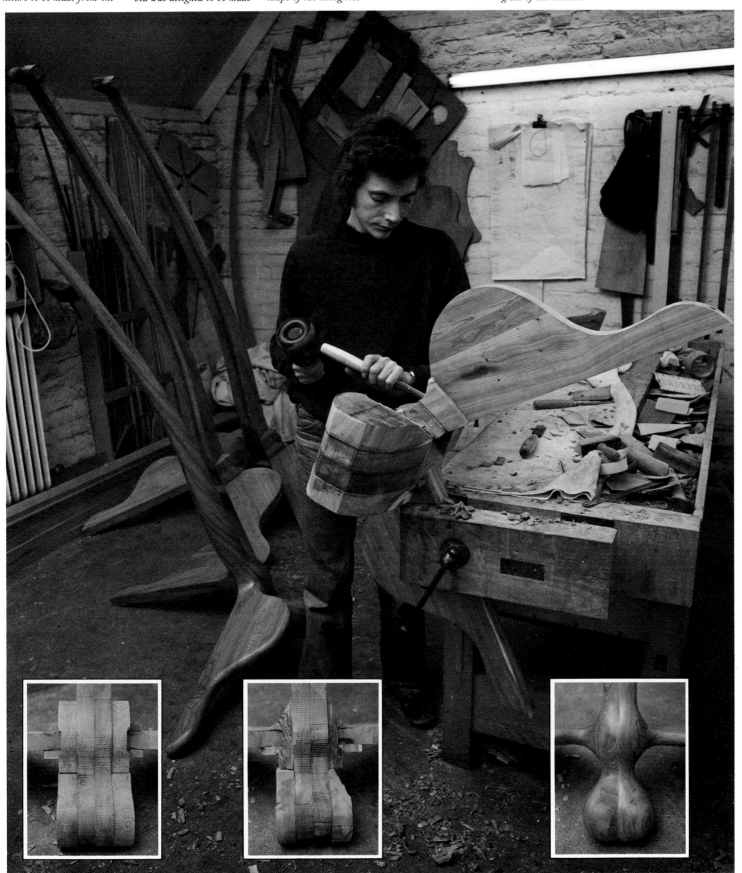

131

Wood in

Photographed in 1926 in the midst of an apparently endless expanse of grassland and scrub, the driver of this South African stage-coach reins in his mule team, slowing them to a walk before fording a stream

Transport

From Runner to Wheel

LABRADOR SLEDGE
*The exceptionally long
runners of the Caribou
Eskimo sledge are joined by
crossmembers lashed with
sealskin thongs. The runners
are coated with peat paste
and are brushed with water
each day to create a layer
of ice over the surface*

ESKIMO SLEDGE
*The lighter, more versatile
sledges of the Mackenzie
Eskimos have a load
platform held above the
runners and have handle-
bars with which the sledge
can be manoeuvred. In
traditional style, joints are
lashed for flexibility*

From time immemorial Man has been forced
by circumstance to devise some means of
carrying his goods. The hunter must take
home the whole carcass of his prey lest the
parts left behind are consumed by scaven-
gers; the trapper in northern lands must
travel quickly and safely between his traps
and the trading stations, and the migrant
plains dweller must be able to move his
family, home and worldly goods easily from
camp to camp.

The use of a vehicle with runners has been
recorded in the carvings and wall paintings
of many ancient cultures. The peoples of
Egypt and the Near East used huge wooden
sledges to transport stone building blocks
and statues – using rollers to ease the passage
of the loaded sledge.

In modern times, sledges are most widely
used in the polar regions. Made from wood,
occasionally from bone, and drawn by rein-
deer or dogs, they form a versatile and effective means of transport over snow and ice.
The sledges are lashed together with no
rigid joints so that over rough terrain the
whole structure can bend and flex without
breaking. Even at very low temperatures
wood retains a remarkable degree of natural
flexibility and resistance to shock.

The Lapp pulka, well adapted to use in
soft snow, resembles a small boat – its wide
keel acting as a simple runner. Canadian
Indians use a similar boat-sledge made of
bark or wood, a simple tray with the front
edge turned up by steaming, and "stone-
boats" have been widely used in Britain and
in pioneer America for dragging rocks and
tree-stumps from fields to be cultivated.

Wheeled vehicles were unknown in the
Americas until the New World was colonized
by Europeans, but the wheel had been used
in the ancient civilizations of Asia and the
Mediterranean for at least 5,000 years before
Christ. The origins of the wheel are unknown, but its development can be traced
from the simple solid wheel with its fixed
axle – a form which still survives in many
parts of the world – to the much lighter and
more efficient dished wheels, which reached
their greatest refinement in the carriages and
coaches of the nineteenth century.

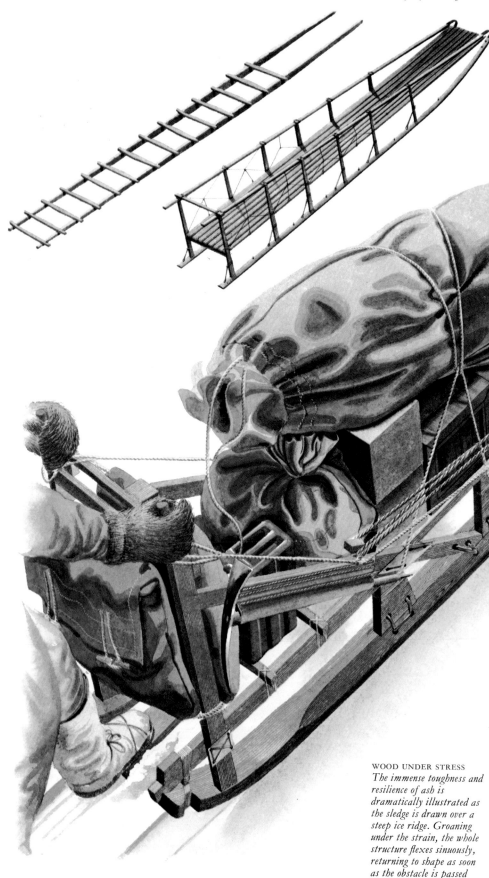

WOOD UNDER STRESS
*The immense toughness and
resilience of ash is
dramatically illustrated as
the sledge is drawn over a
steep ice ridge. Groaning
under the strain, the whole
structure flexes sinuously,
returning to shape as soon
as the obstacle is passed*

NANSEN SLEDGE
The modern Nansen sledge, used extensively by polar expeditions, is a refinement of the Eskimo sledge and has been designed for lightness and reliability. The sledge is built entirely of ash – the parts mortised together and lashed with fine cord and rawhide. The runners, of hickory on some early models, are now made of ash with a coating of low-friction "tufnol". A flexible brake-bar, lashed to the bridges supporting the load platform, can be forced on to the ice or snow with the foot

THE "COW-CATCHER"
The towing rope is looped round the first three pillars to absorb sudden loads and is supported clear of the ground by the cow-catcher – a length of steam-bent rattan cane which acts as a shock-absorber and facilitates easy handling

LAPP PULKA
This simple boat-shaped sledge used by the nomadic Lapps slides easily over soft snow, stabilized by the wide keel, which acts like a single broad runner. The shell is made from planks of pine sewn together or pinned with wooden pegs

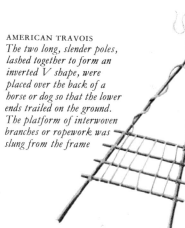

EUROPEAN Y-SLEDGE
The Y-sledge occurs in many forms throughout northern Europe and parts of Africa. The frame of the sledge is simply a natural forked log, trimmed of its bark and any projections and fitted with a plank platform fixed in place with hardwood pegs

AMERICAN TRAVOIS
The two long, slender poles, lashed together to form an inverted V shape, were placed over the back of a horse or dog so that the lower ends trailed on the ground. The platform of interwoven branches or ropework was slung from the frame

DEVELOPMENT OF THE WHEEL

Archaeological finds prove the wheel to have been invented at least 5,000 years ago. The oldest extant wheels are from ox-carts and were either solid-hewn or made from three or more planks dowelled together. The much lighter spoked wheel appears in about 1500 BC in the Near East and was used on fast horse-drawn fighting chariots

*Solid-hewn wheel
Holland: c. 2000 BC*

*Tripartite wheel
Mesopotamia: c. 2500 BC*

*Crossbar-type wheel
Italy: c. 1000 BC*

*Spoked wheel
Norway: c. AD 850*

*Dished cart-wheel
Europe: c. AD 1850*

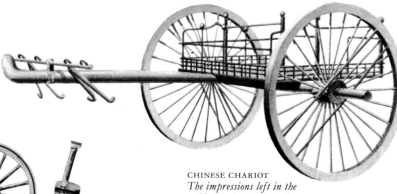

CHINESE CHARIOT
The impressions left in the soil by 19 chariots buried in Hui-hsien made possible a reconstruction proving that the Chinese had both spoked and dished wheels by 250 BC. Later hubs were elm, felloes oak and spokes rosewood

135

The Wheelwright's Craft

The construction of a spoked wheel calls for skilled craftsmanship and a thorough understanding of the intrinsic properties of three contrasting kinds of wood. The central stock, or nave, has always been made from elm, typically *Ulmus procera*, the English elm, or *U.americana*, the American species. It is the toughest timber available and, due to its interlocked grain, the most resistant to splitting. Tapering cylinders of well-seasoned heartwood are first turned on a lathe; the centre of each hub is then bored out with an auger to take the iron axle-box, which revolves on the fixed axle. Mortises are cut to take the spokes, and the nave is then finished and made firm by shrinking on iron bands at either side. Heavy-duty wheels are made concave, or "dished", to help them withstand sudden jolts.

Oak, either English oak, *Quercus robur*, or American white oak, *Q.alba*, is chosen for the spokes. No other timber but ash can be trusted to bear the immense stress as each spoke in turn carries the whole weight borne by each wheel. Spokes are always cleft from the log by hand, to ensure that the grain runs unbroken from end to end. Every spoke must be completely free of knots or any irregularities of grain.

The felloes that make up the rim are cut from well-seasoned, completely knot-free ash, either the English kind, *Fraxinus excelsior*, or American white ash, *F.americana*. Both these timbers have open, porous bands of springwood interleaved with very hard, strong bands of dense summer-wood. This combination enables ash to take sudden jolting shocks without breaking. Each felloe is long enough to take two spokes, as longer sections would necessitate cutting across the grain. It is essential that some annual rings run through the whole length of each felloe. Small round dowels of oak, hidden completely when the wheel is finished, fit into sockets bored in the felloe-ends and bind each felloe to its neighbours.

The iron tyre, expanded by heating, is then fitted on to the wheel and cooled with water – binding the components firmly together and providing a hard-wearing surface over the wheel-rim.

OLD AND NEW
A recently repaired cart wheel, with several felloes and spokes replaced, awaits the addition of a new iron tyre. Behind it rests the broken wheel of an 18th-century gun carriage – its elm stock split across after decades of heavy use

THE SPOKES
Straight-grained oak is the prime timber for the spokes of the wheel. The oak log is cleft, rather than sawn, into billets to preserve the long fibres and is then left to season for a year before being shaped with the draw-knife

THE STOCK OR NAVE
Elm, and only elm, has the toughness and strength to serve as the stock of the wheel. A well-seasoned log, fourteen inches long and twelve in diameter, is turned on the lathe and then mortised to take the ends of the oak spokes

THE FELLOES
Alternating bands of open, porous springwood and dense, tough summer-wood give ash the properties of a natural shock absorber. Here, the shape of the felloe is inscribed on a three-inch plank of ash prior to cutting

FITTING THE WHEEL
Leaning his full weight against the shaft of the "spoke-dog", the wheelwright strains each pair of spokes together as he fits the felloe into place. The final stage before fitting the metal tyre over the rim of the wheel calls upon all the skills of the craftsman. Working over the spokes and rim with draw-knife and spokeshave, the wheelwright pares away excess wood – lightening the wheel but leaving sufficient timber in place where stresses are greatest

137

Farm Carts and Wagons

In many respects the wheelwright's craft is similar to that of the joiner, but whereas the joiner may rely on glue and dowels to strengthen his work, the wheelwright must rely on the tightness of the joint alone to hold his work together. Traditional glues would simply disintegrate under the constant flexing and jarring, wetting and drying that are the lot of the farm vehicle.

So important was the quality of the timber to the wheelwright that very often he alone would supervise the felling of the trees and sawing of the timber. As each piece was delivered to the yard it would be assigned to a specific task; gently curving pieces of beech and ash would be set aside for felloes, straight oak butts for cleaving into spokes, curving ash for the shafts, and straight-grained oak and elm for the frame and boarding of the wagon.

Oak and ash were the principal timbers used in framing the wagon, but different styles developed in Britain and Europe. The four-wheeled farm vehicles of Europe had simple straight-sided bodies of oak which were extremely strong but which restricted the turning circle of the wagon. English wheelwrights, designing their wagons for use in narrow lanes and yards, overcame this drawback by making the wagon body in two parts – the narrow waist allowing the front wheels to turn more sharply.

The flooring of elm planks could be laid either along the length of the wagon or across the width, depending on the loads to be carried. For sand, gravel or manure loads, all of which must be shovelled, long-boarding was preferred; with cross-boarding, an elm slat might curl up in the path of the shovel. For side-members, ash was generally preferred, often in combination with poplar.

Not all country working vehicles, however, are the work of skilled craftsmen and in many parts of the world, particularly in remote upland regions, the farmer must make his own carts from whatever material is available. In China, oak and elm are used but with rosewood taking the place of oak for many purposes, whilst the ever-versatile bamboo is pressed into service for vehicle construction throughout many far-eastern countries.

SIDE BOARDING
The boarding of the sides was usually of elm, particularly when heavy loads were carried, but poplar boards, which were more easily bent, were often used to accommodate the curves of the fore-body

SIDE FRAMING
The main framing members were of ash or oak, with poplar and beech commonly used for the spindles

THE TOP LADE
Ash was generally used for the sweeping curve of the top lade – the outer part of the frame that extended the load-carrying space and protected the wheels

AXLE BED
A massive baulk of oak, or occasionally ash or beech, took the weight of the cart and undercarriage and supported the steel axle arms

SHAFTS
The shafts, or rods, were generally made of ash – a tough, resilient wood which worked well to the long sweeping curve required

AXLE ARM
A forged steel arm, set into the axle bed, carried the elm stock of the wheel

SUSSEX WAGON
The massive wheels of the traditional Sussex wagon were made in a variety of tread widths in different parts of the county. The cart illustrated is of the older broad-wheeled type, its wide felloes shod with heavy iron strakes

NORTH LINCOLNSHIRE
The carts of north Lincolnshire, high-bodied and with no out-raves to break up their austere lines, suggest a close affinity with some Dutch wagons. Each county had its own traditional colours; here an indigo body on a red undercarriage

FLOORBOARDS
The flooring boards of elm or poplar were laid either horizontally or transversely and were nailed in place

RCARRIAGE
carts the main arriage was made of hamfered in places to weight, but un- d at points of um loading

CONESTOGA WAGON
Rugged, versatile and easily repaired, this wagon played a major part in opening up the American West

OSEBERG CART
Made of beech, ash and oak, the Oseberg cart dates from the late ninth century

CHHAKRA
The bow-shaped bullock carts of Bihar in northern India are made of sisu wood and bamboo, held under tension by fibre ropes

CHINESE WHEELBARROW
A wide variety of single-wheeled barrows are used in China for carrying both passengers and merchandise

OXFORDSHIRE WAGON
Weighing no more than 18 hundredweights, the Oxford wagon was among the lightest of the county wagons and, with its sweeping curves, high narrow wheels and deep waist one of the most attractive. The rear part of the body arched into a shallow half-bow to give clearance to the 60-inch-diameter rear wheels. The tail-board is constructed of spindles set into the top and bottom rails, like the sides of the cart, and may be fully panelled or half-panelled. The tail-board was hinged on wooden pins

The Age of Elegance

Heavy coaches were known in Europe from Roman times, but the introduction of sprung suspension in the late seventeenth century revolutionized the craft, and by the early nineteenth century it had developed to produce cabriolets, phaetons, curricles and barouches of great elegance.

The coach-builder's craft had a great deal in common with that of the wheelwright; both used the same tools and very similar techniques, but, while the wheelwright would construct the whole vehicle himself, the coach-builder's workshop might include a great variety of specialized craftsmen, including body-makers, wheelwrights, smiths, carriage-trimmers and body painters.

The woods used in coach-building were hard, well-seasoned, fine-grained timbers and although three-ply panels of cross-grained American whitewood were available in the late nineteenth century, most craftsmen spurned the new material in favour of the traditional oak, ash, beech and mahogany.

The principal wood used in construction was ash – a tough, fibrous wood that could be pressed into shape after boiling or steaming. The wood was stable in use and seldom suffered from twisting. Ash grown on steep hillsides, and home-grown elm, were much sought after for framing, for their unrivalled toughness, but neither could be used for exposed panelling due to their strong, wrinkled grain, which would show through paint even after careful planing.

Body panels were usually made out of Honduras mahogany, a high-quality timber which would plane to a smooth finish and take the high-gloss paintwork required by fashion. Mahogany would also bend easily to the graceful curves of the carriage body. Occasionally, Honduras cedar was used for panels destined to be covered in leather; the wood being far too porous to accept paint.

During the late nineteenth century, the design of vehicles became standardized; the coach builder was able to consult text-books and patterns and even had his own trade journals. His workshop methods, however, remained largely traditional until the internal combustion engine brought about the end of the age of the horse-drawn carriage.

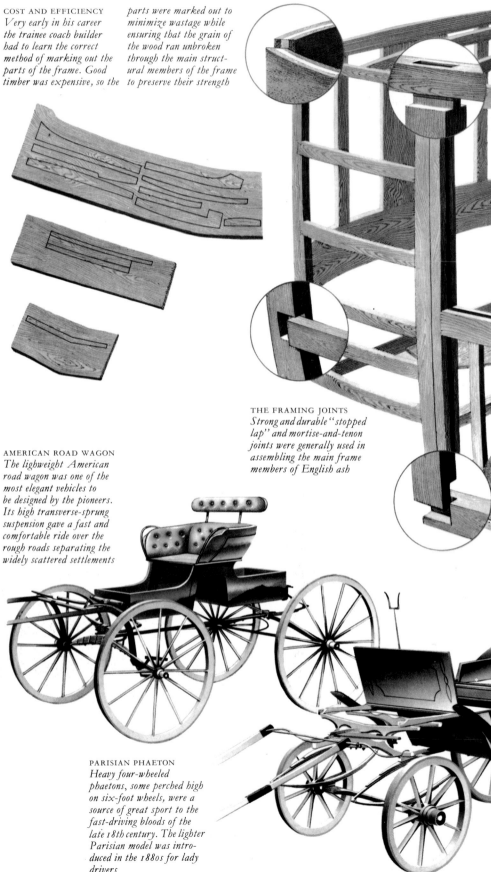

COST AND EFFICIENCY
Very early in his career the trainee coach builder had to learn the correct method of marking out the parts of the frame. Good timber was expensive, so the parts were marked out to minimize wastage while ensuring that the grain of the wood ran unbroken through the main structural members of the frame to preserve their strength

MITRED PANELLING
The top, back and quarter panels were usually mitred together to conceal the end-grain of the wood. Generally of mahogany or, in America, of white-wood, the panels were steamed first to make them more pliable and were then pinned to the frame timbers. Smaller side panels were set in grooves in the framing and secured either by gluing (popular in France and America) or by packing with white lead, commonly used in England. The white lead gave a firm joint and also protected the wood from moisture

THE FRAMING JOINTS
Strong and durable "stopped lap" and mortise-and-tenon joints were generally used in assembling the main frame members of English ash

AMERICAN ROAD WAGON
The lighweight American road wagon was one of the most elegant vehicles to be designed by the pioneers. Its high transverse-sprung suspension gave a fast and comfortable ride over the rough roads separating the widely scattered settlements

PARISIAN PHAETON
Heavy four-wheeled phaetons, some perched high on six-foot wheels, were a source of great sport to the fast-driving bloods of the late 18th century. The lighter Parisian model was introduced in the 1880s for lady drivers

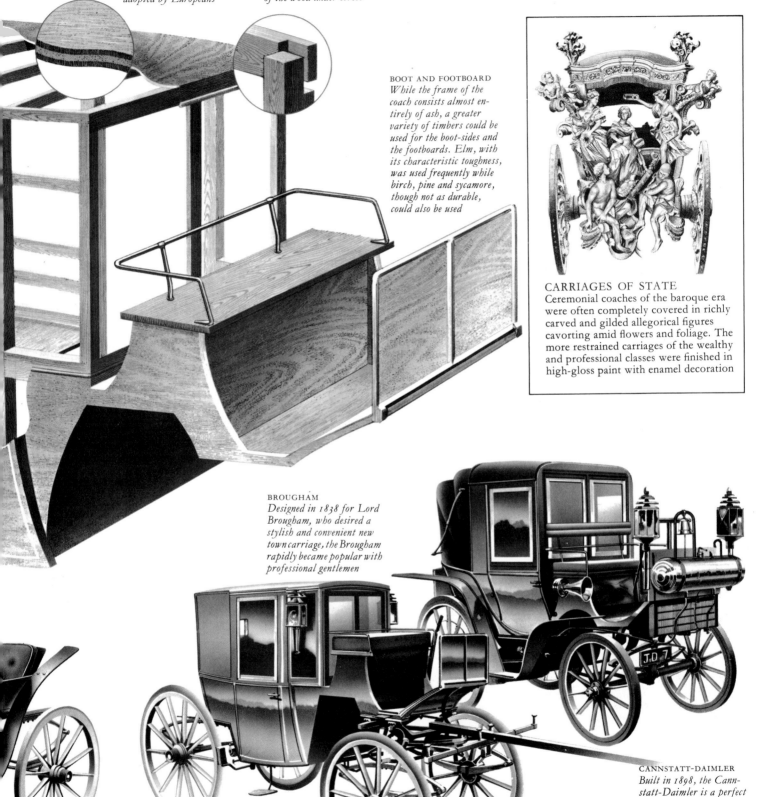

AMERICAN INNOVATION
In the early 19th century one-piece roofing panels of three-ply cross-grained white-wood were imported from America, but, though more convenient than the jointed pine-board roof, the new material was not widely adopted by Europeans

JOINTING TECHNIQUES
Most of the joints used in framing the carriage were of the mortise-and-tenon or lap types – cut from straight-grained timber and pinned with wooden dowels to hold them in place despite the natural movement of the wood under stress

BOOT AND FOOTBOARD
While the frame of the coach consists almost entirely of ash, a greater variety of timbers could be used for the boot-sides and the footboards. Elm, with its characteristic toughness, was used frequently while birch, pine and sycamore, though not as durable, could also be used

CARRIAGES OF STATE
Ceremonial coaches of the baroque era were often completely covered in richly carved and gilded allegorical figures cavorting amid flowers and foliage. The more restrained carriages of the wealthy and professional classes were finished in high-gloss paint with enamel decoration

BROUGHAM
Designed in 1838 for Lord Brougham, who desired a stylish and convenient new town carriage, the Brougham rapidly became popular with professional gentlemen

CANNSTATT-DAIMLER
Built in 1898, the Cannstatt-Daimler is a perfect example of the early "horse-less" carriage. Though powered by a four horse-power engine, the car's chassis and centre-pivoted front axle clearly owe their origins to the brougham

141

Dawn of a New Age

The development of iron and steel during the nineteenth century brought about a rapid decline in the use of timber in the transport industries. Although both Drais's 1818 hobby-horse and Kirkpatrick's pedal-driven bicycle of 1839 had curved longitudinal members of wood, the famous "penny-farthing" was made of iron, and later bicycles were made of tubular steel and a variety of very light alloys. Benz's 1885 tricycle was wooden, and in the same year Daimler mounted his newly developed engine on the rear of a simple wooden cart, but the overwhelming advantages of steel were already apparent and by 1905 it was used almost universally in chassis construction.

Wood continued in use, however, for many years as a material for vehicle bodywork. Ash, the traditional wood of carts and carriages, was used for framing the bodywork, as were oak and mahogany, but teak was also widely used in vehicles destined to be exported to hot countries. Pine, larch and deal were used extensively as board and panelling, but undoubtedly mahogany was the most valued timber where panels were required to follow complex sweeping curves. Birch, hickory, beech and elm were all used where boards were required – for seating structures, footboards and the side panels of boots. Even in the period between the two world wars, timber was more widely used than steel for the bodywork of both motor-cars and commercial vehicles of all types.

In Europe, the first private motor-cars were luxury products sold to gentlemen who expected no less a degree of comfort and elegance than they had enjoyed in their carriages, and for many years the trades of mechanic and coach builder remained quite separate and distinct. Beautifully finished vehicles were built for the wealthy and were exported to many parts of the world, but the end of the wooden motor-car was already in sight. In America, Henry Ford introduced the use of aluminium and achieved both a faster rate of production and a far superior power/weight ratio for his Model "T" than could ever be achieved with wooden construction. In 1913 the age of the modern production-line vehicle began when E. G.

HISPANO SUIZA
This 8-litre Hispano Suiza was custom-built for the millionaire André Dubonnet in 1924. Its rich tones set off by gleaming rivets of copper, the tulipwood body was designed by Nieuport, a Frenchman, who had earlier built aircraft

KNIGHT PROTOTYPE
The English pioneer Knight built a series of motor-car prototypes, of which this 1895 dog-cart was the first. In contrast to the exotic Hispano Suiza, wood was used for the wheel, chassis and body as the cheapest and most familiar material to hand

Budd manufactured, for the Dodge brothers, a car body made entirely of pressed steel.

Though no longer used for production-line car bodies, wood has re-emerged several times in the manufacture of specialist cars, a notable example being the English Marcos sports car of the 1960s, the entire body of which was manufactured from plywood sheets. As a status symbol, wood has retained its popularity and is used frequently for the steering wheels, fascia panels and interior trim of many quality vehicles. Since 1904, the Rolls-Royce dashboard has been made of walnut, embodying, in sadly attenuated

form, a coach-building tradition reaching back through several centuries.

Today, wood has taken on a new role in the world of transport. Plywood sheets are now used to make the box bodies of lorries, containers, rail-cars and vans. The sheets are usually covered with a thin skin of steel or aluminium, protecting the wood from the elements and from excessive abrasion. Once the rectangular box body is complete, with the doors closed, it is immensely strong – even without internal bracing – and will withstand the abuses of heavy commercial use far better than any all-metal body.

STRIKING INNOVATION
Contrary to contemporary practice, Nieuport used wood not for the framing or the panels of this Hispano Suiza, but as plating. A subsequent owner crowned his conception by replacing the aluminium mudguards with tulipwood veneers to match the body

1895 DOG-CART
Like the buggy in America, from which the "runabout" took its form, in Europe the cheap and practical dog-cart was motorized by Knight in England and Benz in Germany. The typical dog-cart had a frame of oak or ash, and deal seats and panels

DAIMLER TRICYCLE
The wooden structure underlying the heavy, cumbersome frame of Daimler's famous 1893 motorcycle had to be reinforced throughout by cast-iron. New stresses and greater weight were soon to curtail the use of wooden members in powered vehicles

BONESHAKER
The wheels and frame of a 19th-century "boneshaker" were wooden, with iron pieces at points of stress. Often the main member terminated over the front wheel in a carved head. For strength, these wheel-spokes are set in two planes

Wood in Flight

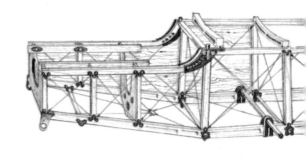

Wood was the basic structural material for most early aircraft. It was cheap, easily worked, had low density but high rigidity and was easy to repair. Before, and after, World War I, most aircraft had a basic framework of high-quality wood, usually ash or spruce, with light wood "formers". The major parts were constructed separately by gluing and pinning, and were joined by large bolts. Thin birch plywood would be used to cover some parts, often the front of the wing, while the rest was covered in lightweight treated fabric.

From the time of Leonardo, flying machines had been proposed with structures of oak, pine, bamboo, whalebone and pliable switches of hazel and willow. By 1910, however, aircraft structures had become fairly precise engineering assemblies, carefully drawn out in advance. Every piece of wood had to be minutely inspected and, by the end of the war, the vast demands of the combatant aircraft industries had caused a great shortage of the best spruce and other choice timbers.

From the 1920s onwards, wood was little used in military aircraft, but still featured in high-quality civil aircraft work. Wood was generally used in the form of a streamlined monocoque, in which the stresses were carried by the outer shell. Used in the 1913 Deperdussin racer, the monocoque design utilized the flexibility of wood and its ability to take on a curved form when pressed in a heated mould.

World War II saw a dramatic revival of wooden aircraft, partly to conserve scarce metal resources and partly to utilize to the full the skills of wood-workers. The outstanding wooden combat aircraft was the Mosquito, a monocoque structure with the skin made of low-density balsa sandwiched between veneers of birch. Mosquitos were still flying at more than 400mph ten years after the war – even in tropical areas, where warping, glue failures and insect damage added to the normal hazards of flying.

Today, many light aircraft and gliders use wood construction for cheapness, fine finish and because of the versatility in use which is increasingly hard to retain in mass-produced metal structures.

LEONARDO ORNITHOPTER
Leonardo da Vinci sketched this man-powered ornithopter (flapping-winged aircraft) in 1490. The wings were to have been of a light, birdlike construction, of pine, raised by the flyer's hands and then pulled down by cords fixed to the feet

WRIGHT FLYER
Success came to the Wright Brothers as a natural result of methodical research. Among their many advances was the first practical lateral (rolling) control, achieved by differential warping of the Flyer's wooden wings

TWO-PART SHELL
The Mosquito fuselage was made in left- and right-hand half-shells, which were much easier to equip with hydraulics and other systems than a one-piece design. Doors and hatches were edged with spruce strips and cut out at a late stage. The main cut-out for the cockpit had reinforced edges to take the metal frame of the transparent canopy

DEPERDUSSIN RACER
This 1913 racer was one of the first aircraft to have a monocoque construction. The streamlined fuselage was fashioned from three layers of tulipwood veneer on a stout hickory frame. Drawbacks were the cost and the difficulty of repair

FOKKER D.VII
This outstanding World War I fighter had a tubular-steel frame with a mixed plywood and fabric skin. The wooden wings had plywood leading edges and fabric skin and were strong enough not to need any additional bracing wires

ROBIN
Built at Dijon, in France, the Robin family of light aircraft are produced in either metal or wood. The wooden Robins are built of selected spruce, shaped by machine to give a precisely repeatable product. The skin is of plywood and fabric

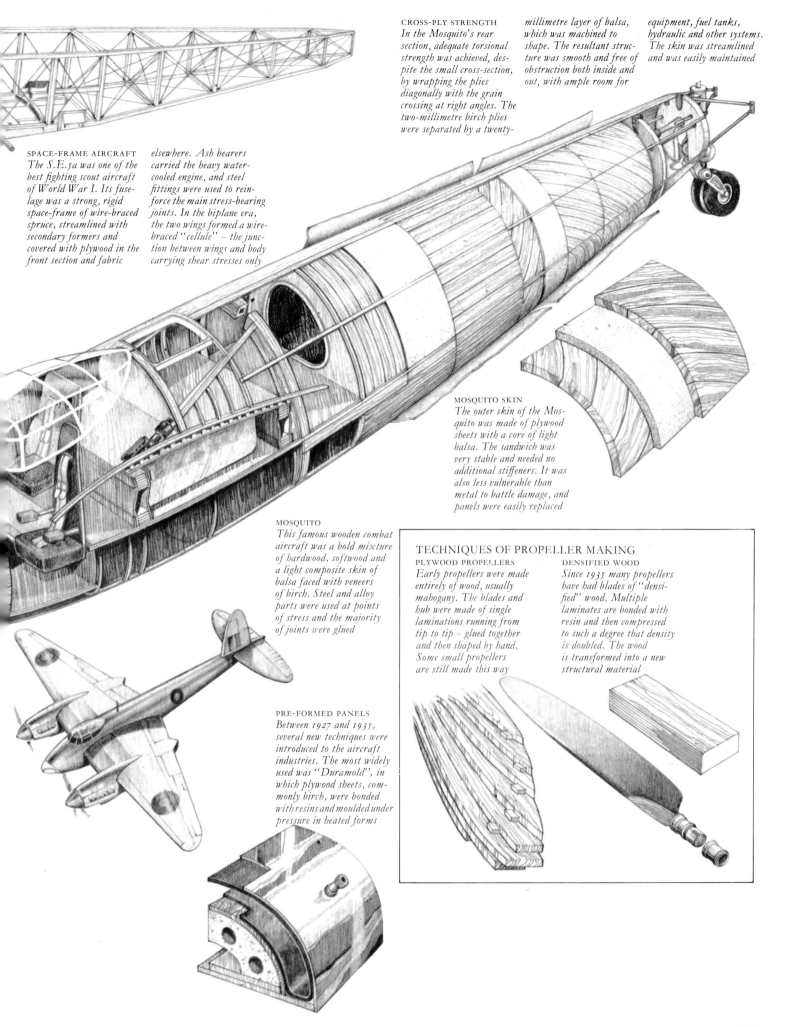

CROSS-PLY STRENGTH
In the Mosquito's rear section, adequate torsional strength was achieved, despite the small cross-section, by wrapping the plies diagonally with the grain crossing at right angles. The two-millimetre birch plies were separated by a twenty-millimetre layer of balsa, which was machined to shape. The resultant structure was smooth and free of obstruction both inside and out, with ample room for equipment, fuel tanks, hydraulic and other systems. The skin was streamlined and was easily maintained

SPACE-FRAME AIRCRAFT
The S.E.5a was one of the best fighting scout aircraft of World War I. Its fuselage was a strong, rigid space-frame of wire-braced spruce, streamlined with secondary formers and covered with plywood in the front section and fabric elsewhere. Ash bearers carried the heavy water-cooled engine, and steel fittings were used to reinforce the main stress-bearing joints. In the biplane era, the two wings formed a wire-braced "cellule" – the junction between wings and body carrying shear stresses only

MOSQUITO SKIN
The outer skin of the Mosquito was made of plywood sheets with a core of light balsa. The sandwich was very stable and needed no additional stiffeners. It was also less vulnerable than metal to battle damage, and panels were easily replaced

MOSQUITO
This famous wooden combat aircraft was a bold mixture of hardwood, softwood and a light composite skin of balsa faced with veneers of birch. Steel and alloy parts were used at points of stress and the majority of joints were glued

PRE-FORMED PANELS
Between 1927 and 1935, several new techniques were introduced to the aircraft industries. The most widely used was "Duramold", in which plywood sheets, commonly birch, were bonded with resins and moulded under pressure in heated forms

TECHNIQUES OF PROPELLER MAKING

PLYWOOD PROPELLERS
Early propellers were made entirely of wood, usually mahogany. The blades and hub were made of single laminations running from tip to tip – glued together and then shaped by hand. Some small propellers are still made this way

DENSIFIED WOOD
Since 1935 many propellers have had blades of "densified" wood. Multiple laminates are bonded with resin and then compressed to such a degree that density is doubled. The wood is transformed into a new structural material

Rural Craft
and
Industry

Leaning against the stump of a tree, the froe, draw-knife, adze, beetle and mallet evoke a bygone age when these simple but effective tools were the standard equipment of every rural craftsman

Cleft and Woven Wood

THE HAZEL CROP
During the spring and autumn, when the sap is flowing, the hurdle-maker collects his raw materials, choosing rods about seven years old and between five and ten feet in length. The wood is used immediately while still flexible

PREPARING THE RODS
Apart from a few rods left in the round for use as the strengthening members, most of the hazel is cleft. The rods are either split from end to end with the bill-hook or by forcing the two halves of a split end to either side of a post

Cleaving is a technique for reducing round timber to plank or paling without using a saw in order to preserve the wood fibres intact. With many timbers, including oak, *Quercus* sp., ash, *Fraxinus* sp. and sweet chestnut, *Castanea* sp., it is possible for a skilled craftsman to cleave a log from end to end by driving an axe blade or wedge into the end-grain on the precise line of one of the rays – the lines of storage cells that radiate from the centre of the log. The two half-cylinders are then cleaved into triangular billets or flat boards using the axe or froe.

Because this method of preparing wood is fast, convenient and requires only the simplest of tools, it was favoured by the American colonists. Fence-posts, palings, gates and bars could be quickly made from trees cut down while clearing new fields; cleft planks were used to clad houses and barns, and roofs were covered with thin cleft shingles of white cedar.

Throughout Europe and America, farm tools of all kinds have been made from cleft timber, particularly from European oak, American ash and the American hickories of the genus *Carya*. All these timbers have a strong grain, which, when preserved uncut, gives the rake- or scythe-handle a toughness and durability never found in sawn wood.

The slender stems of selected trees, often hazels and willows, may be cleft or simply split with a knife or billhook to be woven into hurdles. Used today as portable fences, they were, in the past, also used as the groundwork for clay wall-panels in "wattle-and-daub" buildings. The improvement of hedges to create an effective barrier to grazing livestock calls upon all the ancient skills of cleaving and interweaving wood. The hedger cleaves part way through the growing upright stems, bends, or "lays", them and then plaits them into the adjacent stems to form a tough, living fence.

Baskets of many types may be made from cleft and woven wood. Willow is the traditional wood for light baskets, but cleft and boiled oak laths are used for the strong "spelk" baskets widely used both in America and Britain for carrying heavier goods such as firewood and vegetables.

SAILS AND LATHE
The outer uprights, left in the round, and the intermediate uprights of cleft wood, are placed in holes in a stout plank of wood called the lathe. The upright rods are commonly called the sails. The starting rods along the lower edge of the hurdle are woven in and out of the sails and are also braided together to increase the strength of the hurdle. Each horizontal is twisted round the end upright to bind the hurdle – the rod being itself twisted round its own axis to bind the fibres and prevent splitting

PRACTISED SPEED
*Using no tools other than
the bill-hook and his bare
hands, the hurdle-maker can
make up to ten hurdles a
day. As each rod is "let in"
it is firmly interlocked
with its neighbours; no nails
or wires are used, yet a good
hurdle may last for ten years*

THE SUSSEX TRUG
*The simple, durable trug
is commonly used for
carrying wood, coal or
garden produce. The main
frame is formed by bending
a steamed strip of cleft
chestnut around a stout
wooden form and nailing
the ends together. The
handle, also of chestnut,
is nailed to the frame*

LATHS OF WILLOW
*The basket itself is made
from thin strips of cleft
willow, usually from the
white or crack willow tree.
The laths are pared down
to a slender crescent shape
using a drawknife, and are
then pressed and nailed into
the frame so that they
overlap like the strakes of
a clinker-built boat*

LAYING THE HEDGE
*In order to form a stout
barrier around his field,
the farmer may utilize the
natural hedge. The main
stems are cut part-way
through and then laid over
and interwoven. A hedge
with widely spaced stems
requires additional stakes*

Cooper and Clog-maker

THE STOCK KNIFE
The massive blade of the stock knife, anchored at one end to the cutting bench, slices easily through the soft alder as the sabot takes shape. Though very heavy, the knife can be used with great accuracy by a skilled craftsman

THE SABOT
The traditional clog found in Britain has a wooden sole and leather uppers; the sabot, worn in many parts of Europe, is carved from a solid block. The craftsman uses a series of gouges in hollowing the sabot to fit the customer's foot

Long before men mastered metals, pottery or plastics, they employed wood to make a wide range of cheap and serviceable articles, some of which survive to this day because they are cheaper than alternatives, others because they have some property that cannot be reproduced in man-made materials.

Sherry and whisky can only develop their unique flavours if they are left to mature in oak casks. Though cheaper wooden barrels are quite suitable for storing "dry goods" like butter or flour, the cooper's trade reaches its highest level of craftsmanship in the casks made for liquors. These retain not only water but, more important, the volatile spirits that would quickly evaporate from an inferior vessel. The staves for such high-quality cooperage must be hand-cleft from knot-free logs of oak; the finest woods coming from the European oaks, *Quercus robur* and *Q.petraea*, and the American *Q.alba*.

The wooden-soled clog, or all-wood sabot, is a traditional item of footwear found in many parts of the world. The timber most commonly used in Europe is either beech, *Fagus* sp., or one of the soft, easily carved, yet tough and water-resistant alders, usually *Alnus glutinosa*. The sole blocks are first cleft from the log and then stacked to allow the wood to season. These processes preserve the fibre of the wood and allow the wood to dry and shrink to a stable block. Shaping is done with the huge stock knife, which allows the maker to exert powerful leverage. The craftsman measures the future wearer's feet, observes his gait and stride, and shapes the block into a comfortable and durable shoe.

Water-pipes have been made from wood for centuries, the most popular timber being elm, *Ulmus* sp., which resists decay if kept permanently wet below ground. A long auger is used to drill out the central bore, and each log is tapered at one end to fit into a conical recess in the end of the next section, to create a water-tight union. To make a wooden water pump, an elm pipe is set upright over the well and a movable valve, or "bucket", also carved from elm and equipped with flap-valves of leather, is set inside the pipe, where it can be moved up and down by levers attached to the pump handle.

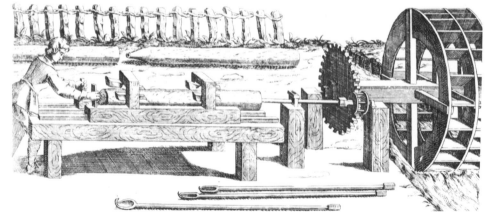

PUMPS AND WATER PIPES
Very few woods have the durability of elm when kept constantly damp. Elm water-pipes were installed in London's New River scheme in 1613—and were found to be perfectly sound when unearthed in 1930. The huge elm logs used in pump-making, some ten feet long and twelve inches square, were drilled with an auger which had a blunt spoon-shaped tip and a curved blade like a gouge. The lower pipe had a two-inch bore; the upper section a five-inch bore to take the moving part – the "bucket"

POWERED AUGER
This mechanized auger, used for boring wooden water-pipes, is taken from Isaac de Caus's "Nouvelle Invention de Lever l'Eau Plus Hault", published in 1664. In the foreground lie the unusual spoon-shaped bits with terminal cutters

SHAPING THE STAVES
The stave of well-seasoned cleft oak is held in a foot-operated clamp as the cooper uses a draw-knife to hollow the inner surface, bevel the edges and adjust the thickness. The edges are finished by drawing them across the blade of a huge fixed plane

SETTING UP
Grooves are cut at both ends of the staves to take the top and bottom of the barrel. The sixteen staves are then held at one end by an iron hoop while a wooden hoop is forced over them – drawing them together so that the bevelled edges meet in a perfect joint

FIRING THE BARREL
Before the iron hoop can be placed over the other end of the staves, the wood must be made pliable enough to bend into the final form of the barrel. This is done by lighting a small fire inside the barrel to heat the wood and increase its flexibility

FITTING THE ENDS
The two hoops nearest the end of the barrel are removed so that the staves spring apart. The top of the barrel, formed earlier from oak planks and with the bung-hole bored, is inserted and its tapered edge fitted into the grooves in the staves, caulked with rushes

THE FINISHING TOUCHES
During work on the barrel, the iron hoops become loose as the wood shrinks and becomes compacted. Fresh hoops are heated and then rammed home using a heavy hammer and an iron wedge called a drift. As the hoops cool, the staves are pulled together

The Key to Progress

From the dawn of the Iron Age, right through to the early nineteenth century, when it was superseded by coke, wood charcoal was the one chemical reagent that enabled Man to win metals from their ores. The whole development of culture and civilization hung, for thousands of years, on the process of smelting – a simple chemical process which supplied the raw metals for Man's swords and cannon, his ploughshares and the healing knife of the surgeon.

Today, the charcoal burner is little more than a folk memory in the industrial areas of the world, but his ancient product has survived in a new role. Now mass-produced in huge metal kilns, charcoal is one of the major purifying agents used in the chemical plants of our technological society.

The tannins are a naturally occurring group of chemicals having the valuable property of arresting the natural decay of hides and skins and converting them into tough, supple and durable leather. For many centuries the main source of tannin was the bark of the European and American oaks, *Quercus* spp., and the American eastern hemlock, *Tsuga canadensis*. Tanbark may only be taken from trees just felled, or about to be felled, as the removal of the bark will normally kill any tree. One curious exception is the cork oak, *Quercus ilex*, whose thick bark may be harvested without permanent damage to the tree.

Resin, also known, in America, as "naval stores" from its traditional use as a caulking material for wooden ships, is obtained by tapping members of the pine family. In southern Europe the maritime pine, *Pinus pinaster*, and in America the loblolly pine, *P. taeda*, are the main commercially tapped species. Sloping cuts, reopened each morning during the spring tapping season, allow the thick resin to ooze from the tree and collect in a small cup. The resin is distilled to separate the volatile turpentine for use as a solvent in the paint and varnish industries. The solid constituent, rosin, is used in printing inks and in specialized paper coatings to promote rapid drying. This same tree-product is also used by the violinist to prepare his bow at the start of each performance.

LIGHTING THE STACK
After removing the central stake, the charcoal burner pours red-hot embers down the flue to light the fire. The mound is then sealed so that slow combustion can convert the wood to charcoal during the two- to ten-day burning period

THE HEART OF THE STACK
The fifteen-foot-diameter mound is formed of seasoned logs stacked around a central chimney of split logs. The wood is then covered with a protective layer of grass before being sealed from the air by a thick layer of earth and ashes

THE RESIN HARVEST
Between January and November
coniferous trees may be
tapped for resin. The outer
bark is removed in a two-
foot, four-inch-wide "blaze"
against which the cup is
fixed. A fine cut is made
in the living sapwood to
promote the resin flow

THE TANBARK HARVEST
The bark is stripped from
newly felled oaks in the
spring, when the bark peels
easily into large plates.
The plates are stacked to
dry with the tannin-rich
inner surface protected
from rainfall, which would
dissolve the valuable chemical

WOODLAND RESIDENTS
Fast or uneven burning, due
to strong winds or a break
in the earth cover, could
quickly destroy the whole
stack, so the charcoal
burner lived next to his
work – sleeping in a crude
hut or tent and vigilantly
watching his charge

153

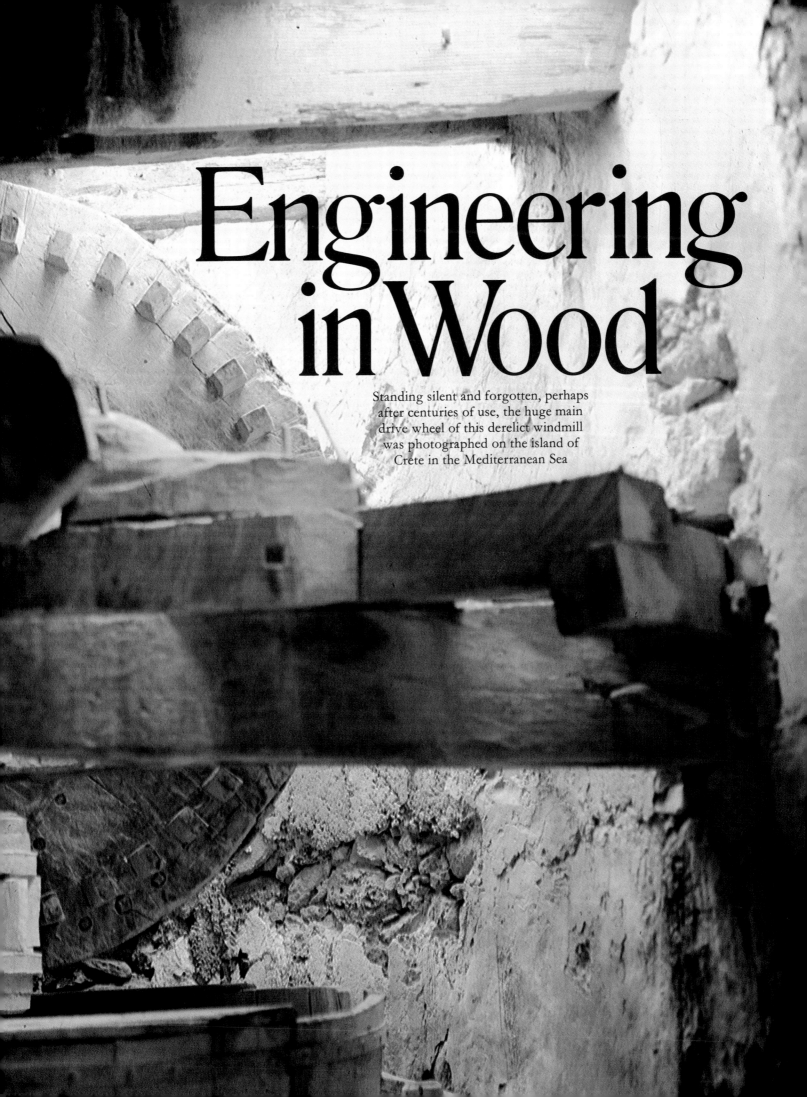

Engineering in Wood

Standing silent and forgotten, perhaps after centuries of use, the huge main drive wheel of this derelict windmill was photographed on the island of Crete in the Mediterranean Sea

Harnessing the Elements

The watermill is first recorded in Greece in the first century before Christ, and in China soon afterwards – both possibly springing from a common source in the Middle East. The simple Greek mill consisted of a horizontal wheel with diagonal blades impelled by water falling from above. It was, however, the vertical wheel set in line with the flow of water that ultimately became the standard throughout the world.

Originally harnessed to grind corn, water power also came to be used to drive pumps and ore-crushers, bellows and forges, textile machinery and gunpowder mills. Until, and indeed long after, the coming of the steam engine, water remained the cheapest and most reliable source of power available.

Windmills of a crude horizontal type originated in Persia in the tenth century, but their main development began in western Europe in the twelfth century – possibly as an independent invention. Of the two main types, the post mill has a wooden body which carries the sail and contains the machinery and can be rotated about a massive centre-post to face into the wind. In the tower, or smock, mill, a small rotatable cap supports the sails on top of the fixed body.

The standard timber for the framework of waterwheels, windmill bodies, drive shafts and gears was always oak, although its high tannic acid content could corrode any iron parts attached to it. Alternative woods for heavy framing were pitchpine, ash, sycamore, yew and elm. The windmill sails were often of fir, for lightness, and, where resistance to wear was a major factor, elm would be used for brake-blocks, bearings and gears.

For centuries, cogs were made of wood; silent when running and, if perfectly cut, remarkably durable. The wheels were often of oak or elm mortised to take teeth made of close-grained hardwoods like apple and hornbeam, beech, holly or hickory.

From the mid-eighteenth century, iron began to replace wood for gears and shafts, but the millwright's traditional wood craftsmanship persisted until, in the present century, alternative power sources brought about the end of the large-scale use of wind and water power.

THE HEAD-WHEEL
The massive oak head-wheel was a craftsman's masterpiece involving the precise cutting of nearly 100 bevelled teeth. Invariably an odd number of teeth were cut to distribute unevenness in wear. Other gears were of oak or elm with inserted peg-teeth of applewood. A brake block operated by weights could be pulled tight against the rim of the head-wheel to control speed or stop the sails

THE SAILS
The light sail-frames of fir are attached to a massive "stock" of pitchpine running from sail-tip to sail-tip. The propeller-like pitch varies from seven degrees at the tip to about 20 degrees at the root, and was traditionally called the "angle of weather"

SUPPORT PYRAMID
The entire weight of the post mill is borne by the central pillar – a massive baulk of oak some 30 inches square. The weight of the mill is transmitted to the stone piers by the four sloping quarter-beams, their upper ends chocked into the post, their lower ends into the ends of the two huge cross trees. No weight is carried by the cross trees where they cross beneath the foot of the post

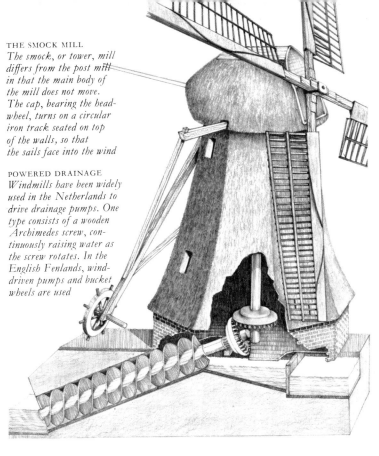

THE SMOCK MILL
The smock, or tower, mill differs from the post mill in that the main body of the mill does not move. The cap, bearing the head-wheel, turns on a circular iron track seated on top of the walls, so that the sails face into the wind

POWERED DRAINAGE
Windmills have been widely used in the Netherlands to drive drainage pumps. One type consists of a wooden Archimedes screw, continuously raising water as the screw rotates. In the English Fenlands, wind-driven pumps and bucket wheels are used

DANISH WATER WHEEL
The wheel is of the overshot type, the water being channelled to strike the wheel at its highest point. The wheel is turned by the weight of water as well as its impact. Undershot wheels rely solely on the pressure of water at the low point of the wheel

THE TUB-MILL
Tub-mills are among the earliest-known forms of water-driven wheels. The water is either directed against the impeller blades by a conduit, or is allowed to fall under gravity on to blades set at an angle to the axis of the wheel

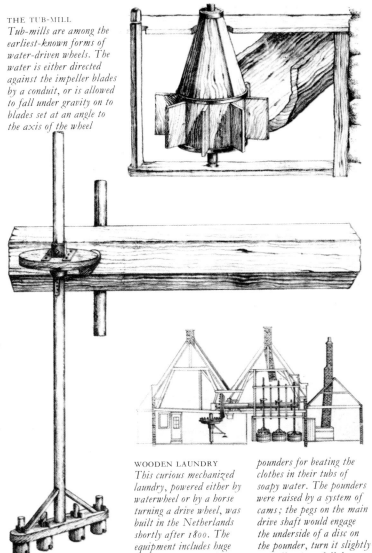

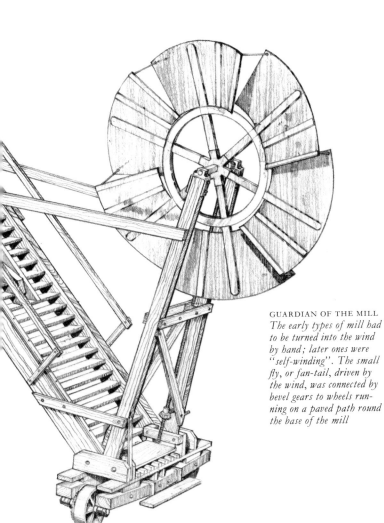

GUARDIAN OF THE MILL
The early types of mill had to be turned into the wind by hand; later ones were "self-winding". The small fly, or fan-tail, driven by the wind, was connected by bevel gears to wheels running on a paved path round the base of the mill

WOODEN LAUNDRY
This curious mechanized laundry, powered either by waterwheel or by a horse turning a drive wheel, was built in the Netherlands shortly after 1800. The equipment includes huge wooden tubs for soaking the linen and massive wood pounders for beating the clothes in their tubs of soapy water. The pounders were raised by a system of cams; the pegs on the main drive shaft would engage the underside of a disc on the pounder, turn it slightly and allow it to fall down under its own weight

The Growth of Engineering

Wood shapes easily, it is tough and provides a firm and solid frame. It has therefore been used since earliest times for wheels, levers, rollers and handles as well as for the apparatus supporting them, and remains the most convenient material for primitive technologies where precision is not important. However, wood is not so hard-wearing, and tends to generate greater friction than metals: already Homer speaks of bronze linchpins on the axles of the Greek chariots before Troy, and bronze gear-wheels dating from before the time of Christ have been unearthed in China.

Nevertheless, until well into the Industrial Revolution, wood remained the chief material used in engineering. Large structures, like the brake-wheels of windmills or the massive gear-wheels of horse-driven "whims", or hoists, could not be cast in iron, and the beams even of the early mining steam-engines were made up of timber baulks. Great wheels were usually of oak, and the cogs were shaped individually before being hammered tight into their sockets. Because oak wears longer and more smoothly when meshed with another timber, the smaller gear-wheels would be of beech or hornbeam, or often one of the fruitwoods.

In certain circumstances, for certain moving parts, wood was actually preferable to metal because of the inefficiency of the only

lubricant then available, animal-fat oil, which soon became too sticky. The pulleys of ships' tackle had blocks of elm and sheaves of lignum vitae, a wood which retains a naturally oily surface even after long use. It seems probable that the Lincolnshire brothers James and John Harrison, responsible for one of the most remarkable mechanical achievements of the eighteenth century, a marine chronometer accurate enough to be used for the determination of longitude, took the idea of using lignum vitae rollers from the pulleys of the ships they had observed in the neighbouring port of Hull.

Wooden clocks had been made since the late seventeenth century in the Black Forest region of Germany, and were made later, perhaps independently but for the same reason – their comparative cheapness – in nineteenth-century Connecticut. Wooden clocks were not as reliable as the more expensive brass clocks, since the wooden wheels, sawn out somewhat crudely from blanks, set up more friction; and as the wood dried with age, the teeth tended to split off along the grain. But the Harrisons also produced a small number of wooden long-case clocks, in which they succeeded, by meticulous workmanship and ingenious design, in overcoming these problems. The Harrisons' clocks were almost entirely of oak; the American ones usually had cleft-oak

plates, wheels of cherry or apple, and pinions of laurel.

It is notable that in Europe ships' tackle, and, in America, wooden clocks, were among the first articles to be mass-produced with interchangeable parts. This is partly attributable to the ease with which wood can be worked. Machines were set up in 1800 in the Royal Navy yard at Portsmouth in England to cut, bore and shape pulleys and pulley-blocks from the tree to the finished article; Eli Terry, the Connecticut clock-maker, mass-produced 4,000 wooden clocks between 1807 and 1810 – in which time a single clock-maker would be fortunate to make 100 brass clocks.

The engravings of Diderot's encyclopedia, which was published from 1751 onwards, show that in virtually every branch of manufacture at that time wood predominated. Arkwright's original spinning machine was made entirely of wood; but as the spate of inventions swelled into the Industrial Revolution, the use of wood dwindled. On the power-looms of the textile industry only the shuttles were of wood (though the supply of boxwood ran short in the 1930s, cornel and persimmon continue to be used for shuttles). Within the space of fifty years, the slow, too tranquil creaking of wooden crown-wheels and spurs had yielded irrevocably to the high-speed clatter of iron and steel.

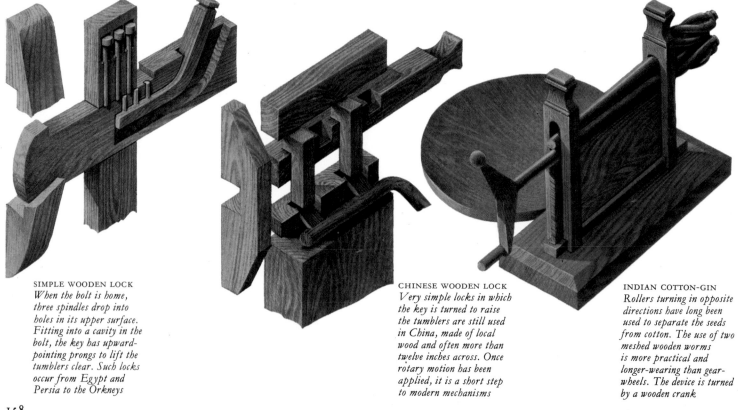

SIMPLE WOODEN LOCK
When the bolt is home, three spindles drop into holes in its upper surface. Fitting into a cavity in the bolt, the key has upward-pointing prongs to lift the tumblers clear. Such locks occur from Egypt and Persia to the Orkneys

CHINESE WOODEN LOCK
Very simple locks in which the key is turned to raise the tumblers are still used in China, made of local wood and often more than twelve inches across. Once rotary motion has been applied, it is a short step to modern mechanisms

INDIAN COTTON-GIN
Rollers turning in opposite directions have long been used to separate the seeds from cotton. The use of two meshed wooden worms is more practical and longer-wearing than gear-wheels. The device is turned by a wooden crank

ANTI-FRICTION DEVICE
To reduce friction in their clock of oak and brass, the Harrisons made ingenious use of the self-lubricating property of lignum vitae. The teeth of the oak wheels meshed with barrel-shaped lignum vitae rollers, held in a cage on brass pins.

Oak wheels were held on brass pins fixed in lignum vitae bushes, but the axle of the brass escape wheel rested on the peripheries of two lignum vitae discs. The escapement of oak was tipped with lignum vitae where it engaged between the escape wheel's teeth

X-RAY OF OAK WHEEL
An X-ray reveals that both the body and teeth of the oak wheels were radially sawn to avoid warping. To increase their strength, the teeth were cut in separate groups of three or four so that the grain should run lengthwise through each

LONGCASE CLOCK
The brothers John and James Harrison, carpenters in a remote English village, made two clocks of great time-keeping precision almost entirely of oak. It was their proud boast that the clocks would not require cleaning for 50 years

Oak escapement

Oak gear segment

Lignum vitae tip

Lignum vitae roller

Brass cage

Lignum vitae bush

Escape wheel

Lignum vitae discs

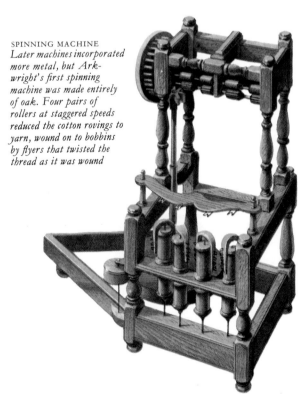

SCREW-CUTTING ENGINE
Reconstructed from a drawing in Leonardo da Vinci's notebooks, this machine was conceived and designed in wood. By gears which are changed to alter the cut, the hand-turned crank moves the cutter over a rotating shank

SPINNING MACHINE
Later machines incorporated more metal, but Ark-wright's first spinning machine was made entirely of oak. Four pairs of rollers at staggered speeds reduced the cotton rovings to yarn, wound on to bobbins by flyers that twisted the thread as it was wound

The Bridge-builders

Since man first crossed a river dry-shod over a fallen tree trunk, the timber bridge has played a major role in his ability to travel, to explore new lands and to keep open the trade routes and communications of his increasingly complex society. Although, since Roman times, stone has been a primary building material for permanent bridges, replaced in later years by iron, steel and concrete, the cheapness of wood, its versatility in use and its availability in areas of difficult terrain have all kept the timber bridge far from oblivion.

Few examples of any great age now exist as punishing wear and exposure to the elements have taken their toll, but carved, drawn, painted and photographed records survive to illustrate the bridges of past ages. The massive wooden structures which carried the early railways across the North American continent, and deep into the hill country of northern India, are a recent chapter in the history of the bridge, while the plaited bamboo suspension bridges of western China are still made to a pattern which caused the traveller Marco Polo to pause and to write of their outstanding strength and beauty of design.

Many different structural types are found, the simplest and most widespread consisting of plain beams supported on piers and planked over to form a deck. A refinement, developed in Britain by Brunel, used splayed struts at the top of high stone piers to spread the load and allow the structure to carry heavy rail traffic. The pioneer railway bridges of America and Africa were often of the trestle type, built up, tier upon tier, of massive timbers braced by diagonal crossmembers.

Simple arched bridges of wood have never been common, but a number of interesting designs have been tried. In 1837–8, the engineer Green developed a laminated structure for viaducts, but the timbers succumbed to internal rot between the laminates.

Truss girder bridges, built in triangular patterns, are far lighter than those made of massive beams and, providing that the joints are sound, just as strong. The design was developed scientifically during the Renaissance by Leonardo and Palladio, but the concept originated with the Romans, and Trajan's great bridge across the Danube is perhaps the most outstanding example. Here the truss was combined with the arch, as it often was in the long tradition of European bridges. Truss bridges built to a variety of patterns proliferated in nineteenth-century America, where, as in Switzerland, they were often roofed over to protect the structure from rain and snow.

Almost every conceivable kind of timber has been used in bridge building. Brunel's viaducts were made of Memel pine treated with mercury sublimate – a forerunner of the creosoting process – which gave the timbers a life of up to 60 years. Elsewhere a wide variety of both hardwoods and softwoods has been pressed into service and, in the countries of Asia, the versatile bamboo – not strictly a wood but one of a highly specialized group of "woody" grasses – has been widely used as a constructional material.

RED SUCKER TRESTLE
The mammoth task of linking the east and west coasts of North America inspired the early railway engineers to build some of the world's largest timber bridges. The eight-tier Red Sucker bridge was built of massive timbers cut from adjacent hillsides

SCHAFFHAUSEN BRIDGE

The long European tradition of combined arch and truss construction reached its peak with Grubenmann's magnificent Schaffhausen Bridge. Two covered spans, each more than 170 feet long, reached out to meet at a stone-built mid-stream pier

CANTILEVER CONSTRUCTION

This simple cantilever bridge over the Spiti River, in the Himalayas, is formed of successive layers of rough-cut timbers extending outwards from a stone pier. The timbers are firmly held by pegged wooden crossmembers to prevent lateral movement

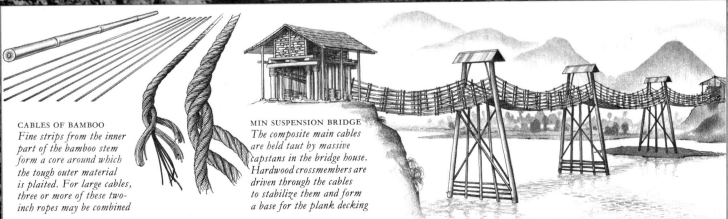

CABLES OF BAMBOO

Fine strips from the inner part of the bamboo stem form a core around which the tough outer material is plaited. For large cables, three or more of these two-inch ropes may be combined

MIN SUSPENSION BRIDGE

The composite main cables are held taut by massive capstans in the bridge house. Hardwood crossmembers are driven through the cables to stabilize them and form a base for the plank decking

Canal and River Locks

The earliest form of lock gate, by which a boat could negotiate a change in water level of a river or canal, was the simple flashlock, or staunch – a single massive vertical gate of oak or elm held in a stout wooden frame embedded in the river banks. The staunch was inefficient and very wasteful of water, but survived until very recently.

The much more efficient pound lock, with double, or single, gates placed at either end of an enclosed chamber, appeared in Europe in the fourteenth century, but remained rare until the Renaissance. The medieval locks were built entirely of wood; later, masonry walls were added to the timber frame to make the structure more durable. The early types had vertical guillotine gates, but the double mitre gate invented by Leonardo da Vinci was far superior and quickly became universally adopted.

To resist the impact of heavily laden boats, and hold back the tremendous mass of impounded water, the frame had to be massive, tightly jointed and built of the finest-quality timbers. Although the sizes of gates may vary considerably, the main timbers are rarely less than a foot square and the counter-balancing beams which serve as levers for opening and closing the gates may be more than twenty-five feet long. Even the gates of a narrow, seven-foot, lock will contain nearly 200 cubic feet of timber.

For the frames, oak was used almost universally until greenheart became available in more recent times. Ordinary deal will serve well as the planking over the frame; it is comparatively cheap and damaged sections are easily replaced. The sills against which the gates close and the buffers, or rubbing blocks, must absorb heavy blows without splitting and elm or oak are preferred.

Along the banks of navigable rivers, erosion may be controlled by walls of timber piles clad with horizontal planking. Bundles of brushwood may be weighted down with rocks or tied to fixed posts driven into the river bed to reduce the damage caused by stones carried along by strong currents or seasonal flood waters, while, on a larger scale, banks and dykes may be stabilized with thick mattresses of willow, birch or hazel.

THE HEEL POST
Greenheart, Ocotea rodiaei, imported from Guyana, is used for heel and mitre posts

The timber is exceptionally hard and strong, is very durable even under harsh conditions of use and is resistant to marine borers

THE HINGE
A section of the massive greenheart heel post is cut away to accommodate the iron straps which hold the

gate securely in its niche in the masonry wall. Even on a narrow, seven-foot, lock each gate may weigh nearly three-quarters of a ton

FOXTON LOCK
Situated on the Grand Union Canal in the English Midlands, the Foxton series will raise a barge through more than 75 feet in a distance of half a mile. The series consists of two groups of five "steps" separated by a wider sec-

tion in which barges may pass and which also serves as a reservoir to maintain the water levels throughout the series. The gates of the upper locks weigh about three-quarters of a ton per pair; those at the bottom, between 1½ and 2 tons per pair

POLDER CONSTRUCTION
Over hundreds of years, the Dutch have perfected techniques of land reclamation by building dykes across major sea inlets. These dykes must be waterproof and resistant enough to withstand the constant battering of the waves. The dyke core of sand is covered with boulder-clay, to waterproof the structure, and thick mattresses of woven willow are overlain to protect the dyke from the scouring action of the waves. The willow mats are resistant to rotting when immersed in sea-water

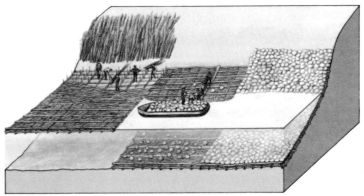

UNDERSEA MATTRESSES
Young willow stems are woven into thick mattresses which are then floated into position and weighted down with rocks discharged from barges. The natural flexibility of the willow allows the mat to follow the contours of the sea-bed

The First Railways

A WORLD BELOW GROUND
*In the early 16th century,
an anonymous artist known
as the Petrarch Master
made a series of woodcuts
illustrating mining activities.
This busy scene shows
a large gallery with hewers,
fire-setters and a "putter"
pushing a small truck*

The railway has its origins in the metal mines of medieval Europe, where small wooden trucks, pushed by hand, carried ore along the underground levels, and occasionally larger surface railways, with horse-drawn wagons, linked the mines with the mills and ports.

For centuries the rails were made of wood, mainly of wear-resistant hardwoods like oak, hornbeam and beech, but even the toughest rail might last for only two or three months. Softwood rails were sometimes used with a replaceable strip of hardwood nailed on to the surface.

The trucks were guided by flanges nailed to the rail edges, by horizontal wheels running between the massive rails or by a fixed peg of hardwood or iron projecting down from the truck into the narrow slot between the broad planks of the track. These trucks generally had softwood bodies, often of fir, with wheels of elm – a dense, compact wood whose fine interlocking grain is exceptionally resistant to splitting. Such simple railways were numerous throughout central Europe with a few outliers in Russia, Sweden, France and Britain.

Railways as we now know them, with flanged wheels, evolved in Britain from the early seventeenth century, generally linking the fast-developing collieries with the inland navigable waterways. The rails were usually of oak, ash, beech or alder and were pinned to the sleepers of ash with oak trenails. Sometimes fir rails were used, capped with a thin layer of beech, which, under constant use, acquired a high gloss, reducing friction between the wheel and rail.

The oak-framed coal wagons were clad with planking of fir and, on the early models, the wheels were made up of two wooden discs of different diameter, to form the tread and flange, or of several pieces held together by S-shaped clamps of iron. Brake-blocks of beech were forced against the wheel rims by a long bar of alder wood.

Iron rails ousted the wooden variety in Europe between 1790 and 1820, but in America wooden rails were widely used well into the 1870s. Indeed, wooden rails are still used on a few lumber railways in America and Australia.

SHIPPING THE LOAD
*As sea-going ships could
not travel far upstream, the
coal was initially carried by
barrow from the rail wagons
to small lighters for transfer to the waiting ships.
Later, staithes allowed the
coal to be loaded directly
into the barges, or "keels"*

STAITHES AND HURRIES
*The staithe was a loading
platform built out from the
river bank and supported on
stout timber piers. Railway
wagons could be run out over
the waiting lighters to
discharge their loads straight
into the hold down chutes
known locally as "hurries"*

THE ENGLISH WAGON
The typical coal wagon of the late 18th century was framed in heavy oak timbers and clad with fir. On steep downhill gradients, the driver would force the beech brake block on to the wheel by leaning, or even sitting, on the alder brake lever

CONTRACTOR'S RAILWAY
The construction and main-tainance of the mine railways was usually contracted out. The contractor was respon-sible for bridges and cuttings and was paid an agreed sum for each mile of track – higher rates being paid for the more difficult sections

WAGON HOIST
Another, less widely adopted, mechanism used a movable counter-balanced platform to lower the whole rail wagon down to the level of the barge. There the load could be discharged by releasing hinged doors in the floor of the wagon

MINING TRUCKS OF EUROPE

TRANSYLVANIAN RIESEN
This crude German flanged-wheel truck, dated early 19th century, runs on a simple track of pine logs

LEITNAGEL HUND
The 1785 "hund", found in a coal mine in the Bernese Oberland, is held on the plank track by the guide-pin (leitnagel) projecting below the body

GUIDE-WHEEL TRUCK
The guide-wheel system was widely used in Hungary and spread to other parts, notably the copper mines of Sweden, in the middle and late 18th century

Wooden Formwork

WOOD-GRAIN FINISH
The exterior concrete of the Hayward Gallery in London has been given a variety of contrasting finishes. Here, rough-sawn boards give a wood-grain effect to the wall of a walk-way – contrasting with the "stone block" finish of the adjacent wall (inset)

Although a number of concreting techniques were known to the ancient civilizations of the world, the use of concrete in large-scale building operations is very much a twentieth-century phenomenon, and, in the developed world, high-rise buildings, motorway flyovers, bridges and dams have become a familiar sign of progress.

Amid this scene of concrete and steel, however, it is easy to underestimate the vast quantity of timber used in a major concrete structure. In addition to the wood used for flooring, staircases, window and door frames and interior finish, and the exterior woodwork used for decorative effect, the whole structure has, at an earlier stage, been "built" in wood in the form of the timber formwork of the concreting process.

The type of board used for the formwork has a direct effect on the appearance of the finished surface; rough-sawn boards and specially textured boards may be used to reproduce on the cast concrete the patterns and textures of wood – often giving a far more pleasing appearance than would be obtained from smooth formwork. Indeed, even using large-area boards with close-fitting tongue-and-groove joints, the wet concrete finds its way into the joints and a smooth finish is only achieved by sanding once the concrete has set. Alternatively, beading may be fixed to the inside of the formwork, over the joints, to create an attractive building-block pattern on the set concrete.

A very high proportion of the wood used in formwork is heavy-grade plywood made from Douglas Fir. The cross-grained construction of the plywood gives it tremendous strength and impact resistance and great stability under conditions of alternate wetting and drying. The composite wood may be nailed close to its edges without splitting and, when treated with an oil before use, can be stripped from the set concrete quickly and efficiently – leaving a clean surface on the work and allowing the board to be reused many times. When the board has reached the end of its useful life as a form, it may still be used as general site lumber or for internal cladding of roofs or walls wherever appearance is not an important consideration.

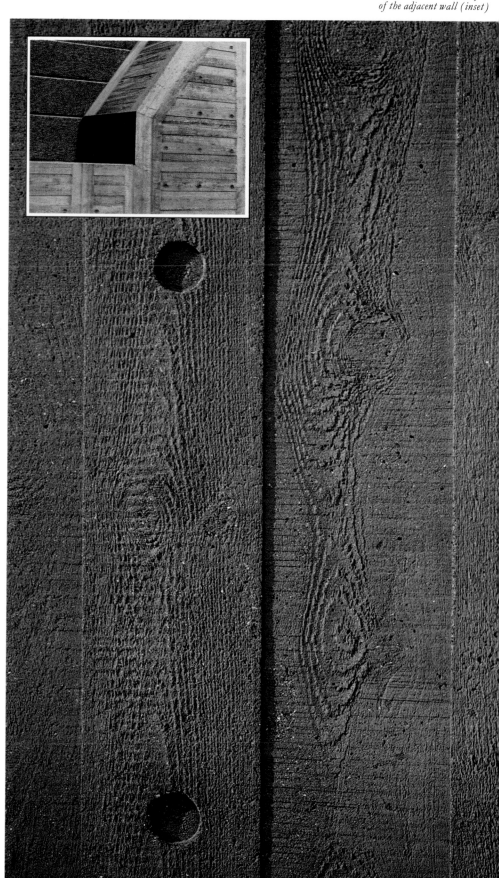

POURING A COLUMN
The timber formwork for an internal column, strengthened with battens and locked in place with iron clamps, has been erected around the steel reinforcing rods. Wet concrete is about to be poured from a crane-mounted skip into the mould

FLYING BUTTRESSES
The Roman Catholic cathedral of Christ the King, in Liverpool, England, is supported by sixteen enormous concrete buttresses each 118ft long. Here, a buttress and its supporting pillars are stripped, revealing the marks of the formwork

MOULDED PINNACLES
Each of the buttresses is linked to its neighbours by concrete tie-rings cast in sections. On completion, the great cone of the cathedral was surmounted by a 70ft-high lantern carrying sixteen pinnacles of cast concrete each more than 60ft high

Forms and Patterns

Unlike so many traditional crafts, the old and little-known craft of pattern-making has increased its importance in the age of technology. Though aided by modern lathes, band-saws and sanding-machines, the pattern-maker still works, one man one job, with the skill of his hands.

Essentially, the task of the pattern-maker is to produce, from the drawings of a designer, a wooden pattern, which forms a vacant shape in the sand of the mould-box into which molten metal is poured to make a casting. Because protuberances and cavities on the pattern would trap or disturb the sand when the pattern is removed, patterns are made usually in two halves or in a number of interlocking pieces. Hollow objects, for instance a pipe, can be shaped only by inserting in the mould-box a block of hardened sand, called a core, for which a separate pattern, a coreprint, must be made. Such complications often have the result that the pattern bears little relation to the completed article.

For three reasons, wood is the ideal material for patterns: it can be easily shaped in any form; it will retain its shape, and it can also be easily altered by paring or by the addition of another piece of wood. The woods most favoured vary from country to country, but the most common are Canadian yellow pine and Western white pine. Besides these two easily worked woods, Honduras mahogany, though still mild, is harder wearing, and gelutong, a knot-free latex wood, has become increasingly popular since the Second World War, especially in Europe. The Japanese make use of katsura, a native timber with the qualities of boxwood.

Pattern-makers still work usually in small, independent shops of about a dozen men, although large concerns support their own shops. The degree of accuracy the pattern-maker must attain is high, and by the time he has served his long apprenticeship, he has learnt enough to make him proficient in several other wood-working trades. It is easy to forget, amid the vast machinery of, for instance, a motor-car production line, that the clattering ranks would soon fall silent without the wood-working skill of the pattern-maker.

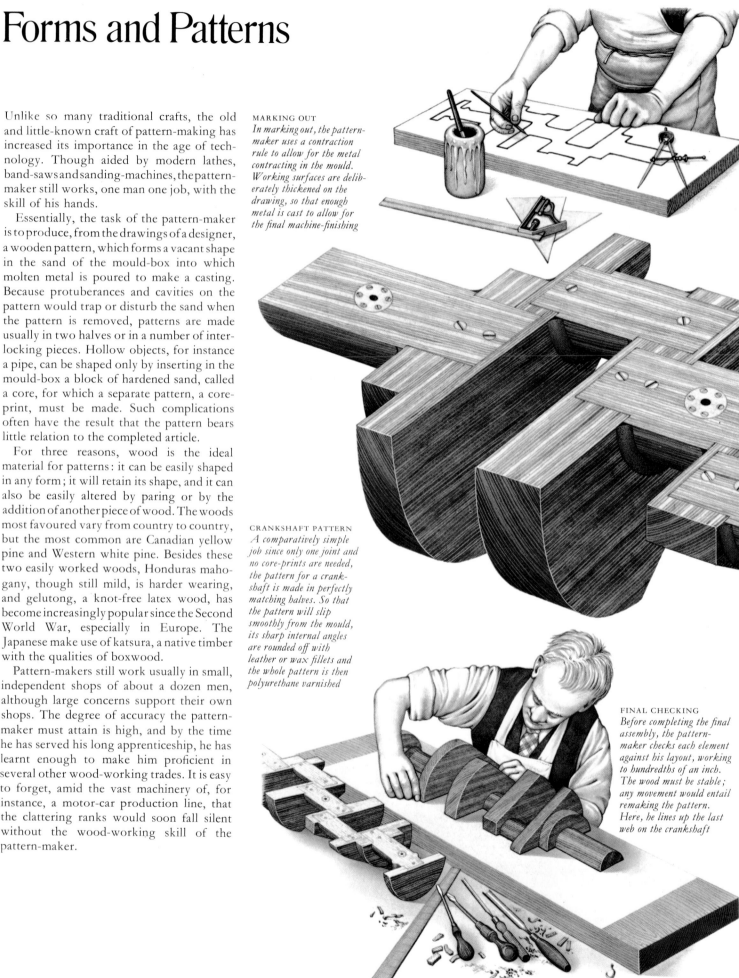

MARKING OUT
In marking out, the pattern-maker uses a contraction rule to allow for the metal contracting in the mould. Working surfaces are deliberately thickened on the drawing, so that enough metal is cast to allow for the final machine-finishing

CRANKSHAFT PATTERN
A comparatively simple job since only one joint and no core-prints are needed, the pattern for a crankshaft is made in perfectly matching halves. So that the pattern will slip smoothly from the mould, its sharp internal angles are rounded off with leather or wax fillets and the whole pattern is then polyurethane varnished

FINAL CHECKING
Before completing the final assembly, the pattern-maker checks each element against his layout, working to hundredths of an inch. The wood must be stable; any movement would entail remaking the pattern. Here, he lines up the last web on the crankshaft

BUILDING UP
*Having prepared and
marked up his wood, the
pattern-maker shapes the
segments. He has at hand
callipers, "pinning dogs",
and a pattern-maker's
cranked gouge. Internal
curves are shaped on a
spinning sanding-bobbin*

DECORATIVE MOULDINGS

Elaborate hand-carved picture frames and
decorative motifs for use on furniture
were prohibitively expensive to produce,
so craftsmen of the 18th and 19th centuries
developed the technique of casting these
items in a paste which, once hardened and
gilded, was indistinguishable from wood.
The moulds were beautifully carved in
boxwood, rosewood or beech – fine-grained
woods which take finely detailed carving and
remain stable almost indefinitely

*A pair of moulds, used in conjunction, produce
a Louis XIV decoration for the front of a table*

*Blocked to prevent the wood from splitting, this
single mould makes all the elements of a frame*

*Worm-resistant and stable, these boxwood moulds
are still used 200 years after they were made*

THE MOULD-BOX
*Like the pattern, the mould-
box is in two halves. Sand
is rammed into each around
the half-pattern and then,
with a sharp tap, the
pattern is removed. When
the casting has been made,
it is trimmed and machined
to its exact dimensions*

Ships and the Shipwright

The ribs of this Scottish fishing boat are cut from oak and hoisted into place over the keel. Once the frame is complete, the two-inch planking of spruce is added – each plank cut by eye alone and steamed so that it will bend to the shape of the hull without splitting

Canoes and Rafts

The one generic factor linking most of the many and varied boats built by Man is the tree – almost invariably the source of the material from which they are made. Wood is tough, buoyant, easily worked, even with the most primitive tools, and readily available.

The simplest form of boat is the raft, consisting usually of a number of large logs lashed together and often braced by transverse members. Many rafts are decked to raise crew and load well clear of the water. Even when fitted with centre-boards, rafts are inefficient under sail and are seldom used as sea-going craft. Notable exceptions are the rafts of ancient Peru, on which Thor Heyerdahl modelled his *Kon-Tiki*.

The basic dug-out, simply a hollowed-out log, is unstable and has very little freeboard. To increase freeboard the sides may be built up with planks, while the stability of the craft is often improved by lashing two hulls together to form a catamaran, or by the addition of an outrigger. Sailing a canoe with a single outrigger requires a unique order of seamanship; the outrigger must be kept to windward at all times and the hull is therefore double-ended to allow tacking.

Lighter boats, using very little wood, may be built by covering a wooden frame with a skin of hide or canvas. The Eskimo kayak, or hunting canoe, and the larger umiak, used for carrying goods and chattels, are among the best-known examples; others are the Irish curragh and the Welsh coracle. The birch-bark canoes of the North American Indians are made of a light cedar frame clad with carefully jointed sheets of bark.

The natives of northern Australia make a remarkable canoe from a single sheet of bark; the sheet is folded and the ends sewn together with strips of the same material. In South America, a similar type of craft is made, but a V-shaped notch is cut in each side and then tightly sewn to raise the open ends of the canoe clear of the water.

Boats made from bundles of giant reeds lashed together are still widely used in coastal and inland Peru, and in the Nile valley. Present-day descendants of the great papyrus boats of Ancient Egypt, they are a direct link with the civilized past.

PAPYRUS RAFT
The mountain scrub around Lake Tana in Ethiopia has little wood, and boats are built with tied bundles of papyrus, Cyperus papyrus, as in Egypt 5,000 years ago. Lasting perhaps only two or three months, these paddled craft may be up to 30ft long and can carry a substantial cargo. Similar reed-built craft sail on Lake Titicaca

BAMBOO RAFT
The concave Formosan raft curves up sharply at the bow and less sharply at the stern and across the beam. Of trapezoidal shape, 30ft long, it is paddled and sailed by three men

GIANT BAMBOOS
Dendrocalamus giganteus or D.strictus, both very tall, thick bamboos, are used in Formosa to construct light, sea-going rafts, the thin ends of the tapering stems being laid at the bow. The mast-step, centre-boards and oars are usually of local pine

Balsa

BALSA TREE
Taking its name from the Peruvian word for raft, the balsa, Ochroma pyramidalis, reaches a height of 75ft. Dry wood is as light as cork, but green logs are used for rafts, as the sap retards water absorption

BALSA RAFT
Sea-going balsa rafts are well attested in ancient Peru, and on rivers and lakes remain in wide use today. The central logs protrude at the stern, on which the steersman stands. Rafts are paddled, or sailed with the aid of centre-boards

BREADFRUIT
Besides food, many species of breadfruit, Artocarpus, provide good boat-building timber that is easy to work. Palm-trees and mango-trees also serve for hulls, and coconut fibre is sometimes used to lash the outriggers and the built-up side-planks

OUTRIGGER CANOE
An outrigger is usually a smaller log attached on one or both sides of the hull by lashed poles, to lend it stability. Often very fast, sometimes 70ft in length, built-up outrigger canoes are found throughout the Pacific and Indian oceans

IROKO
Any thick, tall tree with durable wood is suitable for making a dug-out canoe; iroko, Chlorophora excelsa, is a common tree in West Africa and widely used. The log is trimmed and hollowed by repeatedly charring, then adzing the wood

Iroko

White spruce

DUG-OUT
Unseaworthy and suitable only for lakes and rivers, the simple dug-out is found all over the world in great variety. This West African canoe has a flat stern on which a spear-fisherman can stand, or to which now-adays an outboard motor is commonly fitted

PAPER BIRCH
The bark of the paper birch, Betula papyrifera, is slit vertically and peeled from the trunk in one piece. Cuts are made where necessary to ease the bending of the bark into the shape of the canoe. The ribs and thwarts are usually made of cedar

WHITE SPRUCE
Roots of the white spruce, Sitka glauca, which grows widely across Canada, are used by British Columbian Indians to sew together the birchbark pieces making up their canoes. Pitch from the same tree is used to make the seams watertight

BIRCHBARK CANOE
Found all over the world, bark canoes were perfected in North America. They are normally about 12ft long, but two joined sheets of bark may be used to make longer canoes. These light, paddled craft are ideal for use on fast rivers

Kit-boats of Egypt

ANCIENT TIMBER TRADE
*The cedar forests covering
the mountain slopes of the
Lebanon were for centuries
a source of Canaanite and
Phoenician wealth. A 700
BC Assyrian relief shows
how timber was shipped: the
logs stacked high on deck
and towed in strings astern*

CHEOPS' SHIP
*Cheops' "downstream"
barge was carefully dis-
mantled 4600 years ago and
lay in 1,224 pieces in a pit
beside the Great Pyramid
until discovered in 1954. It is
the oldest ship in the world.
Cheops' sailing barge probably lies hidden near by*

In the fourth millennium BC, the Egyptian ships that plied the 750-mile stretch of the River Nile between the First Cataract and the sea were simple craft made of bundles of reeds lashed together. They rode the river downstream to the coast, returning upstream helped by the prevailing northerly winds. These craft had wooden decks on which cargo was carried and deck-houses to provide shade from the burning sun. They were equipped with paddles and a sail set on a bipod mast positioned well forward. Such simple boats continued to be made throughout Egypt's history, but many stouter vessels were evolved to transport stone from quarry to temple site and to carry priests and pharaohs in fitting style.

The earliest wooden boats copied the hull form of the reed boats and were carefully pieced together, using tenons and dovetailed clamps, from short planks – the only timber locally available. For larger vessels, cedar was imported from Lebanon as early as the 1st dynasty, and one reason for the northward expansion of Egypt's influence was to ensure its supply. In the construction of Cheops' ship, discovered beside the Great Pyramid in 1954, huge planks of cedar up to 75 feet long and five inches thick were used. However, local woods such as acacia, hophornbeam, juniper, siddar and soapberry were used for smaller parts.

The discovery of Cheops' ship proved that even large vessels were built of sewn planks without metal fastenings and without a keel. Instead, like a bow bent by a bowstring, the hull was tensioned at a higher level by a longitudinal member running beneath the deck-beams; in sea-going ships this spine was strengthened by a thick rope truss running above the deck under tension.

Numerous such "hogging trusses" braced the largest ships, like Queen Hatshepsut's barge, built in 1500 BC to transport granite obelisks from Aswan to Thebes. We read that "sycamores from the whole land" were ordered for the ship's construction, but cedar was probably used for all the main beams. This huge ship probably had a displacement of some 7,500 tons, and thirty oar-powered tugs were needed to propel it.

BIPOD MAST
*Though originally designed
for reed-boats, the bipod or
two-legged mast was suitable
also for wooden boats, as it
straddled the hogging truss.
When raised it was sup-
ported by many stays; when
not in use it was lowered to
rest on a gallows. The yard
bore a large square sail made
of cloth or papyrus, stretched
at the foot by a boom. Though
earlier stepped well forward
by 1500 BC it had been moved
to the midships position*

SAHURE'S SHIP
*Reconstructed from a relief
in Pharaoh Sahure's burial
temple, this is a sea-going
vessel of c. 2450 BC. It shows
many elements typical of
Egyptian ship-building:
sewn planking, a bipod mast,
the hogging truss and a girdle
running around the hull*

STEERING OARS
*With its large spreading
sail and lack of stabilizing
keel, the Egyptian ship
needed several pairs of
steering oars to maintain
full control. Smaller craft,
particularly river craft,
required only a single steering oar mounted astern*

THE GIRDLE
*Rope netting girdling the
hull at deck-level counter-
acted any tendency for the
sides of the hull to flare
out under the pressure of
the tightened hogging truss.
Cross-beams supported
the deck and locked together
the upper sides of the shell*

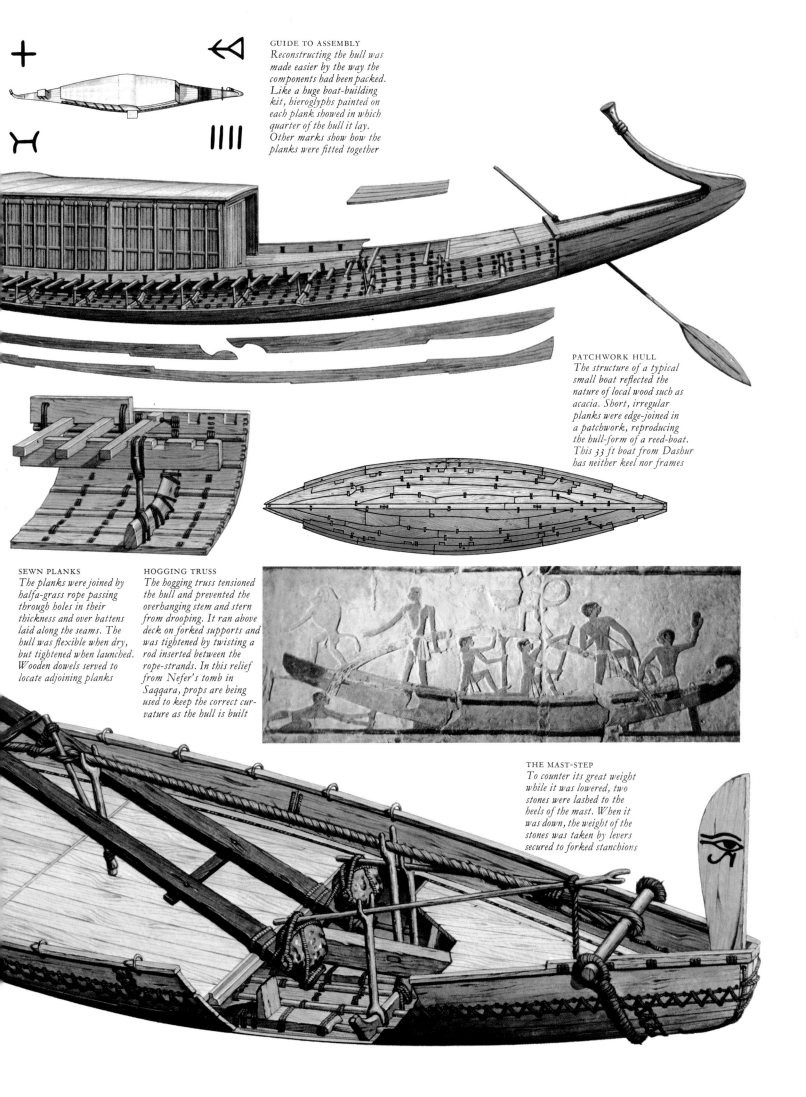

Reconstructing the hull was made easier by the way the components had been packed. Like a huge boat-building kit, hieroglyphs painted on each plank showed in which quarter of the hull it lay. Other marks show how the planks were fitted together

PATCHWORK HULL

The structure of a typical small boat reflected the nature of local wood such as acacia. Short, irregular planks were edge-joined in a patchwork, reproducing the hull-form of a reed-boat. This 33 ft boat from Dashur has neither keel nor frames

SEWN PLANKS

The planks were joined by halfa-grass rope passing through holes in their thickness and over battens laid along the seams. The hull was flexible when dry, but tightened when launched. Wooden dowels served to locate adjoining planks

HOGGING TRUSS

The hogging truss tensioned the hull and prevented the overhanging stem and stern from drooping. It ran above deck on forked supports and was tightened by twisting a rod inserted between the rope-strands. In this relief from Nefer's tomb in Saqqara, props are being used to keep the correct cur-vature as the hull is built

THE MAST-STEP

To counter its great weight while it was lowered, two stones were lashed to the heels of the mast. When it was down, the weight of the stones was taken by levers secured to forked stanchions

The European Traditions

On the fringes of northern Europe, where trees were sparse, as in northern Norway and western Ireland, skin boats were made from very early times. They needed only slender branches for their construction, preferably of willow or hazel, over which a covering of greased animal hides could be tightly stretched.

Elsewhere the tree cover was generally more extensive than it is today and the early ship-builders had little difficulty in finding trees of large size growing reasonably close to the water's edge. At least as early as the beginning of the Bronze Age, cleft slabs from large trees were sewn together to build stout sea-going vessels. In early Bronze Age Denmark there were sea-going vessels with as many as thirty-two thwarts; fighting ships with projecting beaks fore and aft, controlled by a skeg, or steering fin.

One such beaked war-canoe, dating from about 300 BC, was discovered in a bog at Hjörtspring, in Denmark. The shell was constructed from five huge planks of lime, each some fifty feet long and twenty-seven inches broad. When a replica was built in 1971, no limes of this size could be found in the Danish State Forests.

The earliest known boat dates from the Mesolithic period, a time when only the smallest and simplest stone tools were in use. The boat is a dug-out made from a pine trunk ten feet long and twenty inches in diameter; however, oak was generally preferred for all types of dug-out and these continued to be made until the twentieth century in some parts of northern Europe.

It is thought that the clinker-built boat evolved from the simple hollowed-out log boat, to which initially a few planks were added to increase the height of the sides. Like the vessels built in the Mediterranean and in Egypt, they were constructed shell-first. The keel and stem-posts were laid down and scarfed together, after which a series of overlapping strakes was added. The earliest examples were fastened together with root fibres; these later gave way to iron rivets which were themselves superseded by rivets and nails of copper. When the outer shell was completed, frames were inserted to give the hull strength. The secret of the clinker-built boat's lasting success was its flexibility, combined with the great strength and comparative lightness of its radially cleft oak planks. Flexibility and resistance to shock was imparted by the framing, which was lashed to cleats left standing proud of the surface on the inside of the planking.

This boat-building technique culminated in Viking times with the building of supremely beautiful sea-going boats with gracefully curving lines. The finest examples to have survived come from Oseberg and Gokstad in southern Norway and were built in 800 and 850 AD respectively. These too are oak built and the keel of the Gokstad ship was made from a straight-grained oak trunk eighty feet long – the like of which could not be found in Norway today.

As early as the Bronze Age, ships' prows were decorated with carved wooden heads. The Oseberg ship has beautifully executed friezes of entwined mythical beasts at stem and stern, but, as it was a queen's pleasure-boat, it lacks the terrifying dragon-head that surmounted the prow of a warship.

The sail was a comparatively late innovation in Scandinavia, but was widely used in northwestern Europe some five centuries earlier where a different method of boat-building was employed as early as the Iron Age. Although its existence was known from the works of Roman authors, and from contemporary coins, tangible evidence of the ancient Celtic boat-building tradition has only recently been discovered.

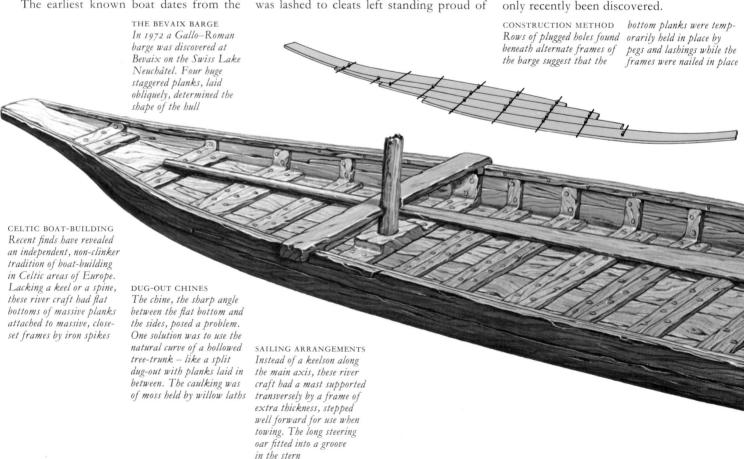

THE BEVAIX BARGE
In 1972 a Gallo–Roman barge was discovered at Bevaix on the Swiss Lake Neuchâtel. Four huge staggered planks, laid obliquely, determined the shape of the hull

CONSTRUCTION METHOD
Rows of plugged holes found beneath alternate frames of the barge suggest that the bottom planks were temporarily held in place by pegs and lashings while the frames were nailed in place

CELTIC BOAT-BUILDING
Recent finds have revealed an independent, non-clinker tradition of boat-building in Celtic areas of Europe. Lacking a keel or a spine, these river craft had flat bottoms of massive planks attached to massive, close-set frames by iron spikes

DUG-OUT CHINES
The chine, the sharp angle between the flat bottom and the sides, posed a problem. One solution was to use the natural curve of a hollowed tree-trunk – like a split dug-out with planks laid in between. The caulking was of moss held by willow laths

SAILING ARRANGEMENTS
Instead of a keelson along the main axis, these river craft had a mast supported transversely by a frame of extra thickness, stepped well forward for use when towing. The long steering oar fitted into a groove in the stern

BRONZE AGE SHIP

Hundreds of ship pictures survive in Scandinavia, engraved on rocks, weapons and utensils. Some seem to represent skin boats, but the largest, with more than 30 thwarts indicated, must have been wooden. The ship carving, right, is in Sweden

NORTH FERRIBY BOATS

On the Humber estuary in Bronze Age England, huge logs were adzed to build hulls up to 50ft long. Their remains show that carving cleats from the solid, for the attachment of internal braces, was already a familiar technique

BOUND JOINTS

The thick oak planks were scarfed together: moss was forced into the seams and held in place by slender oak laths over which passed stitches made from twisted yew roots. These bindings were countersunk to avoid excess wear

THE NYDAM SHIP

Found in Schleswig-Holstein in Germany, the Nydam ship is a pre-Viking, fourth-century open rowing boat, 75ft long. Its outer shell is constructed from only fifteen pieces of wood; each strake is a single curved plank of cleft oak

FLEXIBLE HULL

The strakes were fastened together with iron rivets, but the framing of curved branches was lashed on to pairs of cleats carved in-board from the solid strake. Rowlocks for 30 oars were lashed to the gunwales of the symmetrical hull

THE GOKSTAD SHIP

Many developments mark the four centuries separating the clinker-built Nydam and Gokstad ships. Strakes made of several scarfed planks permit a gracefully curving hull, and oarports allow an increase in the freeboard. Propulsion is no longer by oars alone: the mast, stepped on a longitudinal keelson, is held by a "mast-fish". Below the waterline the flexible system of cleats and lashings is retained, but above it the frames are trenailed directly to the planks of the hull

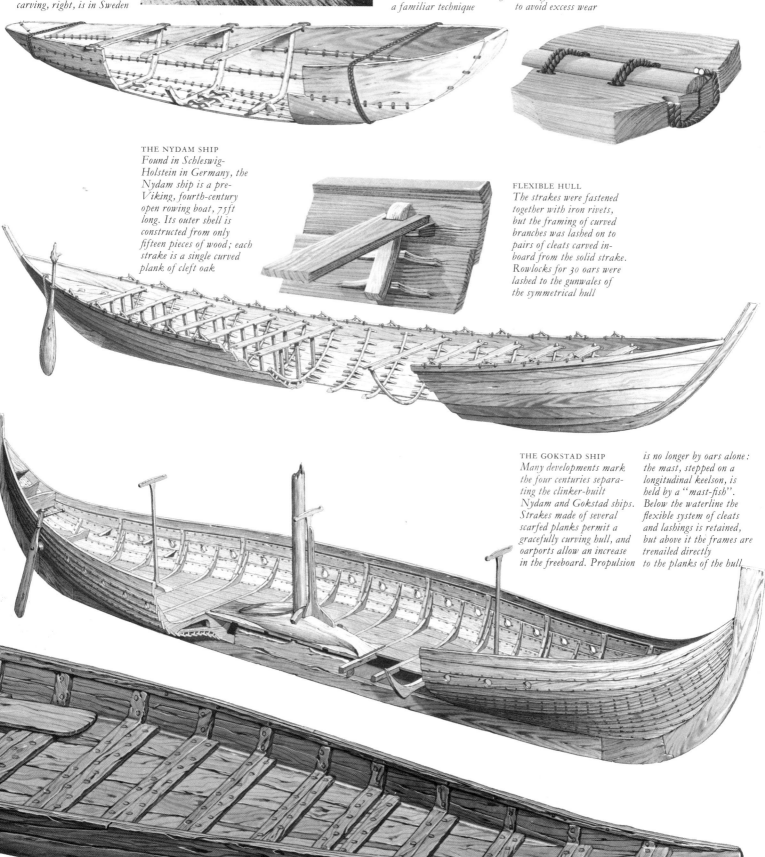

Traders and Galleys

Many contemporary pictures show us what the ships of the Classical world were like and how they were rigged. Twentieth-century discoveries of the underwater remains of the hulls of merchantmen, preserved by the mounds of ballast and cargo piled over them, provide evidence of the way they were built.

The hull was stiffened by a strong keel, which was laid down first, and to which a stem and stern-post were strongly scarfed. To this "backbone" a shell of planks was fitted, each strake being fastened to its neighbour by numerous wooden tenons. Once the shell had been constructed to a point above the turn of the bilge, the heavy frames were marked out, shaped and attached to the planking. At the waterline the first of a series of massive wales was fitted which alternated with the strakes. Heavy beams spanned the hull and projected through the planking to hook on to the wales and complete the box-like construction.

Compared with northern Europe, timber was in short supply in the Mediterranean. Theophrastus, the pupil of Aristotle, tells us what was available: "The ship-building woods are silver fir, fir and cedar. For triremes and other warcraft, silver fir is used for lightness; for merchant ships, fir is used because of its resistance to rot. Some also use it for triremes because silver fir is not available. In Syria and Phoenicia, cedar is used because of the lack of fir. In Cyprus pine is used; this is found on the island and seems to be better than the local fir." The frames were usually made of oak, and trenails and tenons were made of resilient woods like acacia, olive and seasoned straight-grained oak.

Though seasoned timber seems not to have been used, the wood had to be dry enough to expand and tighten all the joints when the ship was launched. The outside of the hull was smeared with pitch and wax or resin, and then sheathed in lead.

Merchantmen were designed to be sailed: the standard rig was a stocky mainmast with one square sail, but the use of more than one mast and of the fore-and-aft sail dates back to before the Hellenistic period. The main deck ran from stem to stern; larger ships had a lower deck and the biggest had two. On the poop-deck beneath a canopy the helmsman operated a pair of steering oars. Special ships such as pleasure cruisers had baths and cabins paved with mosaic and were lavishly furnished. Most of the wrecks of the Roman period so far found are in the range of 150 to 200 tons, though we know that vessels of more than 1,000 tons were sometimes built.

The design of warships differed in that oar, not sail, was the main means of propulsion, although a sail was set when long distances had to be covered. Naval tactics in the Classical period depended on the ability to disable an opponent by means of a projecting ram, often in the form of an animal's head, fitted to the bow and sheathed in bronze. Speed and manoeuvrability were essential: the oared galley had a hull of fir for lightness, and was frequently beached to keep it as dry, and as light, as possible – therefore its keel had to be of a tough wood, often oak. As many oarsmen as possible were accommodated within the water-line length, and many different arrangements of oarsmen were used. The most famous and successful warship was the trireme, which was about 115 feet long, sixteen feet in the beam, including outriggers, and had a freeboard of about six feet. Warships increased in size to accommodate larger contingents of marines, and war engines such as catapults, culminating, in the Hellenistic period, with catamarans, which combined the necessary large deck area with space for many oarsmen.

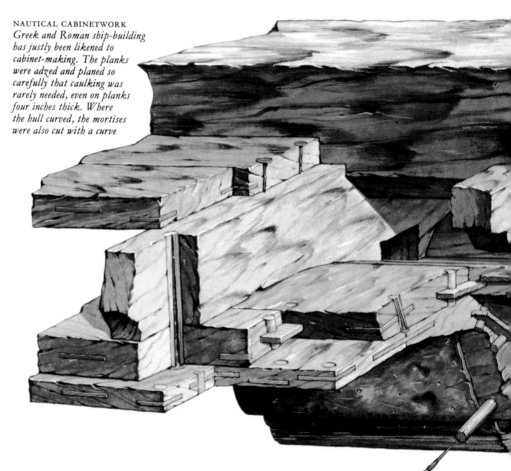

NAUTICAL CABINETWORK
Greek and Roman ship-building has justly been likened to cabinet-making. The planks were adzed and planed so carefully that caulking was rarely needed, even on planks four inches thick. Where the hull curved, the mortises were also cut with a curve

TENON AND MORTISE
The planks of the hulls of Romans ships were fastened together by tenons, fitting into mortises cut in the edge of each plank, and held by wooden dowels. On larger ships, an outer skin covered the seams of the inner, to which it was firmly spiked

LEAD SHEATHING
To protect them against the marine borer Teredo navalis, lead plates laid over tarred fabric and fastened by nails of copper sheathed the hulls of most merchantmen below the waterline. Round the keel a false-keel of beech protected the lead sheathing

The Lenormant relief, c. 400 BC, found on the acropolis of Athens, illustrates the light superstructure of the Greek warship. The tholes of the topmost oars are carried by an outrigger; the lower two tiers emerge through the hull. Two horizontal wales girdle the hull below the outrigger. The men row naked on leather cushions, shielded from the sun by the deck. Before an engagement, sail and gear were often landed, as everything depended on lightness and the ability to manoeuvre. Elaborate tactics were practised in battle

GALLEYS ON MANOEUVRE
As the bristling spears and glinting shields on a Roman fresco from Pompeii suggest, the role of the marines on board gradually became more important. The Roman navy used not only triremes, but small undecked "liburnians", quadriremes and quinquiremes

KEELSON AND BILGES
Above the massive oak frames, and pinned through them to the keel beneath, was a keelson, acting as an inner keel and supporting the mast. Notches cut into the bottoms of the frames let water into the bilges, to be drawn off by pumps of wood and bronze

ROMAN CARGO VESSEL
Broad-beamed, strongly built merchantmen shipped the produce of the Roman Empire. The mainmast carried the square sail and often, above the yard, a triangular topsail. Steering was usually by large quarter-rudders

CONSTRUCTION DETAIL
The main structural members of the keel and hull, left, were located in the area covered by the blade of the rudder in the relief carving, above left

THE STERN-POST
The upcurving stern-post was made in one or several sections, and was rabbeted, like the keel, with mortises and tenons to receive the planking fixed to it. Sometimes an outer sheathing of wood was nailed to it, to protect it from impact

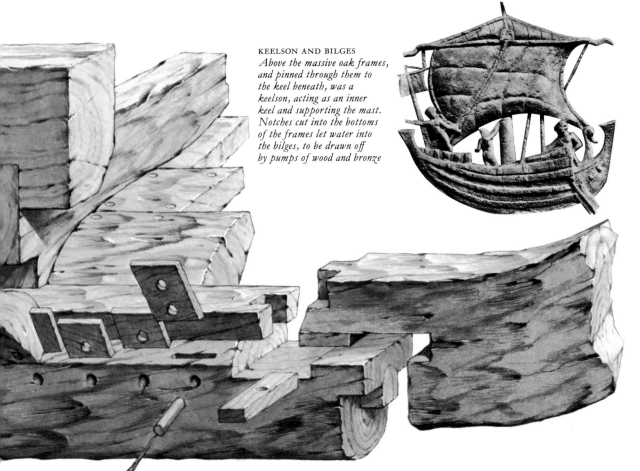

FRAMES OF OAK
To give the hull rigidity, massive frames, usually of oak, were fitted, normally when the first layer of the planking was complete. Trenails, through which long bronze spikes were driven from outside the hull, fixed the frames in position

ATTACHING THE FRAMES
Wooden dowels were inserted through holes bored in both the frames and planking, and large copper spikes hammered through them. By this method the wood was prevented from splitting, and the spike was easily driven in, following the grain of the dowel

STERN-POST SCARF
Fragmentary remains of merchantmen show that their construction could be very complex – notably in the scarfing of the keel and stern-post. Multiple dovetails were often cut at each end of the scarf, and the assembly was locked by a scarf-key

Nautical Archaeology

Nautical archaeology is a relatively young science and, due to the fragile nature of the materials under study, one which, perhaps more than any other, has brought together the skills of the archaeologist and timber expert, laboratory technician and chemist.

The vessels excavated may be of any age; Neolithic dug-out, Viking longship or Dutch East Indiaman, and may be found in a great variety of sites; on land sites such as the ritual burial mound at Sutton Hoo in the English county of Suffolk, in former watercourses as at Graveney, in underwater sites like the Viking Skuldelev wrecks in Denmark, or from tidal foreshore sites like that at North Ferriby in East Yorkshire.

Each site is unique; each presents a new set of problems which the archaeologist must resolve. The finds are seldom complete; topsides and fittings have been dismantled in antiquity, or have been broken and scattered by the action of waves and currents. Fragments are distorted, fastenings lost for all time, and scarfs and other joints forced open. In most cases the surviving wood has been preserved from biological decay by being waterlogged, and this factor, while preserving the wood through the ages, creates a problem for the archaeologist. If the wood is allowed to dry it will shrink and split and may even degenerate so far as to become quite unrecognizable as part of a boat. The wood must therefore be kept sodden throughout the excavation and be treated with the utmost care less vital evidence be lost.

As the timbers are exposed at the site, each piece, no matter how small, is carefully recorded. Scale drawings of the site are made and each find is marked on the plan for reconstruction at a later date. Oblique photographs and photo-mosaics, drawings and detailed notes, even plaster casts are used to ensure that no piece of evidence is overlooked. Associated artefacts are collected, and samples of the surroundings are taken for analysis, before the timbers are lifted and passively conserved in water tanks.

Each boat find, and often each individual piece of timber, requires specialized conservation treatment, for the degree of waterlogging and the degree of degradation of the wood varies not only from site to site but through the cross-section of a single piece of wood. Detailed examination is needed before the conservation schedule can be prescribed with confidence. There are several methods of conserving waterlogged wood, the most widely used at the present time being the replacement of the water in the wood by a water-soluble wax. The treatment is necessarily slow, and five or ten years may elapse before the remains can be mounted for public display.

After excavation, the archaeologist begins the mammoth task of extracting every shred of available evidence from the remains. The species of timber are identified and strength tests are carried out on selected samples. Specimens are taken for age-determination by radiocarbon analysis and by dendrochronology, and samples of the matrix – the sand, mud or earth in which the find was embedded – are analysed to determine the type of environment in which the boat was used.

Each element is carefully cleaned, and great attention is paid to the position of fastenings and the shape of joints and edges. Tool marks are noted, as are any indications of the manner in which the wood was converted from the log. The grain direction of each major timber is examined and its position noted in relation to the living tree. Such evidence enables the archaeologist to determine the criteria used by the ancient boat-builder in selecting his timber, and the sequence of events in the construction of the craft. The reconstruction is first drawn out and then prepared in scale model form; in some cases a full-scale replica may then be built and tested for seaworthiness.

From the direct evidence of the boat finds, and the indirect evidence provided by the laboratory tests and the electron scanning microscope, a comprehensive picture may be built up of the boat-builder and his methods. Many of these research techniques have only recently been developed and few nautical remains have yet been exhaustively examined.

WATERLOGGED WOOD
Waterlogged wood, below, retains its size and shape, but loses strength and hardness. With the consistency of soft cheese, it is very difficult to handle until treated with a consolidant

WATERLOGGED CELLS
This micrograph, right, of waterlogged oak from one of the ships shows how the walls of its fibres are swollen with water, weakening the cell walls and their connecting tissues

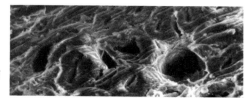

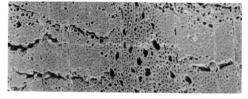

DRIED-OUT OAK *(above) On drying, the individual cells of waterlogged wood distort and severe splitting occurs along the lines of the* weaker, thin-walled cells. *This micrograph (× 100) of desiccated oak shows the torn line of the parenchyma, or food-storage, cells*

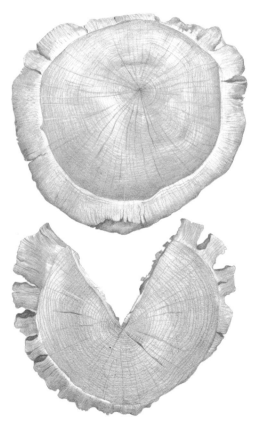

DESTRUCTIVE DRYING
The lower picture shows the same piece of oak after drying in an oven. It has shrunk, split and become totally distorted. To keep the wood in its true shape, the archaeologist must replace the water in the wood with a consolidant

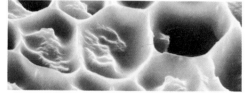

PRESERVATIVE TREATMENT
The principle of all preservative treatment is, by osmosis, to exchange the water in the wood for a hardening and strengthening material. The micrograph, above, shows the cells of oak supported by a layer of tough acetone-rosin

THE SKULDELEV SITE

In a specially constructed coffer-dam, work proceeds on the excavation of the third of five Viking ships found beneath Roskilde Fjord near Skuldelev in Denmark. Sunk to block the channel, the stone-laden vessels were flattened, but many of their timbers were intact. While the mud is washed away, and the site measured and recorded in detail, water-sprinklers prevent the timbers drying out until they can be dismantled, taken up, treated with preservative techniques and finally reconstructed

MARKS OF THE SHIPWRIGHT

This small fragment of pine from a crossmember, photographed from both sides, reveals the shape of the tool used and brings to life the stroke of the craftsman at his work. The wood was worked with a small, narrow-bladed axe and the strokes were directed at angles between 30 and 90 degrees to the grain of the wood. The holes were bored out with a conical bit

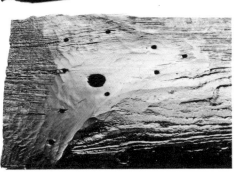

HIDDEN DETAILS

Clearly visible on the surface of this pine plank is the shape of the scarf joint, by which it was attached to its neighbour. Beneath the scarf can be seen the socket of the wooden trenail, which held the frame securely in place. Also visible on this surface are the marks left by the blade of a smoothing plane

CLEFT OAK PLANKS

Oak planks used in the ships were cut from wedge-shaped timbers cleft from the log. The 32 equal-sized billets would be stronger than sawn planks and any knots present would not affect more than one or two of the planks

STRAKES OF OAK

A side-frame decorated with a double swallow-tail secures the seventh and last strakes on the port side of one of the five wrecks. Nailed to it is a cleat which can only have been used for the running rigging. The marks of an axe can be seen on the planks; no trace of a saw was found on any of the ships' timbers

RINGS OF OSIER

The tough fibrous wood of the willow tree was used for trenails and for the twisted rings of osier, or young shoots, which were found still in place along the gunwales of the ship. The mast shrouds were secured to these rings

THE RECONSTRUCTED SHIP

The third, and best preserved of the wrecks, was a half-decked cargo vessel, lightly built of choice timber for speed and strength. The keel was fashioned from a 30ft oak log; the planking was secured by iron nails, and the frames by willow trenails

CONSTRUCTION DETAILS

A close study of surviving timber illustrates details and the sequence of stages in the ship's assembly. Keel, stem, sternpost and the first five strakes were assembled, after which the frames, the internal timbers and the higher strakes were added

The Junks of China

FLOATING REED STACK
*The marshy banks and wide
flats of Lake Tung T'ing
abound with tall reeds of
great use to the Chinese
for mats, blinds and fuel.
The junks that carry the
reeds are almost engulfed
by their cargo, which acts
as a sail before the wind*

From time immemorial the waterways of China have been the arteries of its civilization; shipping its life blood. Myriad wooden rafts, sampans and junks ply the streams and rivers of this vast country, carrying every sort of cargo, ferrying passengers and fishing the rich lakes and coastal waters.

The ubiquitous bamboo is made into ropes and cables of great tensile strength, highly resistant to rot, chafe and stretch, and yet relatively light to handle. It is also woven into screens, roofing mats for the deck-housing, and baskets of all kinds. The bamboo, it would seem, inspired the construction of the junk. The stem is divided by internal diaphragms into a series of water-tight sections, inspiring the water-tight bulkheads of the junk – a feature of ship design that did not appear in the West until comparatively recently.

The junk has no structural keel and few straight lines. There are innumerable types, varying with locality and use, but all are built by eye, without recourse to drawings or moulds, with a few simple tools. Bottom planks are laid on sandbags stacked so that the timbers are pressed into their curved shape by the weight of the ship. The bulkheads are constructed close by and then raised and pinned into place. The heavy curved wales, made from slender trees sawn length-wise, are fixed to the sides to add strength and to protect the hull from abrasion.

The main timbers used are fir and pine. Fir from the southern provinces grows rapidly, is light, tough yet soft, and readily available. The pine *Cunninghamia sinensis*, known as sha-mu, grows widely and varies in colour from red to white. The bulkheads are commonly made from close-grained durable laurel, and from fragrant camphorwood grown in Fukien, Szechwan and the Nan Ling and Nan Shan mountain ranges. In the largest sea-going junks, now no longer found, teak was used extensively both in construction and in making the masts and yards.

To toughen them, the main mast timbers and spars were buried in damp earth for several years, to emerge iron-hard and with great resistance to the ravages of weather and destructive boring insects.

CROOKED STERN JUNK
*The junks which negotiate
the dangerous rapids of
the Upper Yangtze River
are built with irregular
sterns, which leave room
for the helmsmen to work
two stern sweeps to their
full extent without any
risk of fouling*

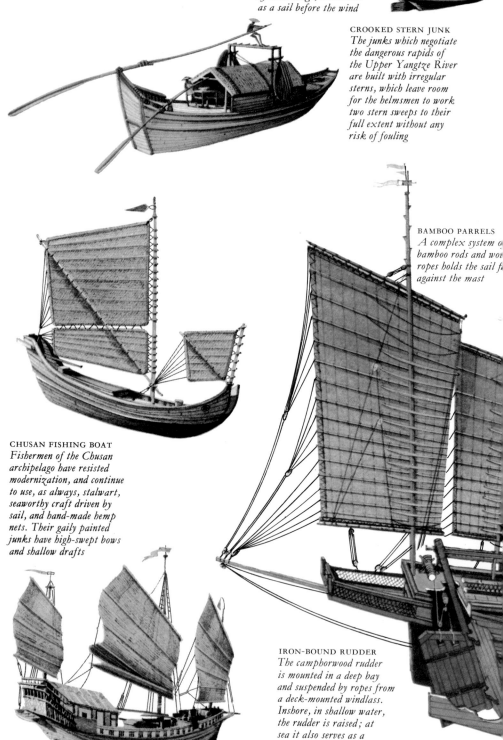

BAMBOO PARRELS
*A complex system of
bamboo rods and wo[...]
ropes holds the sail f[...]
against the mast*

CHUSAN FISHING BOAT
*Fishermen of the Chusan
archipelago have resisted
modernization, and continue
to use, as always, stalwart,
seaworthy craft driven by
sail, and hand-made hemp
nets. Their gaily painted
junks have high-swept bows
and shallow drafts*

IRON-BOUND RUDDER
*The camphorwood rudder
is mounted in a deep bay
and suspended by ropes from
a deck-mounted windlass.
Inshore, in shallow water,
the rudder is raised; at
sea it also serves as a
centre-board and helps
keep the ship stable*

HAINAN JUNK
*Large, three-masted traders
from the island of Hainan
set sail with the winter
monsoon for Indonesia and
Malaya, returning in the
early summer. These junks
had centre-boards forward
of the mainmast, a device
then unknown in the West*

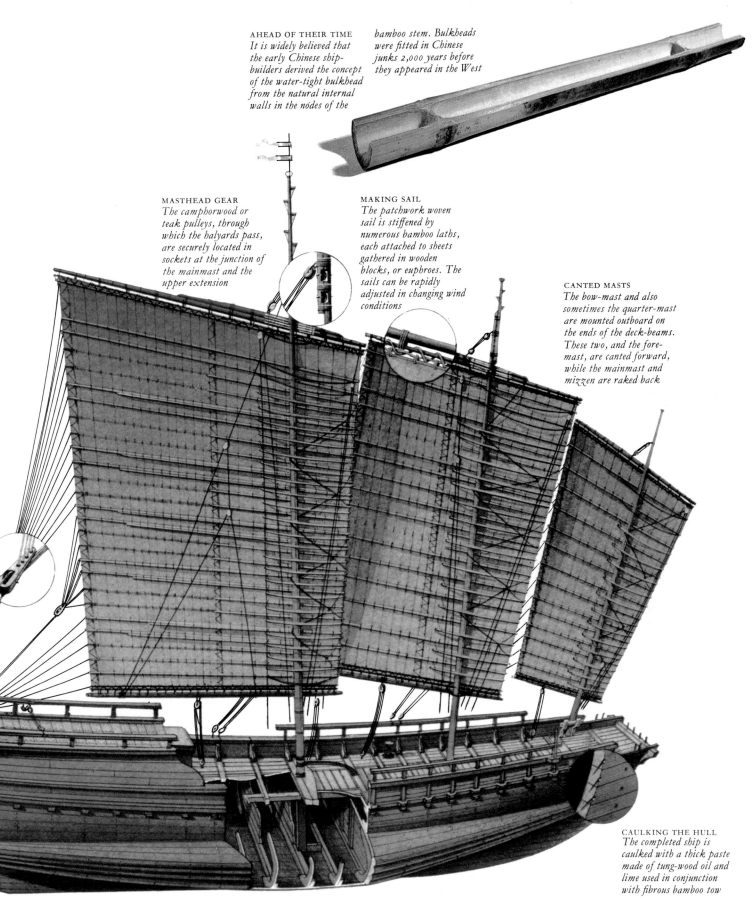

AHEAD OF THEIR TIME
It is widely believed that the early Chinese ship-builders derived the concept of the water-tight bulkhead from the natural internal walls in the nodes of the bamboo stem. Bulkheads were fitted in Chinese junks 2,000 years before they appeared in the West

MASTHEAD GEAR
The camphorwood or teak pulleys, through which the halyards pass, are securely located in sockets at the junction of the mainmast and the upper extension

MAKING SAIL
The patchwork woven sail is stiffened by numerous bamboo laths, each attached to sheets gathered in wooden blocks, or euphroes. The sails can be rapidly adjusted in changing wind conditions

CANTED MASTS
The bow-mast and also sometimes the quarter-mast are mounted outboard on the ends of the deck-beams. These two, and the fore-mast, are canted forward, while the mainmast and mizzen are raked back

CAULKING THE HULL
The completed ship is caulked with a thick paste made of tung-wood oil and lime used in conjunction with fibrous bamboo tow

STRUCTURAL STRENGTH
The mainmast is often a single teak log, rendered strong as iron and immune to boring insects by long burial in damp ground. So stepped as to transfer its weight forward and across the beam, it rakes back – its tenoned heel held in place by a chock bearing against the next bulkhead, while at deck level its thrust is taken to the heavy longitudinals by a massive transverse beam. Heavy vertical timbers join the fourteen bulkheads securely to the sides of the ship

BUILDING TECHNIQUES
First the planks of the flat bottom are joined to-gether with iron pins driven in at a shallow angle; the bulkheads are then raised and fixed in place. Side-planking, sometimes clinkered, more usually laid edge to edge, is attached between the bulkheads and secured by pins. For longi-tudinal strength, five rows of untrimmed half-logs line the mid-region of the sides. Decayed timbers in the pine-wood hull are not removed; new timber is simply over-laid, and an old junk might have as many as six skins

The Politics of Wood

As the scale of maritime war increased in the late seventeenth and eighteenth centuries, the rival nations became less self-sufficient in the materials essential to the existence of their fleets and hence their power. The sources upon which England, France, Holland and even Spain began to depend were the forests of Scandinavia and the vast plains of northern Germany and Russia. From the Baltic, into which flow the rivers which serve this great area, came huge convoys and single ships, with cargoes of oak timber and plank, fir masts, pitch, tar and turpentine. Their destinations and their safe arrival were never far from the centre of the political struggles of northern Europe.

France and England emerged as the most powerful maritime nations as Spain and Holland fell away. At first sight, the great French forests, ruled by authoritarian forest laws drawn up by Colbert, would seem to give the French a great advantage in obtaining her raw materials, for the navy had supreme powers to secure the best timber. England muddled on to final victory in 1815 without much of a domestic timber policy and far less resources. Yet its government had sound credit and had a paramount and permanent interest in the navy; the French kings had neither.

It was at the time of Cromwell's rivalry with the Dutch in the seventeenth century that keeping the Baltic open for timber and stores became a cardinal point of British foreign policy. Between 1658 and 1814 a British fleet was dispatched to the Baltic nearly twenty times, while in every war a number of frigates were always on hand. Admirals Norris, Wager, Gambier, Nelson and Saumarez made reputations there. Copenhagen was bombarded twice, while a helpless Denmark could only bemoan the ruthlessness of British policy. Perhaps the most telling display of British force was Saumarez's command between 1808 and 1813, when, despite Napoleon's continental system, and the hostility of almost every Baltic country, convoys of nearly 600 merchantmen made the yearly journey to England, escorted by British warships.

England, however, would not have come through that crisis if she had not had an alternative supply. For all its nearness, the attested quality of its products and the efficiency of its merchants, the Baltic was not only liable to sudden closure, but was expensive in wartime. To encourage colonial trade, an Act was passed in 1704, granting bounties to merchants – the start of a long effort to make England independent of the Baltic. Since the Baltic region was enveloped in the Great Northern War until 1721, colonial trade was given a good start, but it was never quite to rival the output of the Baltic region.

Very similar problems faced the French. There was less reliance on foreign oak, but from the mid-eighteenth century they had to turn to the Baltic. From the 1730s, Toulon was supplied with oak from Albania, Italy, Corsica and even for a short time from the Black Sea. The great weakness was mast timber. Turning again to the Baltic after early attempts at self-sufficiency, they were outthought and out-organized by the English merchant; frequently they even had to order masts through their rivals. Yet, curiously, when the American Revolution gave France a much-needed new supply, she showed as much conservatism as the English and failed to take advantage of the situation.

Even before 1815 the search for shipbuilding timber had widened, and it was found that the qualities of teak surpassed even those of oak. The coming of steam and iron finally brought the whole problem to an end.

THE AMERICAN SUPPLY
Distant America was never the main supplier of timber to the European navies due to high transport costs and a lack of control over the colonists, who often preferred to sell their timber in more profitable markets. Unreasonably, the British distrusted American oak, but came to rely heavily on New England pine for masts. The loss of the American colonies removed one of Britain's advantages over the French, highlighting the strategic importance of timber in the days of wooden ships

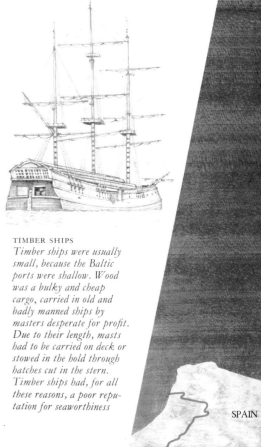

TIMBER SHIPS
Timber ships were usually small, because the Baltic ports were shallow. Wood was a bulky and cheap cargo, carried in old and badly manned ships by masters desperate for profit. Due to their length, masts had to be carried on deck or stowed in the hold through hatches cut in the stern. Timber ships had, for all these reasons, a poor reputation for seaworthiness

🌲 Pine forest

🌳 Oak forest

● Port

⚓ Naval dockyard

⚓ Merchant dockyard

THE ENGLISH SHIPYARDS

The six main English naval dockyards consumed vast quantities of timber in the building and maintenance of warships and many more were built by commercial contractors. By the 18th century, the forests of Kent and Sussex were sadly de- pleted and English oak was taken mainly from the New Forest and the Forest of Dean. Some foreign timber was imported direct to the shipyards, but the majority passed through London, where the navy had to com- pete, on the open market, with the merchant marine

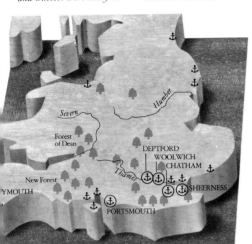

THE POLITICS OF MAST TIMBER

Since the early 16th century, shipwrights had made composite masts from up to seven timbers scarfed to- gether and bound with iron. The French used this long and costly method throughout, but the British came to rely on the cheaper "great" masts supplied by the New England colonies. When the colonies revolted, however, the supply abruptly ended and the British had to make new "made" masts from smaller Baltic timber

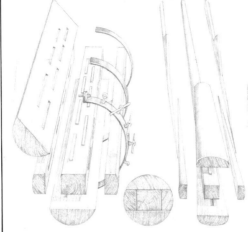

THE MAST POND

While oak must be seasoned until dry, the fir spars and masts had to be kept moist and resinous to retain their strength and resilience. This was achieved by storing the masts in dockyard ponds beneath the mast sheds

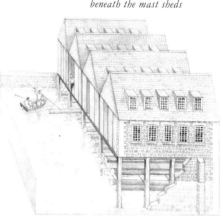

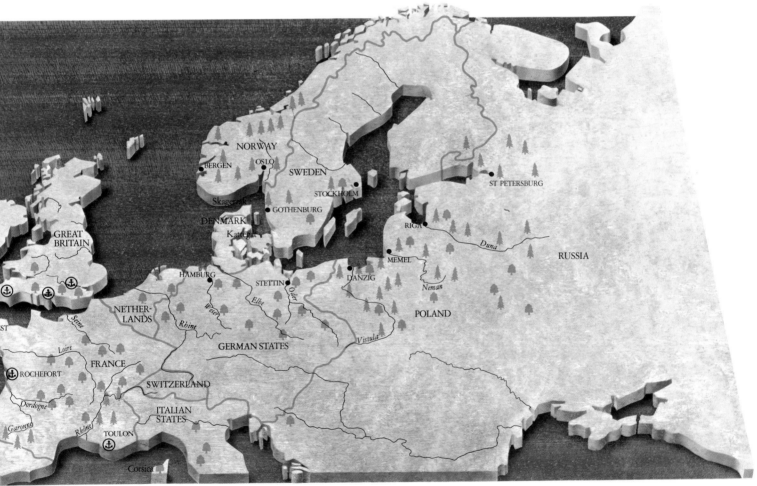

FRENCH TIMBER

Wide rivers flowed through the great French forests of oak, and there was fir in the Dauphiné and Pyrenees. There were also stringent laws to conserve timber for the navy. France was poten- tially self-sufficient, but the laws were difficult to enforce, the fir was not of good quality, and, because none of the rivers needed to transport timber had suitable ship-building sites near their mouths, there were huge difficulties in supply. Only Rochefort could be supplied directly from internal resources

CHANNEL STRATEGY

Dominating the sea-lanes down to Brest and Roche- fort, England had a con- siderable strategic advantage in the struggle for mast supplies from the Baltic. Since the English very often held local superiority, France could not even transport domestic timber by sea. In 1761 Brest had no wood at all because no ships could get through from Le Havre. The positions were reversed in the Ameri- can war, when privateers made convoys essential for the timber-carrying traffic from England

BALTIC TIMBER

The Baltic timber trade was notable for its efficiency. A system of quality control ("bracking") developed, and "Danzig crown" tim- ber was reliably best quality. By the early 17th century the Dutch, with efficient saw-mills and low-cost ship- ping, dominated the trade. Later, their influence declined before Swedish and Russian rivalry. Denmark controlled the only safe entrance, the Sound, which brought them lucrative fees; but it was a small country, and the power of the British fleet guaranteed access

From Forest to Fleet

The building of a wooden warship was a major undertaking, comparable in scale to building a cathedral. Besides requiring a vast number of men and great capital expense, a large ship would consume a small forest of timber. In a mid-eighteenth-century warship of 74 guns, the timber used was equivalent to about 3,700 mature trees; a 100-gun ship like the *Victory* would involve the felling of about eighty acres of oak forest.

Difficulties of supply were apparent even in Henry VIII's reign, and as ships grew in size and numbers these difficulties increased. Great timber, that of very large dimensions, was required for the stern-post, which took some of the greatest stresses in the ship; a stern-post made of jointed timbers would soon fail. Compass timber was timber grown into curved shapes, the grain running unbroken along the curve. It was therefore far stronger than timber sawn to a curve and was used extensively throughout the structure of the ship. Dockyard timber was a perpetual problem in the turbulent eighteenth century. It takes more than a hundred years for an oak to reach its full maturity and even those governments with foresight enough to plan ahead, and those, like the French, with powers of requisition, could easily be caught unprepared.

The English navy had guaranteed but limited sources of supply from the Royal Forests, the largest of which was the New Forest, but the dockyards were still forced to procure much of their timber, cut and carted, from contractors, whose hold on the main sources of supply was never broken. The practice of the great seventeenth-century ship-builder Phineas Pett, who, before commencing work on a ship, travelled the forests choosing his timbers as they stood, was seldom followed. Corruption in the dockyards had been rife since the sixteenth century and, generally, the whole system of timber supply for the navy was costly and inefficient.

Due to shortages, and lack of foresight, unseasoned timber was often used and even sound timber suffered greatly from rot in the damp, enclosed spaces of the ship. Some ships were lost because they simply fell apart in heavy weather. Wartime damage and thoughtless neglect in peacetime lowered the life expectancy of these magnificent floating castles to less than twenty years.

OUTBOARD PLANKING
Thick planks, joined edge to edge, lay across the frames, forming the skin of the hull. From stem to stern, just above the waterline on each side, ran the girdling wales of "thickstuff" – planks not less than eight inches thick. The planking along the waterline, "twixt wind and weather", was a region of weakness, vulnerable to decay and requiring constant attention and repair

GROWN TIMBERS
The ideal trees for curved, or "compass", timbers were the hedgerow oaks: woodland oaks provided the straight "great" timbers. For the knees and clamps, wood was taken from the point at which a stout branch grew from the tree

CAULKING
Oakum – strands picked from old hemp rope and smeared with pitch – was rammed between the joints of the planking. Caulking kept out the water and tightened the joints, so the timbers should not work unduly under stress

INBOARD TIMBERS
Where the deck-beams met the side, clamps beneath, knees above and lodging knees beside, supported them. These were all huge right-angled brackets of oak grown to shape. Strengthening the sides were the thick, vertical riders

THE STEAM-CHEST
Since the outboard planks had to follow the curving lines of the ship, they were often steamed in order to make them more pliable. The heavy plank had to be rushed from the steam-chest into position, and fitted while the timber was hot

FLOOR AND KEELSON
The floor, the lowest part of the main frame, was scarfed to the middle part, the futtock, and fixed to the keel with wrought-iron bolts. Running the length of the ship, the keelson was attached through the floor to the keel

SCARFING THE KEEL
To achieve the strongest possible joint, each timber was tapered to overlap the next, tongued and grooved so that the timber itself bore virtually all the strain. Bolts were passed through the joint – their ends clenched over washers

THE FRAMES
The frames of the ship were set in place in pairs; the main frame crossing the keel and rising to the gunwales while the secondary rib, which did not cross the keel, gave added support. Each of different shape, the frames were laid out from actual-sized moulds of deal battens, "laid-off" on the floor of the mould loft by the shipwright, working from a single 1/48th scale drawing. Until they were finally fixed by the planking, the frames were held by shoring and by longitudinal ties called ribbands

PRIZED TIMBERS
One of the most difficult pieces of compass timber to find was that for the wing transom; it had to come from a large tree with an uncommonly widely forked stem. Less open forks were used inboard to strengthen angled joints. For the stern-post, a tall, straight trunk was required

THE WING TRANSOM
The transoms were heavy transverse timbers joining the stern-post to the aftermost frames. The highest transom – the crucial wing transom – carried a vast amount of strain and was formed from a single baulk of compass timber taken from an unusually widely forked tree. The timber was often 30 feet long and more than two feet thick

THE STERN-POST
As the keel supported the frames, the great timber of the 30ft-tall stern-post linked the planking and the thickstuff wales girdling the ship's sides. For 300 years, shipwrights had complained of the shortage of such great timbers, for the loads could be borne only by a single, unjointed baulk of oak. This alone limited the size of wooden ships

187

The Boat-builder Today

Though far from extinct, traditional wooden boat-building has declined in recent years, not so much because it has been superseded but because the costs of timber and labour have rendered it uncompetitive. Materials such as glass reinforced plastic (GRP) need less maintenance and are more than adequate for recreational use. But where sturdy, hard-wearing seaworthiness is required, as in lifeboats and fishing vessels, boats continue to be clinker-built in the old manner.

New methods of using wood have, however, kept steady pace with the development of replacement materials. Modern preservatives, adhesives and paints have made marine plywood extremely durable and strong and highly versatile. It can be moulded to virtually any shape and built up into a composite member of any size – completely overcoming many of the problems which hindered the ship-builders of old. For the purposes of racing craft, marine plywood can offer a much better strength/weight ratio than many of its competitors.

Difficulties in the supply of timber have to some extent been overcome by the introduction of a number of tropical hardwoods, such as utile, iroko and opepe, which are the equal, if not the superior, of more familiar woods like oak, teak and mahogany. The use of plywood and veneered structures has considerably reduced both the amount of timber required and the degree of waste.

The new boat-building techniques have completely reversed many long-held traditions. The standard European practice since the Viking period, in which the strakes were fitted on to the keel and frames, has given way to the "shell first" method, in which the frames are laid inside the completed hull – a technique practised by the boat-builders of Ancient Rome. Though restricted to use in small boats, wood is more than holding its place in the age of technology.

THE MAIN FRAME
The main frame, comprising stem, keel, sternpost and horn, is built up of ten individually contoured laminates: once the first laminate has been shaped on the mould, the rest are shaped consecutively each over the next. The frame is enclosed over most of its length by the skin; here only one layer on one side is glued and complete. The gap at the stern (below) is closed by a flat, raked transom

THE FIRST SKIN
Honduras mahogany planks of a standard length and width make up the first, inner, skin. They are 3.5 mm thick – the most popular thickness used in the cold-moulding process. From the centre outwards, each plank is stapled on to the mould and marked off. It is removed, trimmed and replaced, and the next one set up beside it. The side-planks meet the main frame at about 45 degrees

THE SECOND SKIN
Over the completed first skin of mahogany, thicker planks of obeche making up second and third skins are laid in the same way. The planked skins run in alternate directions – so obtaining in moulded timber the great strength of ply

COLD-MOULDING PROCESS
Though it uses modern technology to full advantage, the cold-moulding process is none the less very simple, and virtually unmechanized. It makes it both practical and economic to build ocean racers and cruisers with standard wooden planks

FOUR LAMINATIONS
The fourth and final skin is again of Honduras mahogany – laid obliquely like the inner skins, or planked horizontally to render the impression of traditional carvel-jointing. The glued and completed hull is now turned right way up and placed on scaffolding. In this close-up, showing the stern ready to be planed and sanded before fitting the transom, the four laminations can be clearly seen

LAMINATED FRAMES
Bridging the main-frame, which protrudes into the hull like a keelson, these frames strengthen the hull throughout its length. The laminates making them up are shaped in position in the hull, and are then removed for sanding before being glued into place

THE COMPLETED HULL
As the frames go in, the hull nears completion. The next stage is the moulding of the deck, done separately but similarly. The interior joinery is commonly teak. The glue is as strong and as durable as the wood; the boat is wholly watertight

READY TO SAIL
Apart from the aluminium mast and the lead ballast, the boat is almost wholly wooden. With the light "infill" in its skin, it is lighter than GRP-hulled boats of the same class, just as strong, and equal in durability

189

Artistry in Wood

Under the patronage of Charles II of
England, the Dutch-born carver
Grinling Gibbons rose to prominence
as the foremost wood-carver of the late
seventeenth century. Rejecting formal
classicism and ornate Gothic styles, he
brought a new vitality and naturalism
to wood-carving – illustrated by this
detail from the east wall of the Carved
Room of Petworth House in England

Pipes and Reeds

Among the myriad woodwind instruments invented since the dawn of human history there are only three basic types. In all of them, the notes are produced by causing rapid vibrations in the air column in their tubes. In the flutes this is done by blowing against a special lip in the wall of the instrument; in instruments of the oboe type the column is vibrated by a "double reed" – two short pieces of cane set in face to face at the end of the instrument and held in the player's mouth. With clarinet types, a single reed vibrates against the edges of a slot cut in the wall of the tube.

The "woody" sound of the instruments comes from the fact that above the note sounded by the player lies a cluster of upper harmonics. These sympathetic vibrations are too faint to be heard as actual notes, but together they are heard as the tone colour. The different tone colours within the wood-wind family come from the fact that different upper harmonics sound in each case. The wood used and its thickness play a part in determining which of these characteristic tones are heard, but perhaps even more important are the internal shape of the tube and the configuration of the reed or the blowhole.

The flute was "invented", perhaps, by some caveman toying with a hollow bamboo. This strange new voice of nature rapidly became part of religious ceremony and today, in New Guinea and South America, giant endblown flutes, up to seven feet long, are still used in religious ceremonies. The flute used in the West is rather different, but the tones of the woodwind are still among the most evocative in the orchestra.

Very early, fingerholes were bored in the tube so that the player could vary the length of the air column and thus the notes available. However, sharps and flats could only be achieved with tricky "cross fingerings", to cover awkwardly situated holes, and even then the notes were not certain to be in tune. In the West, the increasing complexity of harmony created many problems. Inventors began to add key mechanisms to help the player close combinations of holes by remote control until, today, the simple wooden tube has become festooned in metal.

BALKAN DOUBLE FLUTE
Double flutes are common throughout the Balkans, where there is also a double reed-pipe called the "diple". The arrangement of the holes makes it possible for the player to accompany his melody with a background drone

BAMBOO SHENG
The Vietnamese sheng clearly illustrates one form of the free-reed instruments which were developed in the Orient. At the base of the bamboo tubes, where they enter the wind chest, a bamboo reed is free to vibrate back and forth

CHINESE SHENG
This sophisticated version of the simple peasant instrument illustrated above, is a fine example of the Chinese instrument-maker's craft. It has more pipes than the folk sheng and, more important, the reeds are made of fine strips of metal instead of slivers of bamboo

ALPHORN
Found in Scandinavia, the Pyrenees and Switzerland, this mountain horn is made from a pine log split lengthwise. The bore is cut from the inner surfaces; the outer faces are shaped and the two halves are tightly bound with roots or bark

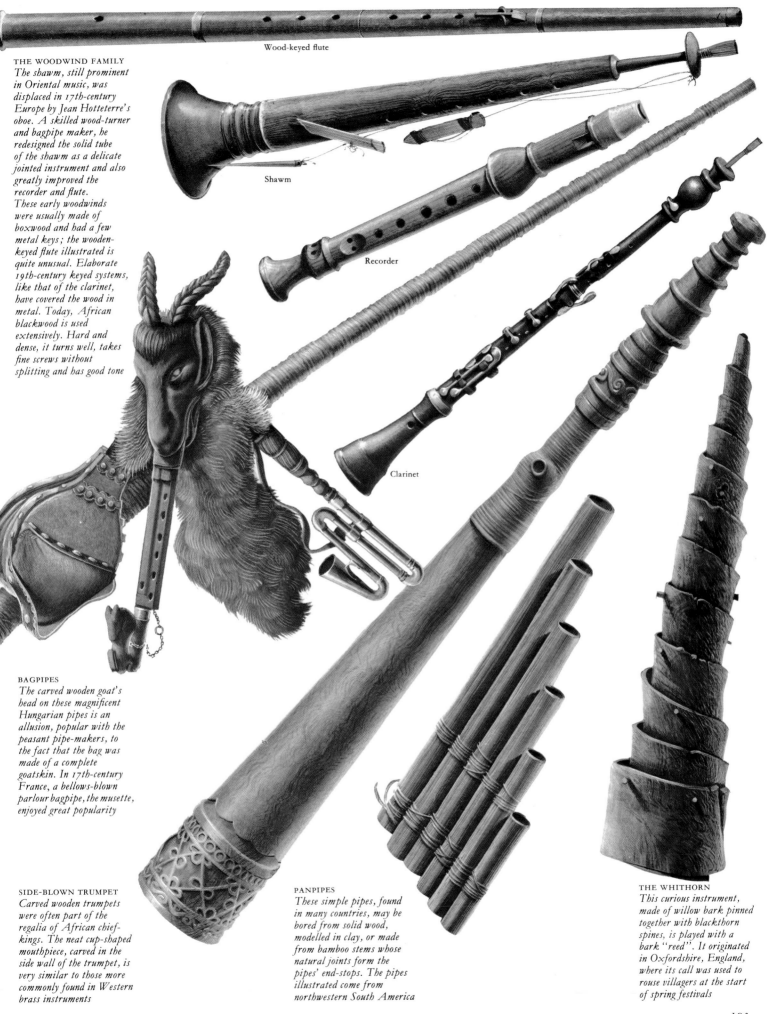

Wood-keyed flute

THE WOODWIND FAMILY
The shawm, still prominent in Oriental music, was displaced in 17th-century Europe by Jean Hotteterre's oboe. A skilled wood-turner and bagpipe maker, he redesigned the solid tube of the shawm as a delicate jointed instrument and also greatly improved the recorder and flute. These early woodwinds were usually made of boxwood and had a few metal keys; the wooden-keyed flute illustrated is quite unusual. Elaborate 19th-century keyed systems, like that of the clarinet, have covered the wood in metal. Today, African blackwood is used extensively. Hard and dense, it turns well, takes fine screws without splitting and has good tone

Shawm

Recorder

Clarinet

BAGPIPES
The carved wooden goat's head on these magnificent Hungarian pipes is an allusion, popular with the peasant pipe-makers, to the fact that the bag was made of a complete goatskin. In 17th-century France, a bellows-blown parlour bagpipe, the musette, enjoyed great popularity

SIDE-BLOWN TRUMPET
Carved wooden trumpets were often part of the regalia of African chief-kings. The neat cup-shaped mouthpiece, carved in the side wall of the trumpet, is very similar to those more commonly found in Western brass instruments

PANPIPES
These simple pipes, found in many countries, may be bored from solid wood, modelled in clay, or made from bamboo stems whose natural joints form the pipes' end-stops. The pipes illustrated come from northwestern South America

THE WHITHORN
This curious instrument, made of willow bark pinned together with blackthorn spines, is played with a bark "reed". It originated in Oxfordshire, England, where its call was used to rouse villagers at the start of spring festivals

193

The Keyboard

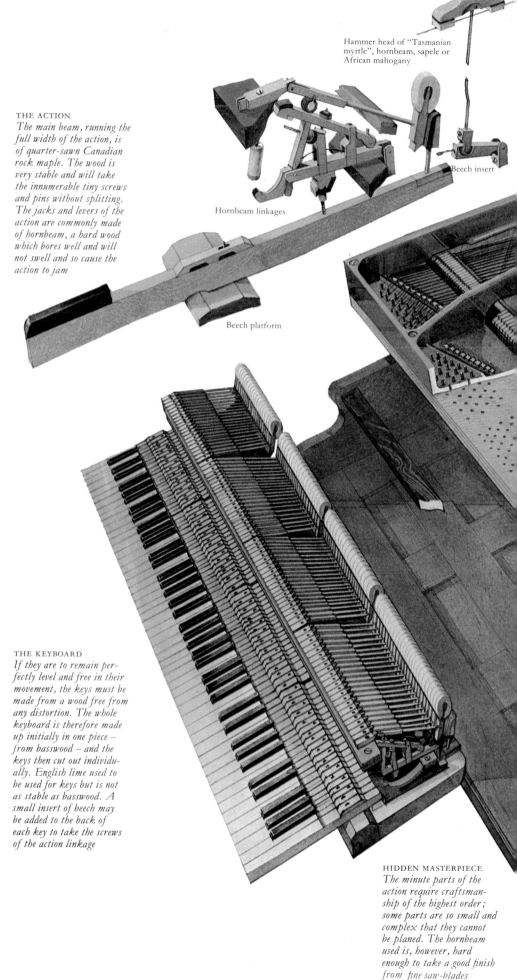

The world's first stringed keyboard instruments emerged in Europe during the late Middle Ages. The earliest was the clavichord, derived from adding a mechanism to the monochord – a single-stringed zither with a movable bridge, or tangent, which was used for teaching the theory of music.

The harpsichord, developed later, combined the keyed principle of the clavichord with the plucking quills used to play the zither-like psaltery. By 1500, its classic winged shape had been established and its basic mechanism worked out. The clavichord's tangent striker was replaced by a small rectangular pillar, called the jack, which carried a plectrum of quill or leather. Depressing the pivoted key forced up the jack and plectrum, which thus plucked the string; it was hinged so that when it fell back it did not pluck the string a second time. A light hair spring brought it back to the vertical. Though ingenious and simple, the mechanism had to be repeated for each of some eighty keys, calling for careful and delicate carpentry skills, which had to be specially developed. In addition, the making of true, responsive and resonant sounding-boards demanded expert joinery in the handling of large areas of thin-gauge timber.

In the early 1700s the harpsichord began to lose favour because it could not provide the subtle gradations from loud to soft, then becoming part of the newly fashionable style. No matter how hard the key was struck, the plectrum always plucked the string with the same force. In 1709, in Florence, Bartolommeo Cristofori unveiled his *gravicembalo col pian e forte*. In this instrument, the string was set in vibration by a hammer, propelled against the string with a momentum determined by the force with which the player struck the key. The faster the hammer travelled, the louder the note produced.

Cristofori's mechanism was almost too effective; the hammers had a greater potential force than the low-tension stringing, necessary on wooden-frame pianos, could withstand. Only in the nineteenth century, when a successful iron frame was developed, did his original concept begin to yield its full potential.

THE ACTION
The main beam, running the full width of the action, is of quarter-sawn Canadian rock maple. The wood is very stable and will take the innumerable tiny screws and pins without splitting. The jacks and levers of the action are commonly made of hornbeam, a hard wood which bores well and will not swell and so cause the action to jam

Hammer head of "Tasmanian myrtle", hornbeam, sapele or African mahogany

Beech insert

Hornbeam linkages

Beech platform

THE KEYBOARD
If they are to remain perfectly level and free in their movement, the keys must be made from a wood free from any distortion. The whole keyboard is therefore made up initially in one piece – from basswood – and the keys then cut out individually. English lime used to be used for keys but is not as stable as basswood. A small insert of beech may be added to the back of each key to take the screws of the action linkage

HIDDEN MASTERPIECE
The minute parts of the action require craftsmanship of the highest order; some parts are so small and complex that they cannot be planed. The hornbeam used is, however, hard enough to take a good finish from fine saw-blades

THE DAMPER
Each time the key is struck, the damper must release its string; as soon as pressure is released it must return to its place. The damper head may be of mahogany, hard "Tasmanian myrtle", or an ebonized wood

OCTAVE SPINET
This beautiful instrument dates from about 1600 and is made of cypress-wood. Due to the shortness of its strings it lacks the rich tone of the full-sized spinet, causing Charles Burney to write, "more wood is heard than wire"

THE IRON FRAME
The steel strings of the modern piano exert a force many times greater than any wood can bear and the introduction of the iron frame dramatically increased the range and compass of this classical instrument

VAUDRY HARPSICHORD
One of only four extant 17th-century French harpsichords, this instrument was made in 1681 by Jean-Antoine Vaudry. It stands on seven spirally turned walnut legs, its walnut frame and case japanned with intricate chinoiserie

THE SOUND-BOARD
The vibrations of the strings are transmitted through the beech bridge on to the slightly arched sound-board, traditionally made of slow-grown "Roumanian" pine. Sound travels along its grain nearly fourteen times as fast as in air. Bars of the same wood, placed beneath the sound-board, distribute vibrations over the surface

THE WREST-PLANK
The wrest-plank of beech, either solid or made up of three to five laminations, must bear the strain of 225 or more strings, each under a tension of more than 150lb. In any other wood, the iron wrest-pin holding the string would crush the fibres at the edge of its hole

CONDUCTOR'S PIANO
This tiny piano, spanning just three octaves, was used by the Victorian conductor Sir George Smart while on tour. Designed to fit into a travelling coach, its elegant casework illustrates the skills of a master cabinet maker

QUEEN ELIZABETH'S VIRGINAL
Though known as virginals like all English keyboard instruments of the period, this is a spinet, dating from the late 1500s. The case bears the coat-of-arms of the Virgin Queen, after whom virginals were thought to have been named

THE CASEWORK
The piano casing is most commonly made up of walnut or mahogany veneer laid on a laminate base. The core veneers are often of agba, obeche or African mahogany. Quality instruments may have parts of the casework of solid wood and, for these, many players feel that mahogany gives the piano its finest tone. The legs, which may support a weight of more than one ton, may be turned from the solid or built up. American mahogany, sapele and beech are used, and the legs are generally veneered

The Violin

The violins, the foundation colour of the classical symphony orchestra, belong to the most recent group of stringed instruments – the bowed lutes. The classic string tone of Indian music is the plucked lute, one of the most familiar being the sitar; that of Japanese music is provided by plucked zithers like the koto or plucked lutes like the shamisen.

In the zithers, each note has a separate string, vibrating freely between two bridges attached to the body of the instrument, which also serves as the resonator. In the lutes, each of a small number of strings can be made to give many different notes by being stopped at different points along a fingerboard so as to give different vibrating lengths.

The invention of the bow seems to have been made sometime about the year 800, perhaps among the steppe peoples of Central Asia, and by the end of the tenth century, bowed lutes were to be found all over Europe and Asia. In Europe, the earliest ones were the rebecs and fiddles, whose close relatives, like the Slav *gusle*, are still prominent in folk music. The rebec had a pear-shaped body and short neck carved from a single piece of wood. The strings stretched between a bridge on the belly of the instrument and tuning pegs set laterally in a peg-box. The fiddle, on the other hand, was built up of separate sections and had a peg disc with the peg holes from top to bottom. The lateral pegs of the rebec made higher string tensions possible, while the segmented build of the fiddle made for a more responsive resonating chamber. During the fifteenth century these elements were combined in a new group of instruments. Known in Spain as *vihuelas*, they gave birth to the guitars and the viols – more refined and elegant than their medieval ancestors. They initiated a tradition of fine carpentry in musical instruments that produced some of the most sophisticated examples of the woodworkers' craft.

The violins, growing from the same origins, began to displace viols in courtly music about 1600. The brash newcomers, first introduced by the Italians, had a more brilliant and incisive tone than the viols, but were soon accepted by musicians and their patrons throughout Europe.

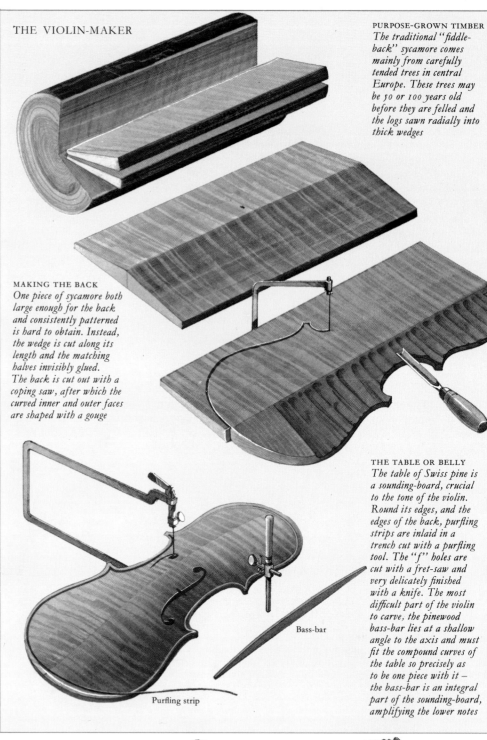

THE VIOLIN-MAKER

PURPOSE-GROWN TIMBER
The traditional "fiddle-back" sycamore comes mainly from carefully tended trees in central Europe. These trees may be 50 or 100 years old before they are felled and the logs sawn radially into thick wedges

MAKING THE BACK
One piece of sycamore both large enough for the back and consistently patterned is hard to obtain. Instead, the wedge is cut along its length and the matching halves invisibly glued. The back is cut out with a coping saw, after which the curved inner and outer faces are shaped with a gouge

THE TABLE OR BELLY
The table of Swiss pine is a sounding-board, crucial to the tone of the violin. Round its edges, and the edges of the back, purfling strips are inlaid in a trench cut with a purfling tool. The "f" holes are cut with a fret-saw and very delicately finished with a knife. The most difficult part of the violin to carve, the pinewood bass-bar lies at a shallow angle to the axis and must fit the compound curves of the table so precisely as to be one piece with it – the bass-bar is an integral part of the sounding-board, amplifying the lower notes

Bass-bar

Purfling strip

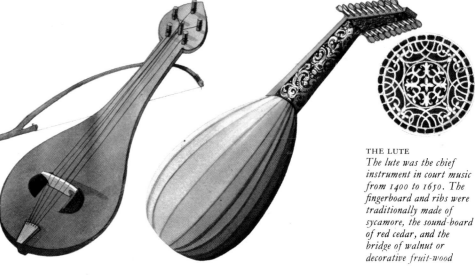

GADULKA
The Bulgarian gadulka is a folk instrument derived from court music. Like its ancestor, the medieval fiddle, its body is carved from a single piece of wood and it has a disc peg-box. The convex bow is still favoured by folk musicians

THE LUTE
The lute was the chief instrument in court music from 1400 to 1650. The fingerboard and ribs were traditionally made of sycamore, the sound-board of red cedar, and the bridge of walnut or decorative fruit-wood

FORMING THE RIBS
The ribs, or sides, are thin strips of sycamore bent to shape round an electrically heated form. Also shaped in this way are the linings – smaller strips of pine providing a base on which to fix the faces with animal glue

THE ASSEMBLY
The ribs, the linings and six carved sycamore blocks round which the violin is built are assembled on a board, then transferred and glued to the back – which is first minutely finished with a tiny thumb-plane

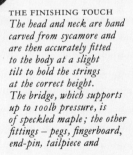

Fingerboard

Pegs

Head

Bridge

Tailpiece

Neck

-rest

Sound-post

nd-pin

ORNAMENTAL CARVING
The old instrument makers often enlivened their work with ornament, like the feminine head of this late-17th-century violin. On the back – a single piece of sycamore – the arms of the Stuart kings of England are carved

THE FINISHING TOUCH
The head and neck are hand carved from sycamore and are then accurately fitted to the body at a slight tilt to hold the strings at the correct height. The bridge, which supports up to 100lb pressure, is of speckled maple; the other fittings – pegs, fingerboard, end-pin, tailpiece and chin-rest – are made usually of ebony, sometimes of rosewood. Once the elements are assembled, the sound-post is inserted through one of the "f" holes and is placed firmly, but not too tightly, below the right foot of the bridge. Finally, the instrument is varnished to protect its finely worked surfaces

INDIAN SITAR
The long wooden neck of the sitar carries seven strings over the arched metal frets and up to thirteen sympathetic strings below. The resonating chamber is a simple gourd – often delicately ornamented with wood and ivory inlay

HURDY-GURDY
The strings of the hurdy-gurdy are vibrated by a wooden wheel, rubbed with wood resin to increase its friction. The melody strings are stopped not by the player's fingers but by wooden slides in the case. Once fashionable at court, it later became a mainstay of folk music

KOTO
The Japanese koto is a type of zither with movable frets. Derived from the Chinese ch'in, it was, above all, the instrument of the intellectuals and was often played by court ladies. The body consists of a convex wooden board, up to six feet long, commonly embellished with laquer-work. Thirteen waxed silk strings are stretched across the base over frets or bridges which can be moved to give different notes, while the instrument is played with plectra attached to the fingers

ARTISTRY IN WOOD

The Sounds of Wood

The sound of wood on wood, rare in our concert halls, has been the very basis of music in some cultures and a subject of religious awe in many others. The gamelan orchestras of Bali consist principally of wood-keyed instruments which are the forerunners of the orchestral xylophone. From Southeast Asia it is believed that the xylophone spread along the trade routes of the Indian Ocean to the shores of Africa; the marimbas of central America have an independent origin. The tribes of Guatemala use only the wood of the female tree to build their marimbas because they believe that the beautiful though mournful tone of the instrument is caused by the tree herself, lamenting the loss of her lover still standing in the forest. Western instrument makers prize the rosewood of neighbouring Honduras for xylophones.

In Africa, slit drums provided the famous bush telegraph; in New Ireland "priest drums" are seen only at moonlight ceremonies, when their voices are taken to be the voices of the fearful monsters carved on the drums. In many traditions drum making has acquired solemn rituals to enlist the co-operation both of the wood itself and of the spirit world. The wood is chosen for such practical qualities as sonority and durability, but the tree must be felled at an auspicious moment and dropped in the auspicious direction. Some societies feel there is a mystic connection between the wood and water, and the drum maker avoids all water while the log is being hollowed out. Offerings are made to the drum in times of drought; and in the Solomon Islands it is believed that the primeval ocean arose from a slit drum sounding of its own accord. The huge ceremonial slit drums of the Nagas of Assam, like the human members of the tribe, undergo an initiation ceremony, and are sprinkled with blood at their naming ceremonies.

Elsewhere, the simple rattle has potent ritual power, in the ceremonies of the American Indian shaman and even in certain festivals of the medieval church in Europe. Today, the xylophones and "temple" blocks of orchestra and dance band are faint echoes of a remote past, when the sonorities of wood stirred men to awe.

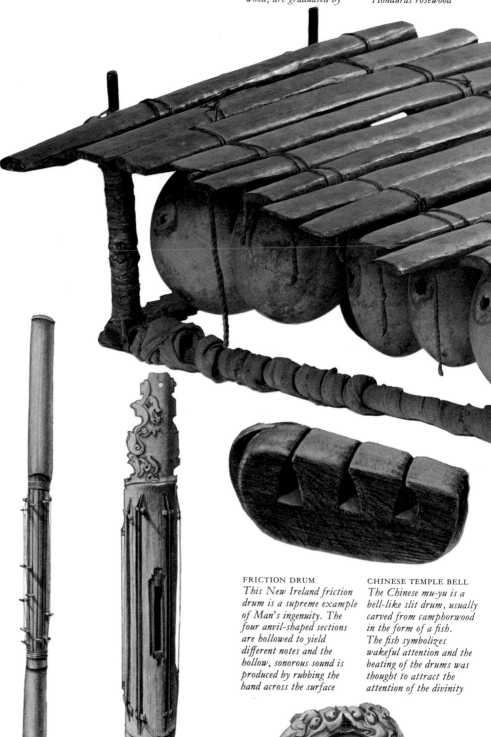

THE MARIMBA
A Portuguese traveller in Ethiopia in the 1580s commented on the "sweet and rhythmical harmony" of the marimba – a gourd-resonated xylophone. The hardwood keys, of resinous mutondo, takula or mwendze wood, are graduated by length and thickness to give different tones. The sound is amplified by gourds beneath each key and, within each gourd, a fine membrane gives a characteristic buzzing tone. Orchestral xylophones use either Malagasy or Honduras rosewood

FRICTION DRUM
This New Ireland friction drum is a supreme example of Man's ingenuity. The four anvil-shaped sections are hollowed to yield different notes and the hollow, sonorous sound is produced by rubbing the hand across the surface

CHINESE TEMPLE BELL
The Chinese mu-yu is a bell-like slit drum, usually carved from camphorwood in the form of a fish. The fish symbolizes wakeful attention and the beating of the drums was thought to attract the attention of the divinity

BAMBOO TUBE ZITHERS
Gently struck with wooden beaters, these tube zithers are wooden instruments in the fullest sense. The "strings" are formed by slivers cut from the wall of the bamboo tube, left attached at the ends and raised over small wooden blocks that form the bridges and allow the strings to vibrate freely. The tube itself provides the resonator. Tube zithers probably originated in Southeast Asia and later were carried to Africa, where they form the principal folk instrument of Madagascar

SHAMAN'S RATTLE
This beautifully carved
North American Indian
rattle, with symbolic frog,
demon and bird, is an
essential part of the
shaman's equipment. The
harsh tones of the instrument
were used to accompany the
chants in religious ceremonies

THE SANSA
The mellifluous tones of the
sansa are used throughout
Africa to accompany both
religious and secular music.
The hard bamboo tongues
are attached to a sounding
board or to a hollow box
resonator – often elaborately
decorated with symbolic
figures. The instrument is
held in the hands and the
keys plucked downwards
with the thumbs and
forefingers

TIGER BOX
Used to announce the end
of a Confucian service, the
stylized tiger is struck
three times, on the head,
with the split bamboo rod,
which is then vigorously
run over the ridges along
the animal's spine, giving
a harsh rattling sound

AFRICAN SLIT DRUM
Used in ceremonies and for
transmitting messages, the
slit drum is laboriously
made by hollowing out the
log through a slit along its
length. One end may be
given a wide bore, producing
a deep "male" note; narrow
bores give lighter tones

THE BONES OF JUDAS
Wooden rattles, like this
triccaballacca from the
south of Italy, were heard
in medieval churches in
Holy Week, when the bells
were traditionally silent.
When shaken, the two outer
hammers rattle against the
fixed central one

199

The Hidden Forms in Wood

To the pioneers of modern sculpture, wood offered the perfect medium in which to break away from the academic and overrefined sculptures of their day. It was the common material of the folk art and primitive sculpture which they greatly admired, and its use entailed carving, in contrast to the work of the academic sculptors, who were trained to model in clay. To carve into a block of wood is an ancient, difficult and irreversible task requiring the artist gradually to release the form he "sees" within the block instead of building it up in soft material.

Wood carving restored some of the common identity of artist and craftsman which had been lost since the Renaissance. In Paris, Brancusi used the economical sawcuts and simple techniques of Romanian country joiners to produce wooden sculptures that were bursting with creative life. In Germany, Barlach used northern hardwoods to create expressive, humanist figures, whose origins lay in the medieval wood carving heritage of his native country.

The idea of deriving form from the material used permeates modern sculpture. The huge wood carvings of the Russian Zadkine take their fibrous, knotty strength from the trunks from which they are made, while the powerful attraction and sensuous quality of the grain and figure of the wood seems to dominate in some of Barbara Hepworth's carvings. In some works the form of the sculpture is the joint offspring of subject and material, as in the reclining wooden figures of Henry Moore.

Others were attracted to wood not for its magnificence, but for its cheapness and accessibility. The refuse of the joiner's shop, odd lengths of lath and batten, were seized on by Picasso to perform his astounding transformation acts, conjuring away the triviality of the material by the force and inventiveness of its arrangement.

To some, the appeal of wood lies in the richness and variety of its grain and colour; to others it is a material full of angles and textures to be explored and exploited; but whether in the form of a man-made sculpture or simply a wave-polished branch thrown up on a beach, it is a material of infinite variety and artistic expression.

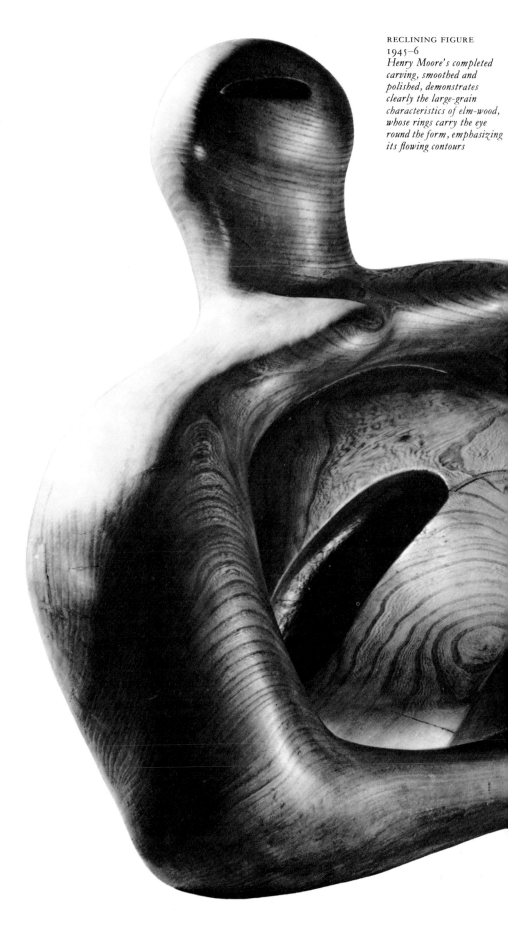

RECLINING FIGURE
1945–6
Henry Moore's completed carving, smoothed and polished, demonstrates clearly the large-grain characteristics of elm-wood, whose rings carry the eye round the form, emphasizing its flowing contours

All Henry Moore's large wood carvings were done in elm-wood because its fast-growing qualities are more suitable for carving than, for example, slow-growing oak, whose twists and turns often cause unforeseen problems.

After several preliminary drawings upon which the sculpture is based, outlines of the reclining figure are drawn on the rectangular block of wood. Initially, a cross-cut saw and an axe are used to rough out the basic form of the figure.

Moore began work soon after the tree was felled. By working on the wood while it was still green and unseasoned, the solid mass was opened out, allowing the wood to season more quickly. Unseasoned wood is much softer and therefore easier to carve.

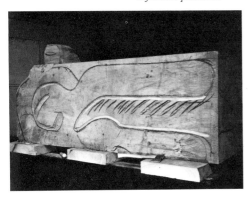

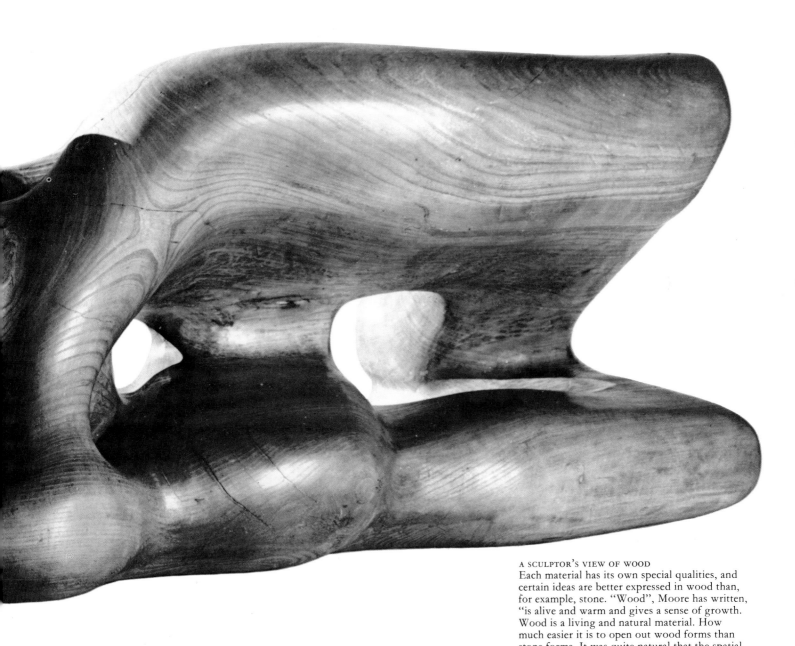

A SCULPTOR'S VIEW OF WOOD

Each material has its own special qualities, and certain ideas are better expressed in wood than, for example, stone. "Wood", Moore has written, "is alive and warm and gives a sense of growth. Wood is a living and natural material. How much easier it is to open out wood forms than stone forms. It was quite natural that the spatial opening-out idea of the reclining figure theme first appeared in wood."

ARTISTRY IN WOOD
Folk Art of America

Though only a by-product of the bustling racial amalgam out of which came America, the folk art of the United States vividly evokes the small towns and homesteads of the eighteenth and nineteenth centuries. It is a world of chalkware and scrimshaw, of homely furniture and fabrics and paintings, a world above all of carved and whittled wood.

Many wooden articles carry clear marks of their creator's country of origin, as do the carved birds of Wilhelm Schimmel, a German immigrant. Others are totally individual, for instance the models of charging Indians carved by a lumberjack during his idle evenings in a remote camp. Nothing but his own eye and an instinctive aptitude have brought to life the crude but affecting figures. Such carvings were usually the work of an artisan or farmer in his spare time, but there were also itinerant carvers or whittlers who for a meal or a drink would, with nothing more than a jack-knife and a piece of broken glass, transform an odd piece of pine or cedar into an ornament for a mantle.

It had long been the practice on both sides of the Atlantic to place outside shops wooden figures advertising their wares, but cigar-store sculpture, in its heyday between 1850 and 1885, is something characteristically American. In those years many ships' carvers found themselves out of work, and not a few tobacco-proffering Indians were carved by axe and chisel out of pine-logs from the spar yards – sometimes even retaining the massive, economical, outward-looking stance of figureheads. But the craft became conventionalized with men like J. T. Melchers, turning out sculptures to paper patterns.

Cedar and white pine were favourite whittling and carving woods. They have a workable softness out of which, on the one hand, craftsmen trained in popular and academic styles could produce circus and carousel animals full of exuberant life and detailed caricature; on the other hand amateur and professional decoy makers could achieve with them an exquisitely naturalistic simplicity. But the true charm of the American folk carving lies in the varied and individual manner in which each untrained hand wins its own struggle to achieve form.

MASONIC INN-SIGN
Inn-keeper Beemer was a Mason and wanted the fact known – no doubt it was good for business. The painted sign hangs like a mirror between the two turned uprights. The sign is as carefully jointed as a piece of furniture

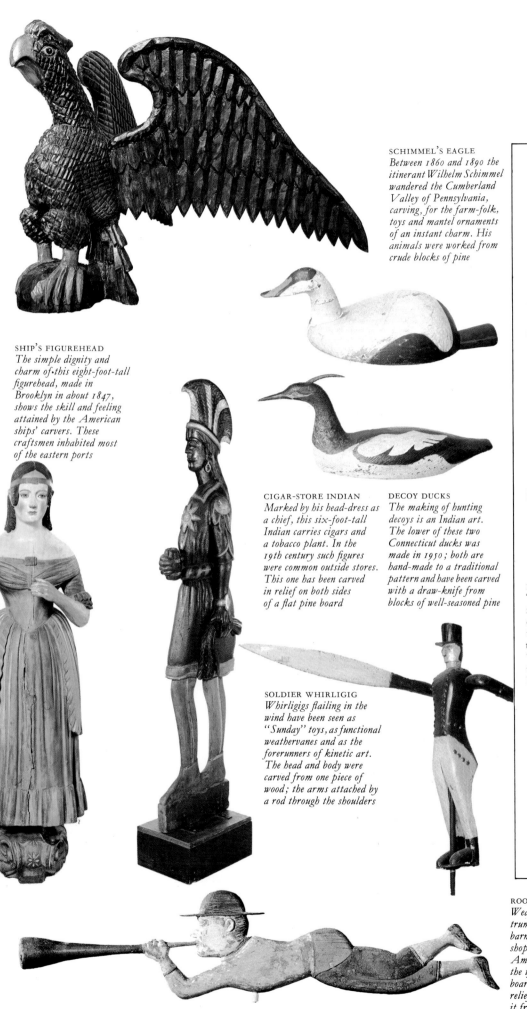

SCHIMMEL'S EAGLE
Between 1860 and 1890 the itinerant Wilhelm Schimmel wandered the Cumberland Valley of Pennsylvania, carving, for the farm-folk, toys and mantel ornaments of an instant charm. His animals were worked from crude blocks of pine

SHIP'S FIGUREHEAD
The simple dignity and charm of·this eight-foot-tall figurehead, made in Brooklyn in about 1847, shows the skill and feeling attained by the American ships' carvers. These craftsmen inhabited most of the eastern ports

CIGAR-STORE INDIAN
Marked by his head-dress as a chief, this six-foot-tall Indian carries cigars and a tobacco plant. In the 19th century such figures were common outside stores. This one has been carved in relief on both sides of a flat pine board

DECOY DUCKS
The making of hunting decoys is an Indian art. The lower of these two Connecticut ducks was made in 1950; both are hand-made to a traditional pattern and have been carved with a draw-knife from blocks of well-seasoned pine

SOLDIER WHIRLIGIG
Whirligigs flailing in the wind have been seen as "Sunday" toys, as functional weathervanes and as the forerunners of kinetic art. The head and body were carved from one piece of wood; the arms attached by a rod through the shoulders

ROOF-TOP WEATHERVANE
Weathervanes like this trumpet blower were set atop barns and houses, and even shops, throughout rural America. This figure has the typical form of a flat board cut out in light relief, painted to protect it from the elements

THE ART OF WHITTLING
Today, wood-carving is a popular hobby, to which the old image of the idler whittling on the front porch is ill-suited. Whittling is an activity requiring only a sharp knife, imagination, and exemplary patience; the wood must be worked with – not dictated to – and the scope is endless. Toys, figures, puzzles and models are made, including the traditional ship assembled in a bottle

FLIGHTS OF THE IMAGINATION
The preliminary stages in making a chain need very careful marking out; the most difficult task is the delicate removal of the tiny pieces of wood remaining between the links once they have been carved out. The advanced student may then progress to caged spheres, balls-within-balls and to bizarre expressions of sheer fancy

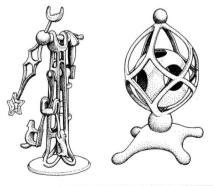

The Blockmaker's Art

Wood has been used as a material for printing blocks for at least two thousand years, its earliest application being for designs on textiles. The art of textile printing was widely practised in China, India and Egypt, spreading into Europe during the early Middle Ages and reaching a climax of perfection in eighteenth-century France and England. For machine production in the nineteenth century the designs were cut on rollers, but for hand printing, blocks were laminated From layers of fruit-wood (usually pear) and plane or sycamore.

It was the invention of paper which led to the use of woodblocks in printing. The fruit-woods were most commonly used and the designs were cut in relief with a knife, with the grain running along the block. This is termed "on the plank". At the beginning of the fifteenth century, woodcuts were in common use, providing cheap illustrations of religious subjects. Texts were sometimes cut into the same block and the resulting prints were hand coloured.

By the second half of the century woodcuts were beginning to appear as illustrations alongside type in Germany, the Netherlands and Austria. A number of great artists, notably Dürer, Holbein and Cranach, designed woodcuts, though their blocks were probably cut by professional cutters. One of the earliest and most popular uses of the medium was in the production of playing cards.

Woodcut illustrations gave way to metal impressions until, at the end of the eighteenth century, the English artist Thomas Bewick rivalled the fine qualities of metal engraving by using boxwood blocks. With the grain running vertically through the block, remarkably fine and detailed work was possible, and the resilience of the material was ideally suited to the new fast-running printing presses.

In the nineteenth century the main use of wood engraving was for cheap commercial illustrations of all kinds. The major commercial application of softwood as a printing medium this century is for "woodletter" type for display posters, though a revival of wood engraving, stemming from the work of William Morris, reached its climax between the two World Wars.

EARLY WOODCUT
The first woodcuts illustrated religious themes, and this symbolic depiction of the five wounds and the instruments of the passion clearly shows the simple, bold lines used by 15th-century cutters, the design resembling stained glass. Many prints were made for travelling pilgrims

THOMAS BEWICK
The intricate work made possible by the use of finely grained wood cut on the cross was fully exploited by Thomas Bewick in the 18th century. The durability of boxwood allowed up to 900,000 impressions to be made before a replacement block became necessary

THE MASTER OF
WOODCUTS
*Woodcut illustrations
achieved great popularity in
the 15th and 16th centuries
in Germany, where masters
like Dürer, Holbein and
Cranach used the medium.
On the left is the original
1511 Dürer block for his*

*"Christ appears to Mary
Magdalene", from which the
print on the right was made.
The design on paper was
pasted on to the plank, or
drawn directly on to it,
then, as can be seen, the
parts intended to be white
were cut away, leaving the
design in relief*

THE ARTIST'S MATERIALS

Early woodcuts, both in Europe and Japan, were made with knives, chisels and gouges (below left) from a block of soft wood like pear, cherry, beech or sycamore. The block was cut "on the plank" – along the grain of the wood – which allowed only limited detail of design. Wood engravers used a block of hard wood, usually box, cut across the grain (below right). Its compact, strong fibres permitted the engraver to achieve a finer point, more delicate lines and greater accuracy with the graver, or burin.

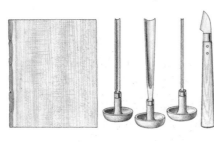

JAPANESE WOODCUTS
*The woodcut was the
traditional printing process
in Japan for centuries, yet
it was only with European
interest in artists like
Utamaro (left) in the
19th century that it was
recognized as an art form
by the Western World*

CHERRYWOOD BLOCKS
*Japanese blocks were
generally cut from a log of
yamazakura wood, a type
of wild cherry. For colour
printing, the hard inner
wood was used for the key
block; the softer outer
wood for the subsequent
colour blocks*

Treen: The Collector's World

The old English word "treen", meaning, literally, "of trees", occurs frequently in old inventories and wills in reference to such items as platters, bowls and drinking vessels. In early days these were made straight from the tree; the main section of the trunk would provide simple furniture, while other parts of the tree were used for domestic vessels and implements. Suitably sized portions were hacked from boughs to make into bowls and platters; smaller pieces of bough were hollowed out for drinking cups and ladles, and still smaller pieces were whittled into spoons. Hard woods like oak, sycamore and maple were used extensively, and from the seventeenth century the heavy lignum vitae, imported from the West Indies, was popular for its reputed medicinal qualities. Other woods traditionally used include beech, birch, walnut and laburnum, holly, yew, box and many fruit-woods.

With the invention of the pole-lathe in the Middle Ages, the manufacture of these domestic items became an important part of the wood-turner's trade, but wood carvers continued to make a wide variety of domestic and agricultural implements. Professional carvers would cut their intricate designs in reverse on butter prints, cheese, biscuit and gingerbread moulds.

Kitchen implements like pestles and mortars, skimmers, washing bats and mangle-boards were made by both professionals and amateurs, but it was the amateur carver who produced a huge proportion of the treen that survives today. In spare moments the husbandman would make clothes-pegs, cheese-scoops or simple nutcrackers for the home, pipe-stoppers for himself and simple toys and whistles for his children. Sailors far from home, French prisoners-of-war, and young men courting, have all expressed their feelings through love-tokens painstakingly whittled from odd pieces of wood.

Only objects handmade before 1830, when machine-made articles began to predominate, are generally accepted as pure treen. Nevertheless, collectors can still find a wide range of domestic ware, smoking or sewing implements like tobacco tampers or lace bobbins, or the ever-popular love-tokens.

DRINKING CUP
The coopers made many small, watertight, staved vessels for use as tubs, buckets, bowls and flasks for spirits and cider. This two-handled drinking cup from Scotland is made of alternate light and dark staves, tightly bound with strips of willow to give a watertight fit

CHILD'S PLATTER
Made from oak cut straight from the bough and hollowed out on one side, this very early platter held meat or stews in the "bowl" side, was then wiped clean with bread and turned over so that the underside could be used for dry foods like bread and cheese. The tough wood is almost unbreakable

WASHING BAT
This English oak bat was used before the days of mangles to beat the excess water out of newly washed clothes and sheets. Leaning against it are an 18th-century fruit-wood ladle, a sugar ladle carved out of boxwood and a 17th-century child's carved walnut spoon

VEGETABLE CHOPPER
The chopping blade was fixed at one end of this efficient 18th-century kitchen implement. Raw vegetables were cut on the raised block near the handle. When the receptacle was filled, the vegetables were tipped into the pot — the side walls preventing spillage

STAY BUSKS

These cumbersome objects were worn in pockets at the front and back of the bodice when a straight contour was fashionable. They are up to fifteen inches long, triangular, wedge-shaped, and made of wood or whalebone. They were probably extremely uncomfortable in use, but once accepted as love-tokens their position close to the heart was perhaps the most important consideration. They are chip-carved with hearts and other love motifs and may also carry carved poems and love messages

THE LOVER'S SKILL

The messages of love conveyed by these tokens were emphasized by the virtuosity of the carving. From a single block of wood the carver might create complex interlocking chains, free-moving spheres held within delicate cages, or a set of spoons (left) held together by a ring

SNUFFBOXES

Snuffboxes carved in the form of shoes were often presented as wedding gifts. Their shape is often an indication of their age – those with square toes dating from the early 19th century. The tiny box is made of laburnum wood, the light sapwood streak cleverly used here for the sole and heel of the shoe. The snuffbox on the right is probably French, carved from a solid piece of walnut and fitted with a hinged lid. The rare bellows shape with the sliding lid is carved from a cherrywood block

207

Wood in Action

The greater the demands the sportsman makes upon himself, his skill and his physical fitness, the higher the standard he expects of his equipment. Once wood must have been used in sport because it was the most convenient material to hand: now the sportsman can bend to his ambition all the resources of modern technology.

Nevertheless, from the tens of thousands of species of timber in the world, it is surprisingly difficult to find more than a few with the right combination of properties, and little advance has been made today over the trial-and-error findings of the village green. Ash and hickory remain the woods used predominantly in sport, superior in their combination of properties not only to other woods but also to man-made materials. These woods were first chosen, and continue to be used, because they resist repeated impact without fracturing, and absorb shock without transmitting it to the arm. They have enough weight to lend force to a stroke, but not enough to make the stroke tiring or unwieldy.

In other games other factors are critical: the heads of golf clubs must be very hard and not wear, split or splinter; fishing rods must be light, and both strong and flexible; on the other hand, billiard cues must be straight-grained and free from movement or distortion. But for most sports ash and hickory are supreme because they possess, with their toughness, the important capacity to be bent to a small radius without appreciably losing their strength.

For handles, the strength and resilience of cane is often utilized, but here its ability to absorb shock is supplemented by the insertion of rubber strips. In other cases, wood has been wholly supplanted: fibreglass vaulting-poles have a strength and resilience with which wood cannot compete, even laminated wood, which has been so successful for skis, archery bows and tennis racquets. Although in the struggle to improve performance traditional woods can continue to be used only if they show themselves equal or superior to modern substitutes, many of them are so much a part of their sport that it would be inconceivable to abandon them.

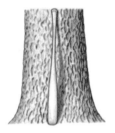

BASEBALL
The bat should be made of ash, of best-quality wood taken from the bole of the tree. Other hardwoods, or laminates, may be used as there is no restriction on the weight of the bat, but their performance cannot match that of ash. As the strike is made, the bat sweeps round in a wide arc – its mass and momentum imparting tremendous speed to the tightly stitched ball

CROWN GREEN BOWLS
The skill and tactics of traditional bowls are a function of the slope of the green and the weight and bias of the bowls – made from the very heavy, slow-growing lignum vitae, which comes from the Caribbean. No two sets of bowls are quite alike and the player must get to know their individuality. Today, a shortage of lignum vitae has necessitated the wide use of composition materials

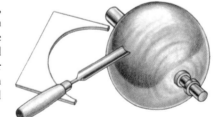

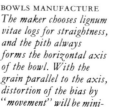

BOWLS MANUFACTURE
The maker chooses lignum vitae logs for straightness, and the pith always forms the horizontal axis of the bowl. With the grain parallel to the axis, distortion of the bias by "movement" will be minimal. Bowls are made in

sets of four, all from a single log, so that their specific gravity, and their behaviour on the green, should not differ. On the selected log, the sapwood is cleft away and the wood varnished and left to season. Any cracks that have appeared are then plugged,

and the wood rough-turned, then more accurately shaped to the desired bias with a template. Lastly, the protuberant ends are removed, and sockets bored in which ivory pieces are set. Bowls must be kept polished and cool, otherwise cracks develop

BOWLS IN PLAY
The bowl is a weighty object which is set on course rather than rolled. To deliver it cleanly and without wobble requires skill. To the sportsman, the heavy clunk of bowl dislodging bowl is a very satisfying sound

THE HANDLE

The handle consists of sixteen pieces of Sarawak cane, planed square and glued into four "slips", between which rubbers are inserted. The lower end is spliced into the blade, and the handle wrapped in twine and fitted with a rubber grip.

CRICKET

Cricket is a game of force and power as well as skill. The willow of the bat blade absorbs the punishing impact of the ball, travelling at up to 80mph, while the springy cane handle-pieces protect the hands from the sudden shock

THE BLADE

Quick-growing cricket bat willow is cleft into wedges, which are roughly shaped and left to season. The craftsman then cuts and shaves the blade, balancing it by feel, and compresses the face and edges to strengthen them

HOCKEY STICK

The head is of ash, steam-bent into as tight a bend as the wood will allow, so that the grain follows the curve and strengthens it. The handle is of cane

LACROSSE STICK

To withstand the strain of hard play – especially at the narrow, sharp bend of the head – hickory has always been used, steam-bent in one length so as to retain its toughness

GOLF WOOD

Persimmon, the traditional wood for the head, is still preferred, but is in short supply. Laminated maple makes a good substitute. Shafts, once of hickory, are now steel

POLO STICK

Weight is critical, and the head must be of bamboo root or of sycamore or ash – these with their growth rings across its width, if they are to last. The shaft is cane

TABLE TENNIS BAT

The light ball delivers no shock, so the bat can be made of plywood. The criss-cross grain will resist warping better than solid wood

FISHING ROD

Rods were once greenheart tipped with degamé, but lengths of the lighter Tonkin cane are now used, or metal alloys. Hickory, ash, hazel and lancewood have all served as rods

SKIS

Originally made of solid ash, skis are now mainly laminated, preferably of hickory, but often still of ash, and of birch sometimes in Scandinavia. The wood must be light, tough and able to withstand continual flexing without splitting

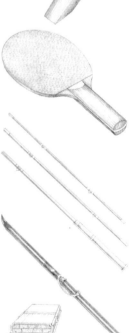

THE TENNIS RACQUET

In the early days of sphairistikè, now known as lawn tennis, the racquet used was of solid ash and weighed 1lb or more. Lamination, introduced in the 1930s, has reduced its weight, its expense, and its tendency to distort

THE HANDLE

Where the laminations of the frame sweep down to the end of the handle, the oval is closed by a glued strip of ash. The wedge behind may be a sycamore, a walnut or a mahogany. The handle-pieces are usually made of sycamore

THE CHOICE OF WOOD

Though ash is a constant, the supplementary woods may vary according to opinion and expense. However, heavier woods used for strength, and coloured ones for decoration, must ultimately contribute to the racquet's balance and feel

THE FRAME

Seven or eight laminations are the optimum, of which typically four are ash. Ash may be supplemented by maroti, a South American wood of similar properties; by beech, to strengthen the string-holes; by fibre, and by decorative woods

Bowyer and Fletcher

The modern sport of archery disguises a deadly intent older than history, for stones and throwing sticks, javelins and arrows, Man's earliest projectiles, served but a single purpose – that of killing.

Any wood, and even bone and horn, may be used to make a bow, but the classic weapon is of yew; its pale sapwood combining with the dark heartwood to unleash up to one hundred pounds pressure each time the bow is drawn and released.

In the Ancient World, archers and slingers were auxiliaries to the main force of cavalry and infantry by which battles were won or lost. The eighty-five-pound longbow came into its own during the Middle Ages, when it was solely responsible for a series of crushing defeats inflicted on the French by the English. At Crécy, in 1346, it struck with all the force of a major technological innovation; the French army, outnumbering the English three to one, charged sixteen times, only to be decimated by the hail of arrows from the ranks of English bowmen more than two hundred yards ahead.

The crossbow, though well developed at this time and extremely effective against armour, was slow to use; in winding up the mechanism, the archer's hands would travel more than 150 feet compared with two feet for the longbow. For every bolt shot from the crossbow, the longbow could release eight arrows; in one minute, 7,000 archers could unleash a lethal rain of 100,000 arrows.

The bow of the Victorian sportsman differed from the medieval bow in having two separate limbs spliced together in the centre. The yew tree is often gnarled and the wood very knotty so that single lengths, suitable for bows, were rare. Even in the Middle Ages, good yew was a prized commodity and considerable diplomacy and subterfuge were employed in ensuring that supplies, mostly imported from Spain, were not interrupted.

Steel, laminates and glass-fibre have carried the bow into the twentieth century. Traditional bowmen, however, demand one thing in particular; unstrung, the bow should return to the straight. That unique property is shared only by the yew and by the American hedgerow tree, osage.

CLEAVING THE LOG
Even a long, straight yew log from the bole of the tree may conceal rotten wood, knots, or twisted grain. The bowyer cannot know until the log has been cleaved open. Using mallet and wedges, he works, as in every stage, with the grain of the wood. Logs suitable for "self-bows" are rare and most bows are made in two separate limbs, or from laminated woods

Self-bow: English y

SEASONING
Yew for a bow should be seasoned for at least five years, but may be stored indefinitely in the form of staves or billets. The end-grain, and sometimes the whole billet, is sealed to prevent the wood from drying quickly and splitting

THE BOWYER'S CHOICE
A remarkable number of different woods have been used in the manufacture of bows. For "self" bows, those made from a single wood, the finest woods are yew and osage, followed by lancewood and degamé. Laminates of these woods, and of ash, are used to back other woods, from cornel and crab-apple to the more exotic ameranth and snake-wood, but few composite bows are sweet; many kick badly in use

TAPERING THE LIMB
The great stresses at the crown, or centre, of the bow require a greater thickness of wood. The bowyer works with impressive accuracy – using the axe to cut away fine shavings along the grain until the limb takes on its initial tapered form

THE ESSENTIAL SAPWOOD
The bowyer uses spokeshave and float to reduce the billet to its rounded form. An even layer of sapwood is retained over the outer surface of the bow and knots are left standing, for to cut through them would seriously weaken the bow

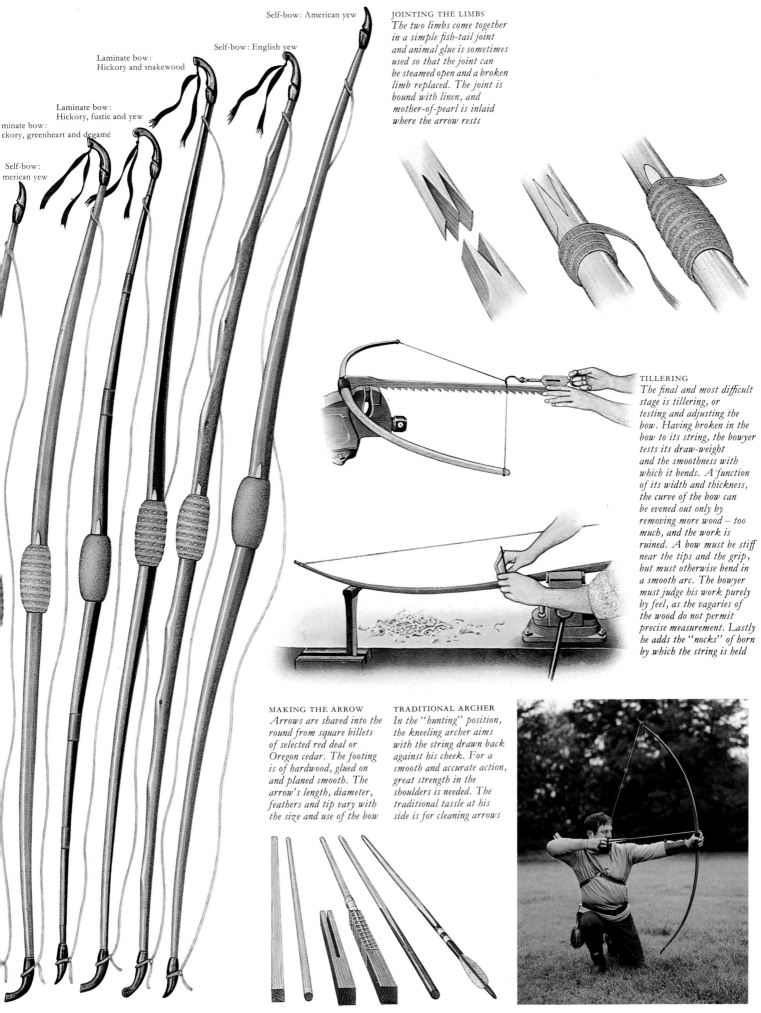

Self-bow: American yew

Laminate bow:
Hickory and snakewood

Self-bow: English yew

Laminate bow:
Hickory, fustic and yew

minate bow:
ckory, greenheart and degamé

Self-bow:
merican yew

JOINTING THE LIMBS
The two limbs come together in a simple fish-tail joint and animal glue is sometimes used so that the joint can be steamed open and a broken limb replaced. The joint is bound with linen, and mother-of-pearl is inlaid where the arrow rests

TILLERING
The final and most difficult stage is tillering, or testing and adjusting the bow. Having broken in the bow to its string, the bowyer tests its draw-weight and the smoothness with which it bends. A function of its width and thickness, the curve of the bow can be evened out only by removing more wood – too much, and the work is ruined. A bow must be stiff near the tips and the grip, but must otherwise bend in a smooth arc. The bowyer must judge his work purely by feel, as the vagaries of the wood do not permit precise measurement. Lastly he adds the "nocks" of horn by which the string is held

MAKING THE ARROW
Arrows are shaved into the round from square billets of selected red deal or Oregon cedar. The footing is of hardwood, glued on and planed smooth. The arrow's length, diameter, feathers and tip vary with the size and use of the bow

TRADITIONAL ARCHER
In the "hunting" position, the kneeling archer aims with the string drawn back against his cheek. For a smooth and accurate action, great strength in the shoulders is needed. The traditional tassle at his side is for cleaning arrows

Toys and Dolls

The great mixed forests of Europe, rich in both hardwoods and softwoods and with abundant water power from fast-flowing mountain streams, naturally became some of the world's great centres of toy-making. Wood-carvers, joiners and carpenters, at their busiest in the summer months, often turned to toy-making during the winter – selling their products either locally or to travelling fairs and merchants.

Patterns of toy-making have evolved through history and makers have repeatedly borrowed ideas from the past. The crudely cut horse on wheels given to a Roman child is not very far removed from the purposely simplified model of a progressive modern toy, and simple lever-operated wooden models invented in Ancient Egypt reappear almost unchanged in the carved spruce fairground toys of Russia and Bohemia. Toy-making centres sprang up wherever fast alpine streams provided power for lathes and, from the fourteenth century, Nuremberg in southern Germany was famous for its toys, with a rigorous system of guilds enforcing the type of work done by each craftsman. Forts and soldiers, farms and animals, constructional toys and a wide variety of games and puzzles were made in the early 1800s, and whole families in the surrounding countryside were involved in their manufacture. Spruce and larch were used for cheap toys; lime, beech and ash for carved models, and boxwood for items which faced a great deal of wear.

European techniques were imported to America by colonists and by the eighteenth century toy-making centres were well established in the eastern states – based on the fine forest of basswood, chestnut, hickory and rock maple. American toys of this period were strong and robust: wagons, sledges, building bricks and ships were common; toy guns were made from hollow elder branches, kites from supple hazel and fishing rods from ash with tips of lancewood.

In Russia, Czechoslovakia and the Saxon Erzegebirge, small communities produced wooden folk dolls like the Russian *matryushka*, nest of dolls, while areas with a tradition of religious carving, like Oberammergau, Berchtesgaden and Nuremberg, specialized in the production of fine fashion dolls.

In eighteenth-century England a customs embargo prohibited the import of painted wooden dolls to encourage the home-based industry. Simple dolls had a turned skittle-shaped body made from a single piece of wood, whittled arms and legs, and painted features. During the nineteenth century, many alternative methods, including papier mâché, wax and china, were perfected for the manufacture of dolls' heads, but the bodies were still made of wood or of fabric packed with wood dust from the sawmills.

In the 1700s, a European fashion of collecting and displaying small treasures soon led to the development of the dolls' house, in which crudely made spruce items stocked the servants' quarters while finely detailed work in oak, walnut and mahogany, often by master craftsmen like Chippendale, reproduced the elegant furniture and fittings of "above stairs".

ANIMAL RINGS
In Saxony, one of Europe's busiest centres of toy-making in the 19th century, an ingenious form of mass-production was employed, particularly for the construction of Noah's Ark animals. A cross-section of a spruce log was lathe-turned in a ring, with suitable grooves cut so that a slice or section roughly resembled a bird or animal, different sizes of ring being used for different animals. Refinements such as tails, ears or horns were later hand-carved, and each animal was then painted in its natural colours

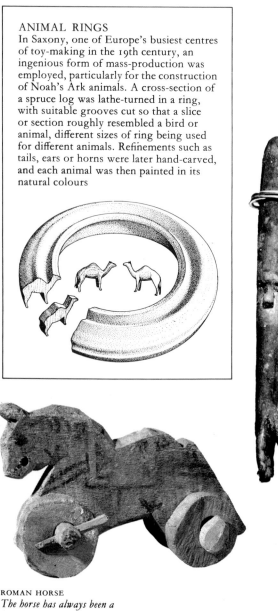

ROMAN HORSE
The horse has always been a favourite toy, whether in miniature or big enough to ride on. This tiny wooden horse, made in Egypt in AD 200, imitates contemporary chariot construction with its solid wheels secured by lynch-pins on the outside

ROMAN DOLL
Found in a sarcophagus of the 1st century AD, this jointed doll is fashioned from oak. The limbs are articulated at the elbow and knee with pivoted mortise and tenon joints, and even the fingers and toes are movable

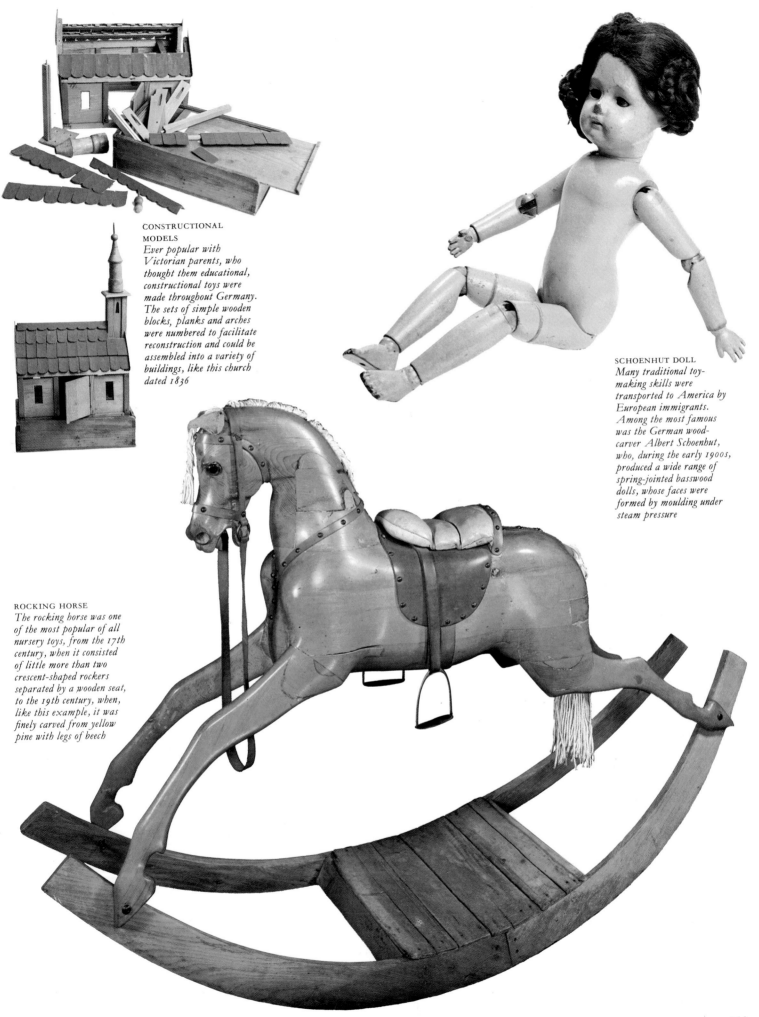

CONSTRUCTIONAL MODELS
Ever popular with Victorian parents, who thought them educational, constructional toys were made throughout Germany. The sets of simple wooden blocks, planks and arches were numbered to facilitate reconstruction and could be assembled into a variety of buildings, like this church dated 1836

SCHOENHUT DOLL
Many traditional toy-making skills were transported to America by European immigrants. Among the most famous was the German wood-carver Albert Schoenhut, who, during the early 1900s, produced a wide range of spring-jointed basswood dolls, whose faces were formed by moulding under steam pressure

ROCKING HORSE
The rocking horse was one of the most popular of all nursery toys, from the 17th century, when it consisted of little more than two crescent-shaped rockers separated by a wooden seat, to the 19th century, when, like this example, it was finely carved from yellow pine with legs of beech

Lore, Legend
and Belief

The awesome face of this Singhalese
ritual mask depicts one of many
"sickness demons" whose anger must be
subdued in elaborate exorcism ceremonies
in order that the afflicted person
may be returned to good health

Instruments of Belief

Protected by all the trappings of civilization, Western Man has largely ceased to fear anything more than his own potential for destruction. But outside this material cocoon, the world still appears a dangerous and even hostile place. Lacking the strong teeth or sharp claws which other animals have developed, Man relies on tools to control his world. The immediate availability of wood, and the comparative ease with which it can be shaped, has made it an obvious material for this purpose. When someone is using a mallet it is obvious what is happening, but some "tools" have a less obvious function.

Mechanical manipulation of the visible world represents only one of the ways in which Man has sought to control events and his surroundings. For most of history, and for most men still, the forces from which Man has sought to protect himself, and finally to use for his own purposes, are as much mystical as mechanical. According to this view, the cosmos is controlled by abstract and personal forces – strength, good, evil and fertility – and the gods, spirits, devils and demons who represent them. These spiritual agencies are in varying degrees like Man himself, and if they can be correctly identified and communication established with them, then they may be persuaded to exercise their powers for Man's benefit and at his request.

As one material from which instruments are fashioned for these purposes, wood has represented a wide range of beliefs and has acquired a whole mythology of its own. Specific woods may be prescribed by tribal traditions, methods of felling trees follow strict patterns, and ways of carving remain faithful to ancient designs. The Iroquois Indians of North America insist that their False Face masks be cut from within the tree, usually an ash; the Maoris use soft woods, which are easily carved; while the Marquesans, anxious that their carvings may be preserved, will use hard, close-grained woods, which give their article the desired permanent status.

For the individual, wooden instruments of belief have three main functions, perhaps best described as "definition", "communication"

and "protection". Definition of relationships and roles is part of the process of manipulating abstract forces, so the individual human must first define himself. This may be done by putting on a mask which claims, in effect, a particular identity. It states what the individual is doing, what social group he belongs to, what function he is performing in a particular ceremony, or whom or what he represents. The mask may thus be of an ancestor or of a particular spirit or god, and, by putting it on the individual also puts on, for the time being, that personality and performs that role. Just as a mask defines a character in the Japanese Nō theatre, elsewhere too it defines the role, and usually defines it as part of the sacred sphere of action. This sphere concerns abstract and invisible forces, which are also made perceptible by the use of other ritual objects, like flutes and drums, altars, wands, sacrificial fires, carvings and paintings. Wood is employed for all such objects; they are tools for communicating with the invisible world.

In the Niger Delta, the Ijaw people carve small wooden figures as spirit companions, through whom they speak to their ancestors. It is said that when pleas for help are heard, the figures tremble. Wood is also used to communicate with particular spirits when it is employed in divination. The Ifa board of the Yoruba people in West Africa helps not only determine present facts but to guide future actions, as does the rubbing-board oracle of the Azande farther east.

Communication in this sense often overlaps with the desire to seek help or protection. Wooden amulets are designed to do this directly. The Chinese would carry pieces of peachwood to ward off evil spirits, Hopi Indians kept bits of petrified wood as lucky charms, while the Iroquois carried miniature wooden canoes to save them from drowning. To "touch wood" as a safeguard against tempting fate originates from the many sacred and magic properties attributed to different woods.

WOODEN AMULETS
Personal charms, imbued with magical properties, were worn by the Dayaks of Borneo. In times of illness or distress, a witch-doctor would take small chips of wood from the amulet and prepare a remedy with them

ANTELOPE HEAD-DRESS
Representing the spirit who introduced agriculture to the Bambara people in Mali, this stylized carved head-dress is worn by young men during dances of the secret society celebrations held before planting and harvesting of crops

TWIN FIGURES
When twins are born to Yoruba parents in Nigeria, the event is commemorated by the carving of a pair of wooden figures. Should one of the twins die, the figure of the dead twin is ritually fed at the same time as the survivor

BABEMBE MASK
Carved roughly from a piece of very light wood, this curiously and suitably owl-like mask represents the genie of the forests, and is worn at the "kalunga" celebrations by members of the Bembe tribes in eastern Zaire

GELEDE MASK
Aimed at appeasing the witches in their community, the masks of the Gelede society in the Yoruba kingdoms of Nigeria are carved with a variety of subjects. The masks acquire a patina when steeped in resin, oil and charcoal

JAPANESE MASK
A single piece of wood with the base of a twig branching from it has produced this terrifying mask, probably worn in the kyogen farce performed in the interval of a Nō play. The grain of the wood clearly emphasizes the grotesque nose

IFA DIVINING BOARD
Wooden objects are held to have magical properties all over the world. They are used particularly in divination, whether by a witch-doctor using a wooden rubbing-board oracle, or throwing small wooden chips, or a Western water-diviner clutching an uncontrollably twitching forked hazel twig. In Nigeria, round wooden trays with carved raised edges are used as divining boards. These are sprinkled with sand, or special wood dust, on which the diviner marks out various signs. He then throws palm nuts, pebbles or wooden chips on to the board, and interprets the position of fall

HEAD-REST
With elaborate hair-styles like those shown in the carving, many Africans use small wooden head-rests to protect their coiffure, usually specially created for festivities. This head-rest was carved by a member of the Luba tribe in Zaire

WOODEN CULT VESSEL
Usually dated soon after the Spanish conquest, the Peruvian kerus are painted and lacquered wooden beakers whose designs represent mythological or ritual scenes. This one illustrates a scene from the bird cult

A KING'S STOOL
Like all African carving, the Ashanti chief's stool is carved from a single piece of wood. Made from osese wood, each stool has an individual design, and is held to be a repository for the souls of ancestors of the tribe, and is thus sacred

The Tribal Image

BURIAL POLE
A form of totem pole, this carved symbol of a killer whale holds the body of a chief of the Tlingit Indians on Wrangel Island, Alaska. When the ground was too hard for interment, the dead were often placed on scaffolds or in trees

HAIDA TOTEM POLE
For the Northwest coast Indians the wood from the red cedar was the most easily worked, inexhaustible raw material readily at hand. Its size made it a perfect wood for the totem pole, a symbol of the group, like a family crest

Just as wooden objects can lend psychological support to the individual who uses them in rituals or ceremonies, so in a similar way they can be used to further social ends by furnishing visible symbols of the social unity of the group. A sense of loyalty is encouraged by the knowledge, for example, that people are descended from a common ancestor, and a representation of the family tree, showing different ancestral origins in its many branches, is usually an object of pride to any member of that family.

Where a family, or a clan, or even a whole tribe sees common descent as the qualifying characteristic for membership of the group, then figures of the ancestor who "founded" the group are often used as focal points in rituals which make this coherence tangible.

Figures of ancestors are venerated among many peoples – particularly in West Africa, Melanesia and Indonesia. These societies believe that their ancestors' spirits travel to another world, where they are in contact with the spirits who control the world, so here the ancestor worship also involves a degree of appeasement. Where it is essential to maintain physical contact with the ancestors in order constantly to draw strength from them, parts of the body, most often the skull, may be preserved in miniature wooden coffins or shrines. The Fang people of Gabon and many tribes in Melanesia do this; the Fang people keeping the skulls in circular bark containers guarded by a carved wooden figure, which, by protecting the skull, symbolically protects the whole community. The efficacy of their spiritual power is maintained by periodical cleaning and oiling – both of the skulls and of the wooden figures.

Spiritual power is not necessarily always present in wooden objects, sometimes it must be persuaded to enter the object, specifically carved, through traditional ceremonies. Other objects may be activated by carving or painting, like the churingas, or bull-roarers, of Oceania, or Aborigine shields and Maori godsticks, whose designs have both totemic and spiritual significance. All totemic objects, like the tall poles of the Haida Amerindians, both symbolize and focus the mystical unity of the society.

GRAVE EFFIGIES
Perched high on a cliff in the highlands of the Indonesian island of Sulawesi are the Toradja grave effigies. Standing lifelike in a wooden gallery, staring out over their village, they guard the graves which have been hacked out of the rock behind them. Carved realistically from teak, they are dressed with the clothes and jewellery of the dead owner, for the Toradja believe that the afterlife is an extension of the earthly one, so the dead man must be accompanied by his possessions

MAORI CARVING
Carved in the traditional manner, before New Zealand was colonized by Europeans, this lintel shows the "clan mother", her eyes made from abalone shell

BONE GUARDIAN
Like many ancestor-worshipping tribes, the Fang people of Gabon preserve the bones or skulls of their ancestors in bark boxes. On top of each box stands a carved guardian figure, itself representing a primeval ancestor

ANCESTOR FIGURES
In the belief that ancestor spirits could be kept in a carving, people in New Guinea produced softwood figures. This composite figure was recovered intact from Lake Sentani, where missionaries had persuaded tribesmen to throw it

HOUSE POSTS
A simply carved tree trunk is the centre post of a Iatmul cult house in New Guinea, thought to be of considerable age. Though it gives little architectural support, it functions as a guardian spirit protecting the occupants

The Power of Wood

Trees have played a central role in Man's rituals from time immemorial and sacred groves appear at all times. Man's earliest temples were simply clumps of trees, whether hidden in the depths of forests or growing prominently on some significant hillside. Prehistoric circular temples like Woodhenge in southern England may have been built in imitation of these natural woodland temples, and it is possible that the lofty pillars and vaulted roofs of later churches owe their origins to the natural forms of these distant forerunners.

Throughout the long history of religions, wood and trees appear constantly as sacred objects. This sacred character sometimes derives from a symbolic relationship which the trees bear to a spiritual force or being, but sometimes it expresses a sacredness belonging directly to the tree itself.

Spirits, whether dryads, the attendants of the goddess Artemis, or other nature spirits like the Nagas, the sacred cobras of India, who bestow rain and fertility, are believed to dwell in trees and, in a natural development of the mythology, the tree has become so identified with the spirit that the two are almost inseparable.

Specific trees are worshipped as gods all over the world. The oak tree, sacred to the Druids, symbolized strength and offered protection. The cedar was revered in ancient Lebanon by Christians, Hebrews and Muslims alike, each for their own different reasons. Carried in the exodus from Egypt, the wood from the acacia tree was used by the Israelites to build their holy objects, the Tabernacle and the Ark of the Covenant. The Bo tree is sacred to Buddhists, because Gautama Buddha sat meditating beneath it until he found nirvana, and the Banyan tree is revered by Hindus, who believe that Brahma was transformed into one.

Rituals held on special occasions, particularly those marking climaxes in the annual cycle, like the spring equinox or the harvest, often employed, and still do employ, wooden objects, partly because wood is itself organic; its own cycle of growth makes it particularly appropriate to occasions which are linked to the natural seasonal cycle. The maypole and its attendant festivities – a familiar sight on English village greens even today – are remnants of the ancient spring fertility rites of pre-Christian Europe. In Ancient Rome, in spring, a pine tree symbolizing Attis, the lover of fertility goddess Cybele (himself born of the almond tree), was carried to her temple on the Palatine Hill. Similarly, throughout northern Europe, youths would go out to the woods and bring home decorated tree branches, phallic symbols of many fertility rites, around which they would dance.

While many trees, and woods, are believed to have sacred powers in their own right, many hundreds more have had powers attributed to them through the complex web of mythology and belief which ties Man to his natural environment.

THE SACRED OAK
Throughout the northern hemisphere the oak tree, and oak groves, enjoy a rich and varied mythology. Oak was the Norse tree of thunder, sacred to the god Thor, and gave protection to those sheltering under its boughs. If struck by lightning, pieces of the shattered wood were kept as protective amulets. The oak was also held in reverence by the Druids, as a symbol in its own right and by association with the sacred mistletoe – the guardian of the tree's life during the winter

THE FOLKLORE OF WOOD

ROWAN is a charm against witchcraft in European folklore. Protective rowan crosses, made without using a knife, are tied to cows' tails in the Isle of Man on May Eve to protect them from evil influences

ASH is renowned for its protective and curative properties. Its wood is thought to cure warts, and a child, if passed through a split sapling which is then bound up, will be cured of rickets or ruptures

ELDER, despite its medicinal properties, is unlucky perhaps because Judas hanged himself on its branches. Witches transformed themselves into elder and, if cut, the wood is said to bleed

WILLOW, one of the best divining woods along with hazel and birch, is a symbol of mourning and forsaken love. Full of magical properties, it brings luck in childbirth and is thought to cure the old and sick

HAWTHORN is ambivalent in country lore: to bring its blossom into the house foretells a death in the family, while sitting under the tree on Midsummer Eve or Hallowe'en can cause fairy enchantment

YEW trees may symbolize both life and death. Once sacred to Hecate, queen of the underworld, they are found in almost every English graveyard, their evergreen quality representing the triumph of everlasting life

The Tree of Life

THE CRUCIFIX
The tree of life, rejected by Adam in the garden of Eden in favour of the tree of knowledge, renewed its promise of immortality when it reappeared, symbolized by the crucifix, offering redemption and everlasting life

THE WORLD TREE
From the legendary ash tree, Yggdrasil, stems all Norse mythology. Its branches hold the heavens, home of the gods, its trunk the middle earth. Its three great roots descend to the underworld, watered from the sacred fountain of destiny

With their roots in the ground, their trunks rising through the plane on which we live, and their branches and leaves towering above our heads, trees span the three levels of Man's experience. Their symbolic importance is a recurring theme in cultures throughout the world, representing birth, regeneration and life itself.

Trees feature in many myths about the creation of the world. The Scandinavian mythological ash tree Yggdrasil represents the cosmos and provides the bridge between the heavenly and terrestrial spheres, by which the gods descended to go about their work of creation. In Dahomey and in Haitian voodoo the tree is the symbol of the backbone and also a bridge by which the messenger of the gods, Legba, descends to earth and into the bodies of the initiates – who then become possessed.

The tree is a kind of skeleton holding up the body of the world. It is also the symbol of growth, of death and rebirth, and of fertility. Its shape makes it an obvious symbol for the erect phallus, and it has been widely represented and worshipped as such. The water it draws up from the ground and the sap which permeates the trunk are similarly symbols of semen, and the explosion of leaves in spring is the very image of birth and life. In Tonga, in the South Pacific, men are said to have grown as sprouting shoots from the "world tree", while in the Admiralty and Banks islands it is thought that Man was first carved from the trunk of a tree. Egyptian tomb-paintings often portray the earth-goddess leaning out of a tree of which she is a part, offering her breast as a source of life.

The tree also represents death, the tree of hanging in the West, and of the crucifixion. The Cross and the Tree of Life have been mystically linked by Christians. According to medieval legend, Eve planted a branch from the Tree of Knowledge on Adam's grave, where it grew into a tree whose wood was finally used to build the Cross.

The tree is in fact a symbol of the whole cosmos, not only in physical but also in moral terms. It is the tree of knowledge, of both good and evil; it is the tree of life and of death. It is the world.

THE DIVINE TREE
The Hindu god Krishna, renowned for his loves with milkmaids and shepherdesses in the forest of Vrindavana, stands at the centre under the sacred kdamba tree, the call of his flute attracting mortals to unite with their god beneath the tree

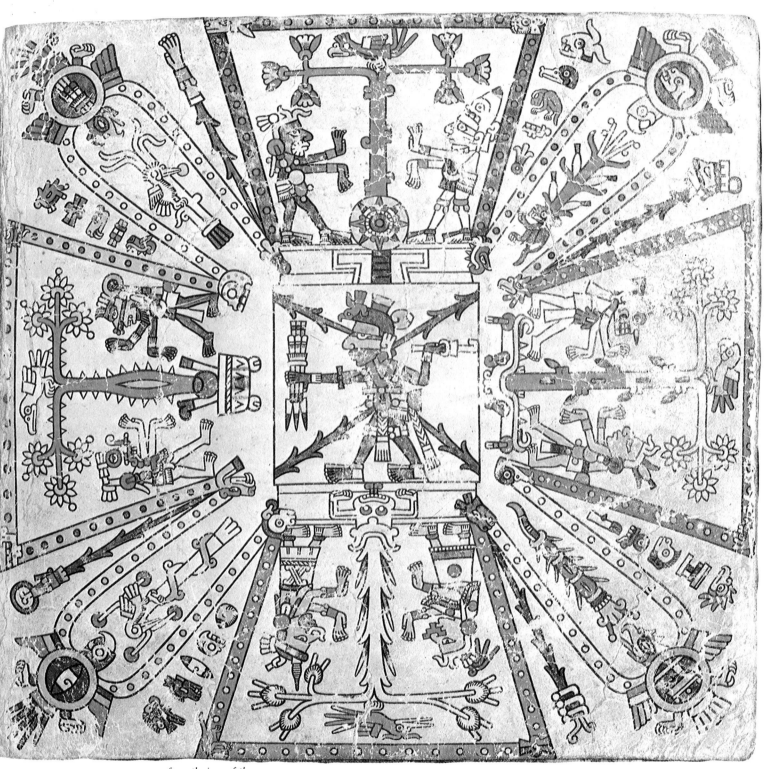

AZTEC SYMBOLS
A painted calendar chart represents the five regions of the world, the earth at the centre and the four trees rising to heaven. In the east (top) is the tree rising from a sun image, symbolizing birth; in the south the tree of sacrifice grows from the jaws of the earth; the western tree rises from the body of the dragon of the eclipse, symbol of the death of the sun, while the tree of the north, growing from a dish filled with the emblems of expiation, represents rebirth and continuing life

World Timbers

A number of terms used in describing timbers are explained briefly below for the reader's convenience in using the following catalogue of the world's major commercial timbers.

GRAIN refers to the alignment of the fibres relative to the long axis of the log. Often it is more or less straight, but occasionally wavy or sometimes irregular. In many tropical woods the grain spirals in alternate directions in successive growth increments, producing a stripe figure on quarter-cut surfaces.

FIGURE is the decorative appearance of wood caused by structural features such as rays, growth rings, grain and colour variation. It varies according to the orientation of the saw-cut used to expose the wood surface.

TEXTURE defines the "evenness" of the wood surface: in softwoods it is influenced by the ring-width and the contrast between early- and latewood; in hardwoods it is determined by the size and distribution of the pores. Where these are large, as in many elms, the texture is coarse, but where there are many fine pores, as in box-wood, the texture is very fine.

FLAT SAWN surfaces are defined as those cut so that the growth rings meet the face at an angle of less than 45 degrees; QUARTER SAWN, or RADIAL SAWN surfaces are produced when the angle is greater than 45 degrees. ROTARY-CUT wood is peeled from the log.

WEIGHT varies considerably between species; it is affected by internal structure and by the amount of water retained. Comparisons are made at uniform moisture content – usually that attained in wood used indoors. Hardwoods vary enormously from very light balsa to the very heavy lignum vitae. Familiar woods range through mahogany (light) and beech (medium) to rosewood (heavy). Softwoods cover a more limited weight range than hardwoods.

MOVEMENT is defined as the dimensional changes occurring in dry wood exposed to varying humidity conditions. It takes place over the width and thickness of the wood, but longitu-dinal dimensions remain stable.

DURABILITY refers to the ability of the wood to resist decay in outdoor use without need for preservative treatment.

The Trade in Timber

Of the 2,500 million cubic metres of wood used every year by Man, a staggering 46 per cent is burnt as fuel within a few miles of the site of felling. Of the remaining 54 per cent, used for building and other industrial uses, all but 8 per cent is used in the country of its origin: this 8 per cent, some 200 million cubic metres of timber, accounts for the entire world export trade in roundwood, sawn wood and wood-based panels.

Commercial timber is extracted from three main types of forest. The coniferous forests, source of the world's softwoods, extend across the arctic and subarctic zone of the northern hemisphere, and also occur at lower latitudes in East Africa, the southeastern United States and in Central America. Temperate hardwoods are also widespread in the north where they merge with the softwoods; in the southern hemisphere temperate hardwoods occur in Chile and New Zealand and in Australia, where eucalypt forests form a major resource. The majority of the world's tropical hardwoods come from the rain forests of S. America, Africa and SE Asia.

Canada to Japan/Korea
1.457 million cu.m

Canada to USA
20.75 million cu.m

Canada to Australia
4.371 million cu.m

Canada to Eur
1.816 million c

USA to Canada
3.481 million cu.m

USA to Japan/Korea
13.81 million cu.m

USA to Europe
1.589 million cu.m

USA to Australia
0.424 million cu.m

Brazil to Europe
0.373 million cu.m

Brazil to USA
0.297 million cu.m

WORLD TIMBER TRADE
The volumes of exported round and sawn wood, shown on the map in units of million cubic metres, represent 158 million cubic metres out of an estimated world total of 187 million cubic metres. The remaining 29 million cubic metres consists of small-volume and localized trading

Brazil

WORLD UTILIZATION OF TIMBER
Annual world production; 2500 million cubic metres

WOOD UTILIZATION

Burnt as fuel 46%
1150 million cu.m

Used in industry 54%

Export roundwood 4.6%
115.4 million cu.m

Export sawnwood 2.9%
71.9 million cu.m

Export wood panels 0.6%
14.0 million cu.m

Used at source 45.9%
1148.7 million cu.m

TYPE OF TIMBER

Softwood 44.8%
1120 million cu.m

Tropical hardwood 38.0%
950 million cu.m

Temperate hardwood 17.2%
430 million cu.m

EXPORTED TIMBER PRODUCTS
The volumes of timber products exported from the major producing areas are given in units of million cubic metres

Plywood and veneer

Korea 1.322
China 1.188
Canada 0.636
Finland 0.597
Philippines 0.509
Singapore 0.469
USA 0.455
W. Malaysia 0.349
USSR 0.316
France 0.242
Japan 0.155
Romania 0.123
Italy 0.100
Congo 0.099
Germany 0.093

Particle board

Belgium 0.755
Finland 0.480
Austria 0.408
Germany 0.364
Sweden 0.332
France 0.267
USSR 0.168
Romania 0.164
Norway 0.163
USA 0.162

Sawn softwood

Canada 23.22
Sweden 9.379
USSR 8.202
Finland 5.195
USA 4.122
Austria 3.340
Romania 1.176
Poland 0.765
Brazil 0.688
Czechoslovakia 0.653
Portugal 0.643
Norway 0.466
Yugoslavia 0.395
Germany 0.353

Sawn hardwood

W. Malaysia 2.046
Singapore 1.110
Yugoslavia 0.740
France 0.734
USA 0.670
Romania 0.562
Philippines 0.427
Canada 0.402
Brazil 0.367
Germany 0.285
Sarawak 0.251
Ghana 0.240
Ivory Coast 0.238

Pulpwood

USSR 6.662
Sweden 2.533
France 1.508
Czechoslovakia 1.310
Canada 1.267
Poland 0.715
Hungary 0.687
Germany 0.333
Romania 0.328

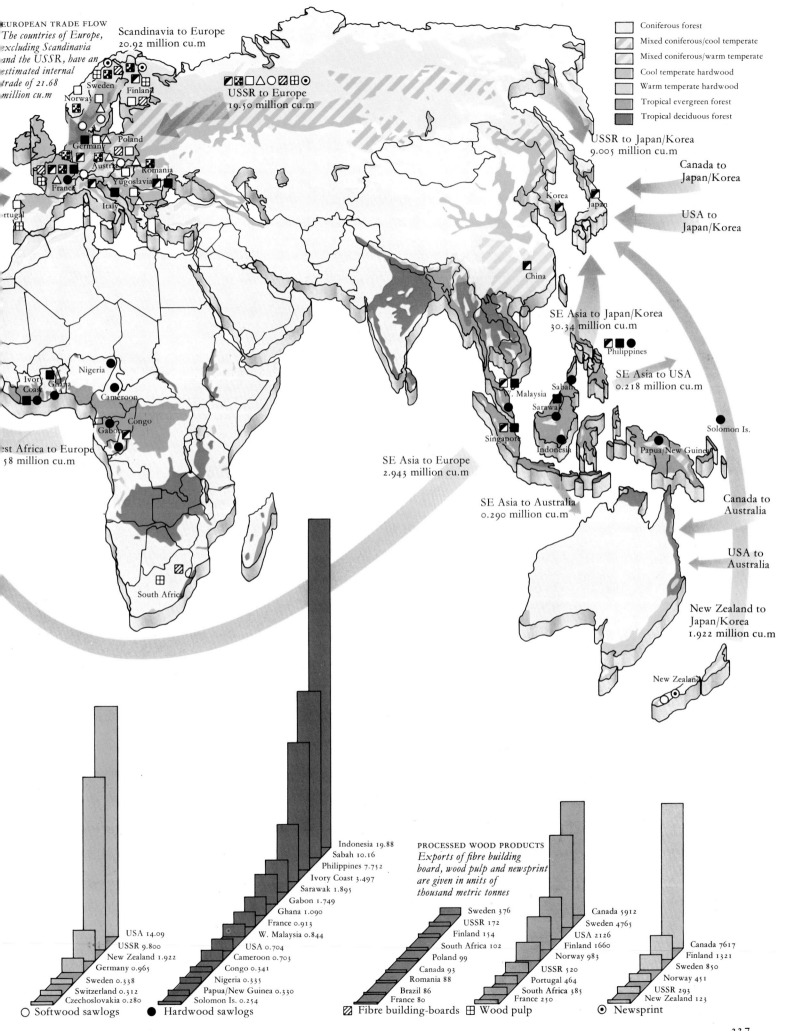

EUROPEAN TRADE FLOW
The countries of Europe, excluding Scandinavia and the USSR, have an estimated internal trade of 21.68 million cu.m

Scandinavia to Europe
20.92 million cu.m

USSR to Europe
19.50 million cu.m

Sweden
Finland
Norway
Germany
Poland
Austria
Romania
Yugoslavia
France
Italy
Portugal

USSR to Japan/Korea
9.005 million cu.m

Canada to Japan/Korea

USA to Japan/Korea

Korea
Japan
China

SE Asia to Japan/Korea
30.34 million cu.m

Philippines

SE Asia to USA
0.218 million cu.m

Nigeria
Ivory Coast
Ghana
Cameroon
Congo
Gabon

West Africa to Europe
58 million cu.m

W. Malaysia
Sabah
Sarawak
Singapore
Indonesia

Solomon Is.

Papua/New Guinea

SE Asia to Europe
2.943 million cu.m

SE Asia to Australia
0.290 million cu.m

Canada to Australia

USA to Australia

New Zealand to Japan/Korea
1.922 million cu.m

South Africa

New Zealand

Coniferous forest
Mixed coniferous/cool temperate
Mixed coniferous/warm temperate
Cool temperate hardwood
Warm temperate hardwood
Tropical evergreen forest
Tropical deciduous forest

Indonesia 19.88
Sabah 10.16
Philippines 7.752
Ivory Coast 3.497
Sarawak 1.895
Gabon 1.749
Ghana 1.090
France 0.913
W. Malaysia 0.844

USA 14.09
USSR 9.800
New Zealand 1.922
Germany 0.965
Switzerland 0.312
Sweden 0.338
Czechoslovakia 0.280

USA 0.704
Cameroon 0.703
Congo 0.341
Nigeria 0.335
Papua/New Guinea 0.330
Solomon Is. 0.254

○ Softwood sawlogs   ● Hardwood sawlogs

PROCESSED WOOD PRODUCTS
Exports of fibre building board, wood pulp and newsprint are given in units of thousand metric tonnes

Sweden 376
USSR 172
South Africa 102
Poland 99
Canada 93
Romania 88
Brazil 86
France 80

Canada 5912
Sweden 4765
Finland 1660
Norway 983
USSR 520
Portugal 464
South Africa 385
France 250

Canada 7617
Finland 1321
Sweden 850
Norway 451
USSR 293
New Zealand 123

USA 2126

▨ Fibre building-boards   ⊞ Wood pulp   ⊙ Newsprint

What Wood is That?

The size, form and distribution of different types of cell are distinctive for each variety of wood. They account for many of the characteristic physical and mechanical properties of woods and also make it possible to identify and classify the thousands of different wood species. Identification of a wood by its unique cell composition and arrangement can often be achieved by examining, either with the unaided eye or using a low-power hand lens, a cleanly cut end-grain surface. A sample only three millimetres wide and about ten millimetres long (taken in the radial direction) is usually sufficient. For some hardwoods, and for most softwoods, a lens examination is not adequate for an accurate identification and it is then necessary to prepare a thin section for closer study under the microscope.

Identification is made by comparing the unknown sample with known specimens – by matching the cellular pattern. Various aids to identification have been devised, the most familiar being the dichotomous key in which a series of questions is answered, each question forming an elimination stage until the identity of the wood sample is reached.

The chart on these pages illustrates the working of a dichotomous key; it has been worked out for twenty-two woods featured in the reference section of the book and works only to identify these particular woods. When large numbers of woods are to be identified the same principle is used, but punched cards and mechanical sorting techniques are employed to speed the processing. Using the micrographs and the additional information supplied below, follow the key to arrive at the identity of the twenty-two samples.

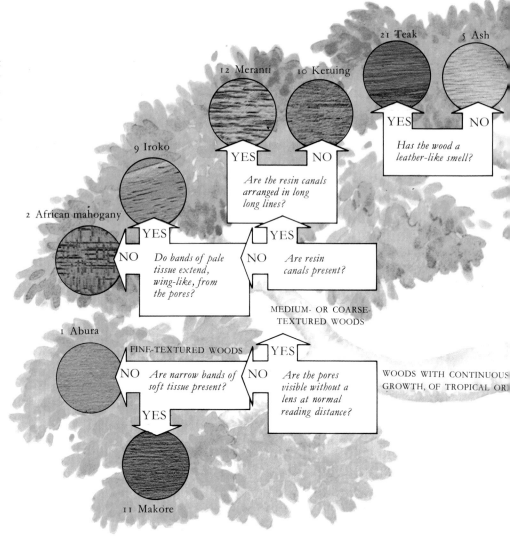

21 Teak
5 Ash
12 Meranti
10 Keruing
9 Iroko
2 African mahogany
1 Abura
11 Makore

YES — NO
Has the wood a leather-like smell?

YES — NO
Are the resin canals arranged in long long lines?

YES — NO
Do bands of pale tissue extend, wing-like, from the pores?

YES — NO
Are resin canals present?

MEDIUM- OR COARSE-TEXTURED WOODS

FINE-TEXTURED WOODS — YES

NO — *Are narrow bands of soft tissue present?* — NO — *Are the pores visible without a lens at normal reading distance?*

YES

WOODS WITH CONTINUOUS GROWTH, OF TROPICAL OR

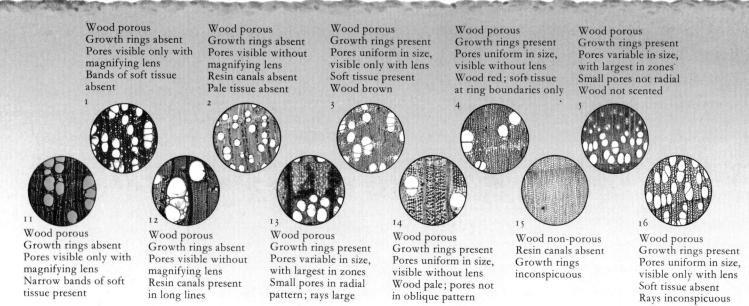

Wood porous
Growth rings absent
Pores visible only with magnifying lens
Bands of soft tissue absent
1

Wood porous
Growth rings absent
Pores visible without magnifying lens
Resin canals absent
Pale tissue absent
2

Wood porous
Growth rings present
Pores uniform in size, visible only with lens
Soft tissue present
Wood brown
3

Wood porous
Growth rings present
Pores uniform in size, visible without lens
Wood red; soft tissue at ring boundaries only
4

Wood porous
Growth rings present
Pores variable in size, with largest in zones
Small pores not radial
Wood not scented
5

Wood porous
Growth rings absent
Pores visible only with magnifying lens
Narrow bands of soft tissue present
11

Wood porous
Growth rings absent
Pores visible without magnifying lens
Resin canals present in long lines
12

Wood porous
Growth rings present
Pores variable in size, with largest in zones
Small pores in radial pattern; rays large
13

Wood porous
Growth rings present
Pores uniform in size, visible without lens
Wood pale; pores not in oblique pattern
14

Wood non-porous
Resin canals absent
Growth rings inconspicuous
15

Wood porous
Growth rings present
Pores uniform in size, visible only with lens
Soft tissue absent
Rays inconspicuous
16

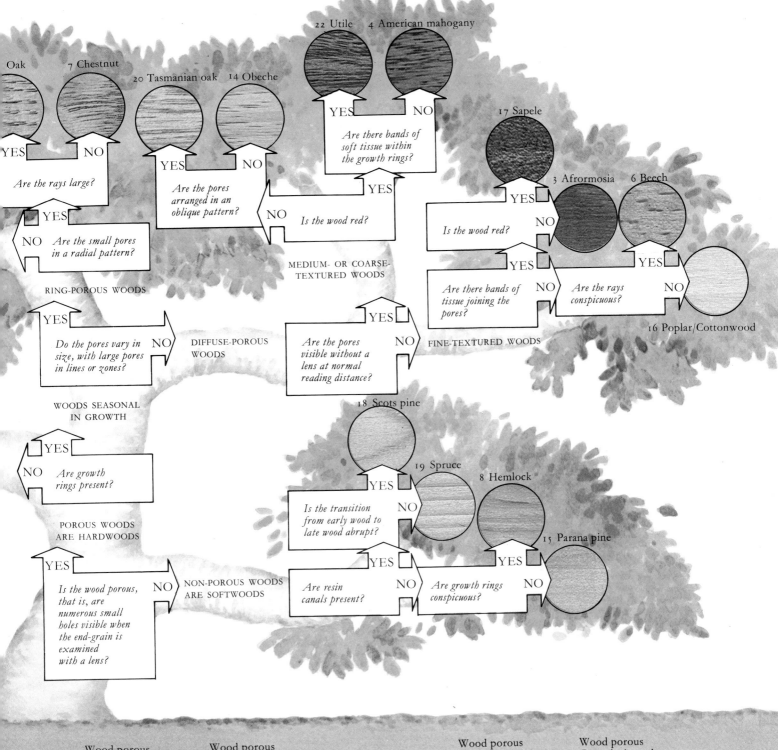

Oak · 7 Chestnut · 20 Tasmanian oak · 14 Obeche · 22 Utile · 4 American mahogany · 17 Sapele · 3 Afrormosia · 6 Beech

YES / **NO** — Are the rays large?

YES / **NO** — Are the small pores in a radial pattern?

RING-POROUS WOODS

YES / **NO** — Are the pores arranged in an oblique pattern?

YES / **NO** — Is the wood red?

MEDIUM- OR COARSE-TEXTURED WOODS

YES / **NO** — Are there bands of soft tissue within the growth rings?

YES / **NO** — Is the wood red?

YES / **NO** — Are there bands of tissue joining the pores?

YES / **NO** — Are the rays conspicuous?

FINE-TEXTURED WOODS

16 Poplar/Cottonwood

YES / **NO** — Do the pores vary in size, with large pores in lines or zones?

DIFFUSE-POROUS WOODS

YES / **NO** — Are the pores visible without a lens at normal reading distance?

WOODS SEASONAL IN GROWTH

YES / **NO** — Are growth rings present?

POROUS WOODS ARE HARDWOODS

YES / **NO** — Is the wood porous, that is, are numerous small holes visible when the end-grain is examined with a lens?

NON-POROUS WOODS ARE SOFTWOODS

18 Scots pine · 19 Spruce · 8 Hemlock · 15 Parana pine

YES / **NO** — Is the transition from early wood to late wood abrupt?

YES / **NO** — Are resin canals present?

YES / **NO** — Are growth rings conspicuous?

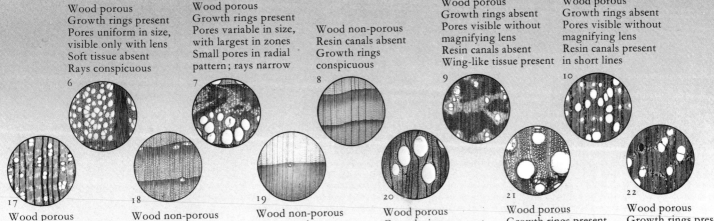

Wood porous
Growth rings present
Pores uniform in size, visible only with lens
Soft tissue absent
Rays conspicuous
6

Wood porous
Growth rings present
Pores variable in size, with largest in zones
Small pores in radial pattern; rays narrow
7

Wood non-porous
Resin canals absent
Growth rings conspicuous
8

Wood porous
Growth rings absent
Pores visible without magnifying lens
Resin canals absent
Wing-like tissue present
9

Wood porous
Growth rings absent
Pores visible without magnifying lens
Resin canals present in short lines
10

17
Wood porous
Growth rings present
Pores uniform in size, visible only with lens
Bands of soft tissue joining pores; wood red

18
Wood non-porous
Resin canals present
Transition from early to late wood abrupt

19
Wood non-porous
Resin canals present
Transition from early to late wood gradual

20
Wood porous
Growth rings present
Pores uniform in size, visible without lens
Wood pale; pores in oblique pattern

21
Wood porous
Growth rings present
Pores variable in size, with largest in zones
Small pores not radial
Has leather-like smell

22
Wood porous
Growth rings present
Pores uniform in size, visible without lens
Wood red; bands of soft tissue within rings

Commercial Hardwoods

Australian Blackwood   Sen

AUSTRALIAN BLACKWOOD
Acacia melanoxylon

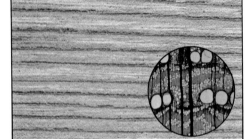

(x25)

THE TREE The several hundred species of *Acacia* produce woods with a wide range of colour, weight and appearance. One of the most attractive is Australian blackwood, which comes from a tree of SE Australia, particularly important in Tasmania, where it produces logs up to 1m in diameter from trees 30m in height. It is also planted outside Australia, notably in South Africa, where it supplements the depleted stocks of native hardwoods.

THE WOOD Though called blackwood, it is a golden to chocolate brown with darker markings. The grain is generally fairly straight, but is sometimes wavy, when it gives a cross-banded or fiddle-back figure which, combined with a high natural lustre, results in a very decorative ap-

pearance; the texture is medium, but even. The wood is moderately heavy.

TECHNICAL PROPERTIES A timber with good properties, it dries without trouble and saws easily. It is a strong wood, comparing favourably with beech. It machines to an excellent finish, takes a high polish, and bends well after steaming.

USES Blackwood is one of Australia's most attractive woods, even plain timber having a fine appearance. It is used mainly for decorative effect for panelling, interior joinery in prestige buildings and for furniture, but it is also an important wood for bent work, e.g. coachwork and boat-building. Logs, especially those with a wavy grain, are often sliced and, outside the countries where it is grown, it is seen mainly as veneer.

SEN
Acanthopanax ricinifolius

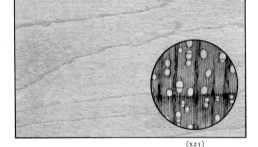

(x10)

THE TREE The small supplies of sen available commercially come from Japan, though the tree also grows in China and Korea. Its wood is easily mistaken for ash. Sen is a large tree, up to 25m high, yielding good-quality logs up to 1m in diameter.

THE WOOD A combination of pale colour, straight grain and ring-porous structure gives sen its remarkable likeness to ash. The wood is almost white, sometimes with a greyish tint, and is often, like some Japanese ash, rather slowly grown. Some 20 per cent lighter than ash, it can be readily distinguished if a cleanly cut end-surface is examined, because, unlike ash, it has wavy lines of tissue between the zones of its large pores.

TECHNICAL PROPERTIES Sen is a mild wood

which presents little difficulty in handling, though shrinkage on drying is fairly high and it does tend to surface-split. Lighter in weight than ash, and appreciably weaker, it especially lacks the outstanding toughness of ash and is particularly brittle when slow grown. The timber saws easily and well, machines to a good finish, and can be sliced to give an attractive veneer. Sen tends to split when nailed and for outside use should be treated with preservative. The wood is moderately stable in use.

USES Sen is used in Japan for many purposes – for furniture, decorative surfaces, lacquer work, handles, combs, etc. It is commonly made into veneer and plywood – the form in which it is usually seen outside Japan.

MAPLE/SYCAMORE
Acer spp.

(x25)

THE TREE Maple is a northern temperate timber important in North America, Europe and Japan. Two main types are recognized in America, rock, or hard, maple (mainly *A.saccharum*) and soft maple (*A.saccharinum* and other spp.). In Europe, the principal species is *A.pseudoplatanus*, known in England as sycamore. Japanese maple is mainly *A.mono*.

THE WOOD Maple is a pale wood, normally straight-grained, but the European sycamore occasionally has a wavy grain, giving a fiddle-back figure which is much sought after. Rock maple, which sometimes produces a bird's-eye figure, is somewhat heavier than beech, while the soft maples are about 25 per cent lighter. Japanese maple approaches rock maple in weight.

TECHNICAL PROPERTIES Maples dry well, if slowly, and are moderately stable in use. Though more easily worked than rock maple, soft maple lacks the strength and the outstanding resistance to abrasion of the heavier wood. Maple should not be used out of doors unless it has been effectively treated.

USES A high density and fine, even texture give the heavier maples an exceptional resistance to abrasion. Rock maple is an outstanding flooring wood, used in industry and dance halls, bowling alleys and gymnasia, and for escalator treads. It is also preferred for shoe lasts and for parts of piano actions. Sycamore and soft maples are used for kitchen utensils and for rollers, while wavy-grained sycamore is the traditional wood for violins.

HORSE-CHESTNUT
Aesculus spp.

(x25)

THE TREE Several types of horse-chestnut occur in the warmer regions of the northern temperate zone, but they are all of only minor importance as a source of commercial timber. European horse-chestnut (*A.hippocastanum*) is widely grown in Europe as an ornamental tree; it often has a short bole, branching low down. American timber (*A.octandra*), known as buckeye, occurs in the central and Atlantic states, and other timber species occur in India and Japan.

THE WOOD Horse-chestnut is a lightweight, fine-textured, creamy-white wood. Though usually straight, the grain may be irregular, especially in old or misshapen trees. In character it is very similar to poplar or American cottonwood, but is a little heavier.

TECHNICAL PROPERTIES Horse-chestnut dries well with only moderate shrinkage and, when dry, is stable in use. It works easily, but does not finish well unless tools are kept very sharp. It is rather a weak wood, soft and tending to be brittle. It has a low resistance to fungal attack, but can be readily treated with preservatives.

USES A plain wood of low strength, horse-chestnut is used for a number of general purposes, though it is in limited supply and sometimes of poor quality. It is milled to sawn boards and is used for turned items such as brush handles and backs and kitchen utensils, often interchangeably with similar hardwoods. It is suitable for use in the manufacture of lightweight boxes and crates, and, when straight-grained, for engineering patterns.

    Flat-sawn timber   Radial-sawn timber

Maple/Sycamore

Horse-chestnut

Afzelia

Alder

Aningeria

Mersawa

AFZELIA
Afzelia spp.

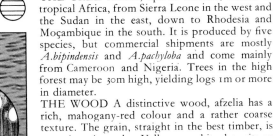

(×10)

THE TREE Afzelia, apa or doussié, occurs in tropical Africa, from Sierra Leone in the west and the Sudan in the east, down to Rhodesia and Moçambique in the south. It is produced by five species, but commercial shipments are mostly *A.bipindensis* and *A.pachyloba* and come mainly from Cameroon and Nigeria. Trees in the high forest may be 30m high, yielding logs 1m or more in diameter.

THE WOOD A distinctive wood, afzelia has a rich, mahogany-red colour and a rather coarse texture. The grain, straight in the best timber, is sometimes irregular. Yellow or white deposits or flecks are often a feature and these may cause staining of the wood in damp conditions. It is a dense wood, some 10 to 15 per cent heavier than oak.

TECHNICAL PROPERTIES Afzelia combines high strength with outstanding durability and stability. It is slow to dry, but shrinks very little; once dry, it is superior even to teak in stability. Generally stronger than oak and rather more difficult to work, it can blunt tools if chalky deposits are present. It is extremely resistant to fungal attack, and has some natural resistance to termites.

USES A high-class wood, it is used in joinery where a distinguished appearance combined with ease of maintenance is required, e.g. for window frames, doors and door surrounds in prestige buildings. It is used for bench and counter tops, and makes an excellent flooring for domestic and public buildings. Afzelia is noted for its acid resistance and is used in chemical works for vats and presses.

ALDER
Alnus spp.

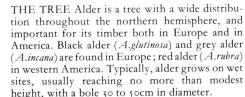

(×25)

THE TREE Alder is a tree with a wide distribution throughout the northern hemisphere, and important for its timber both in Europe and in America. Black alder (*A.glutinosa*) and grey alder (*A.incana*) are found in Europe; red alder (*A.rubra*) in western America. Typically, alder grows on wet sites, usually reaching no more than modest height, with a bole 30 to 50cm in diameter.

THE WOOD Freshly cut alder-wood is pale, but turns to a bright orange-brown on exposure. It has a fine texture, but lacks a distinctive figure. The wood is of medium density; European alder is about the same weight as mahogany.

TECHNICAL PROPERTIES The wood dries readily and well, but European alder shrinks more than red alder. It saws easily and takes a good

finish, provided that the cutting tools are sharp, and it rotary-peels to a good veneer. It is not durable.

USES European alder makes a good general-purpose plywood, used especially in Russia to supplement supplies of birch, though it lacks the clean, white appearance of birch and is softer and weaker. Red alder is the most common commercial hardwood of the Pacific seaboard of America, where it is widely used for furniture. In both Europe and America alder tends to be used for craft and small industry purposes, such as brush and broom backs, handles, toys, etc. It is a traditional wood for clog-making and has been used for artificial limbs, rollers for textile mills, and in the manufacture of charcoal for gunpowder.

ANINGERIA
Aningeria spp.

(×25)

THE TREE Aningeria came into use only in the late 1960s. This African wood was first introduced as Tanzanian walnut, misleadingly, since it is neither a walnut nor did it come from Tanzania, though it occurs there. Commercial shipments are mainly from the Ivory Coast, where it is known as anegré.

THE WOOD Aningeria is a blond wood with a slightly pinkish tint, and about the same weight as African mahogany. It is somewhat plain, but lustrous and fine and even in texture. It is usually straight-grained, but occasionally a wavy grain gives it a mottle figure.

TECHNICAL PROPERTIES The siliceous content of aningeria, though very small, is enough to make it abrasive to cutting tools. Other, more

amenable, timbers were available for sawn production, so aningeria attracted little interest until a demand arose for a plain, fine-textured veneer. Aningeria was then found to slice well and its even texture meant that veneer as thin as 0.6mm could be produced and handled without difficulty.

USES Seen almost entirely in the form of veneer, aningeria has an attractive, if plain, appearance. However, its fine, uniform texture provides a particularly suitable base on which to print finishes that simulate highly decorative, and more expensive, woods. For this purpose it is superior to paper, since its true wood grain shows through. As printed veneer it is used for furniture and panel surfaces, and it may well become popular for these same uses in its natural colour.

MERSAWA
Anisoptera spp.

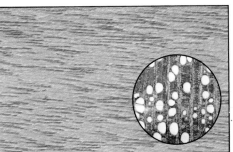

(×10)

THE TREE Mersawa is the Malay name for a timber which occurs from Bangladesh to New Guinea and the Philippines. Some dozen or more species provide commercial timber, which comes mainly from western Malaysia, from Thailand (known as krabak) and the Philippines (known as palosapis). Typically a tall tree with a long, straight stem, mersawa produces large and cylindrical logs.

THE WOOD Different species have woods which vary in character, but are typically pale-yellow, moderately coarse in texture and straight-grained. A plain wood lacking any decorative feature (unless it is accurately quarter-cut to give a ray fleck), mersawa varies in density from species to species. It is about the same weight as teak, but palosapis is somewhat heavier.

TECHNICAL PROPERTIES Sawn mersawa is difficult to dry and, because it contains silica, though in very small amounts, it is abrasive to saws and cutting tools. These two properties limit its use as sawn wood. However, it can be rotary-peeled to give a good veneer. It has some resistance to decay, but should not be exposed to fungus or termite attack. The wood is difficult to treat effectively, even by pressure methods.

USES Though sometimes sawn, e.g. for flooring, mersawa is now a popular timber for plywood production, to which it is more suited. Logs have been exported to Japan, but plywood is being made in increasing volume in SE Asia. The outer plies of white-faced Malaysian plywood are usually either of mersawa or white meranti.

Commercial Hardwoods

Conçalo Alves　　　　Gaboon

CONCALO ALVES
Astronium spp.

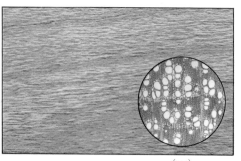

(x25)

THE TREE Gonçalo alves has a variety of names, including zebrawood and, in America, tigerwood and kingwood; however, these are not distinctive and some are more often and better used for other woods. The timber occurs widely through Central and tropical South America, but comes mainly from Brazil, where it is known as gonçalo alves. It is a medium-sized tree, yielding round logs up to 90cm in diameter.

THE WOOD A very distinctive wood, medium red-brown with almost black bands giving a highly decorative figure, which varies in appearance according to the way the wood is cut. It is further enhanced on quartered surfaces by an interlocked, or more rarely wavy, grain giving a stripe or fiddle-back figure. Gonçalo alves is a very

dense wood, generally being about 30 per cent heavier than beech.

TECHNICAL PROPERTIES. A wood noted in tropical America for its combination of high strength and good durability. Although heavy, it is not unduly difficult to saw and can be machined to give a smooth finish, but care is needed if the grain is irregular. It can be cut to produce highly decorative veneer.

USES A timber sought in Brazil for structural use and, because of its appearance, for furniture and cabinetwork. Its weight and limited availability preclude its general use outside tropical America and it is usually seen as veneer. In the solid, it is of interest for high-value items, such as brush backs, knife handles and billiard cue butts.

GABOON
Aucoumea klaineana

(x25)

THE TREE The usual English name for this timber is gaboon, deriving from the West African republic of Gabon; but it is known as okoumé in many countries. The single species producing gaboon is a large tree, up to 40m in height, found only in Gabon, Equatorial Guinea and the Congo, but where it occurs it is very abundant and more gaboon is exported than any other African wood. Shipped to continental Europe very largely as logs, gaboon is one of the main timbers used by the plywood industry.

THE WOOD Gaboon is a pale-pink, fairly fine-textured and rather characterless wood. It is occasionally figured, but more usually has a straight grain. It is similar in weight to spruce or whitewood.

TECHNICAL PROPERTIES Because its siliceous content makes it abrasive to saws and cutting tools, gaboon is rarely sawn. Instead it is peeled, yielding veneer which dries well, glues well and makes an excellent plywood, though not suitable for use in conditions favouring decay.

USES Gaboon was the first tropical hardwood extensively exploited by the plywood industry in France and in other European countries. Gaboon ply is used for a range of general purposes, for furniture, door skins, partitions, etc., and because of its light weight it is also used in the construction of small boats, though its low durability makes careful maintenance necessary. In addition to plywood, gaboon is also used in the manufacture of blockboard and laminboard.

RHODESIAN TEAK
Baikiaea plurijuga

THE TREE Rhodesian teak, from a tree which grows in Zambia and Rhodesia, has some of the properties of true teak, but differs in appearance. The tree is of modest size and may reach a height of 20m, yielding high-quality logs, usually 3 to 7m long.

THE WOOD Rhodesian teak is an attractive wood. It is a rich red-brown, sometimes with darker flecks, and any figure shown derives more from variations in colour than from the grain, which is usually straight or interlocked. It has a fine, even texture and a smooth surface. The wood is almost 40 per cent heavier than true teak.

TECHNICAL PROPERTIES Rhodesian teak dries slowly but well and, once dry, is stable in use. It is very heavy, difficult to saw and to work by

hand or by machine, and it cannot be nailed; its high density and very fine texture make it highly resistant to abrasion. It is very durable, very resistant to fungi, and to a lesser degree to termites. The wood stains badly on contact with iron under damp conditions.

USES Being strong, stable and durable, Rhodesian teak is particularly suitable for structural and outdoor use in Africa. Logs are commonly converted to give large-size pieces for local use, and the offcut timber, often of good quality from the outer heartwood, is exported. Sawn into strips and blocks, it makes excellent flooring; it has an attractive appearance and is durable enough to withstand the most exacting conditions of use in domestic, commercial and industrial environments.

PAU MARFIM
Balfourodendron riedelianum

(x25)

THE TREE Pau marfim, also known as moroti, occurs in southern Brazil, Paraguay and northern Argentina. It comes from a tree of moderate size, up to 25m high, and with a bole sometimes 80cm in diameter; more slender boles are, however, common. It is a recent introduction to the timber market, shipped to the United States and Europe as sawn lumber and squares.

THE WOOD Though somewhat featureless, pau marfim has a characteristic appearance, due to its pale-yellow colour, fine, even texture and its growth rings, visible on flat-sawn surfaces. Its grain is usually straight, but sometimes wavy. It is a heavy wood, almost the weight of hickory.

TECHNICAL PROPERTIES Pau marfim has attracted interest because of its strength proper-

ties, especially toughness and resistance to suddenly applied loads. In these properties, straight-grained pau marfim is superior to ash, though not the equal of hickory. It is believed to dry without trouble and can be sawn and worked easily to a fine, smooth finish. Its durability is uncertain, but it is probably not suitable for outdoor use.

USES Because it is tough and finishes smoothly, pau marfim selected for straight grain is being used for the handles of striking tools. In South America it is used for construction, furniture and turned articles. Its fine texture and yellow colour suggest its use as an alternative to boxwood, e.g. for drawing instruments and rules; similar also to rock maple in density and texture, it should make a hard-wearing floor.

Rhodesian Teak

Pau Marfim

Birch

Muhuhu

Boxwood

Brazilwood

BIRCH
Betula spp.

(x25)

THE TREE Birch is a northern hemisphere tree, particularly important in Canada (especially yellow birch, *B.alleghaniensis*) and Europe (*B.pendula* and *B.pubescens*). It is often a very common tree, but not, as a rule, very large; in Europe, it is at its best in the northern countries, where it has a straight, cylindrical stem. Birch rotary peels well and provides high-grade logs for the important Finnish and Russian plywood industries.

THE WOOD Birch is fine-textured and almost white, though yellow birch has a brownish heart. Normally it is straight-grained with little figure, but occasional logs give figured wood, e.g. flame birch and masur birch. It is a fairly heavy wood, yellow birch being about the weight of oak and European birch a little lighter.

TECHNICAL PROPERTIES Birch compares favourably with ash for toughness and is generally better in other strength properties. It works well by hand or machine, turns well, and rotary-peels to give an excellent veneer. It is not durable in conditions favouring decay.

USES Birch is more familiar as plywood than as solid timber; because of its strength properties it makes an excellent structural plywood, and the British Mosquito aircraft of World War II was built of birch ply. Today it is used in building, for wooden components and flooring. Solid birch is used in furniture, especially for the frames of upholstered chairs, and is turned for brush backs and handles. The pulp is important for the making of writing paper.

MUHUHU
Brachylaena hutchinsii

(x25)

THE TREE Muhuhu is an East African wood shipped mainly from Tanzania almost entirely in the form of strips and blocks. It comes from a small to medium-sized tree which averages 50cm in diameter, and is often of poor stem form.

THE WOOD Muhuhu is hard, heavy, very fine-textured and varies in colour from yellowish to medium brown. The grain is interlocked and in wood from misshapen stems, often irregular, which contributes to the decorative appearance of the wood. It is a dense wood, some 30 per cent heavier than beech. A distinctive feature, most noticeable when the wood is being worked, is its pleasant, spicy smell: the tree contains an aromatic oil which is distilled and marketed commercially as a substitute for sandalwood oil.

TECHNICAL PROPERTIES Muhuhu, though very heavy, is not unduly troublesome to dry; once dry, it is very stable. It is, however, difficult to work and care is needed to obtain a good finish on wood with an irregular grain. Its hardness, high density and fine texture make it very resistant to abrasion; otherwise its strength properties are only moderate. It is very resistant to fungi and moderately resistant to termite attack.

USES Muhuhu is an outstanding and widely used flooring wood, favoured in public and domestic buildings for its attractive appearance. Because of its excellent wearing properties it is also used in the most exacting industrial conditions. Muhuhu is also commonly used in East African sculpture for the carving of wooden animals.

BOXWOOD
Buxus sempervirens

(x25)

THE TREE The name boxwood is used for several heavy woods with a very fine, even texture and pale-yellow colour. Strictly, it refers to timber of species of *Buxus*. European box, *B.sempervirens*, occurs locally from Britain, through southern Europe and Turkey to Iran. Other true boxwoods occur in Asia and South Africa. European box is at best a small tree, up to 12m high, but is often little more than a shrub. Its timber is marketed in short billets, often only 10 to 20cm in diameter.

THE WOOD Boxwood has a distinctive yellow colour and is amongst the finest-textured of commercial woods. The grain is straight or, more often, irregular – especially in wood from misshapen stems. It is heavy, and even after drying will only just float in water.

TECHNICAL PROPERTIES Boxwood must be dried carefully if splitting is to be avoided, but once dry, its usefulness depends on its working properties. It is an excellent wood for turning and can be carved to give remarkable detail, making it an outstanding wood for engraving.

USES Boxwood has a long history. The Bible and ancient classical literature mention boxwood combs, spinning tops, writing tablets, etc., and its use in decorative inlay. More recently it has been used especially for engravers' blocks, for rulers and for shuttles, particularly in the silk industry. Nowadays, despite the small quantity available, it is familiar from its use in small turned items such as chessmen, corkscrews and occasionally as tool handles.

BRAZILWOOD
Caesalpinia echinata

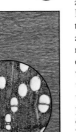

(x25)

THE TREE The name brazil was used in the Middle Ages for plants producing a red dye – notably the East Indian wood *C.sappan*. When the Portuguese colonized South America in the sixteenth century, they discovered a similar dye-wood and, by association, gave the name Brazil to the new land. The wood therefore gave its name to the country and not, surprisingly, vice versa. Brazilwood, *C.echinata*, also known as Pernambuco wood, is found only in the coastal forests of Brazil, where it is usually a small to medium-sized tree, yielding short billets 1m long and up to 20cm in diameter.

THE WOOD Brazilwood is orange when freshly cut but darkens to a deep red. It has a fine and even texture and can be selected for straight grain.

TECHNICAL PROPERTIES Though hard and heavy, brazilwood works fairly easily and can be finished to a very smooth surface.

USES Brazilwood was once so highly prized as a dye-wood that for several centuries the trade was a royal monopoly; by the mid-nineteenth century demand had declined. Today it has one special use, for violin bows, for which it is said to have the right combination of weight, flexibility and strength. The closely related partridgewood, *C.granadillo*, is brown to almost black in colour and has been used for truncheons and umbrella handles. Pau ferro, or ironwood, *C.ferrea*, is also related botanically and is another heavy but decorative wood used as veneer and for the manufacture of furniture.

Commercial Hardwoods

Silky-oak

Hornbeam

SILKY-OAK
Cardwellia sublimis

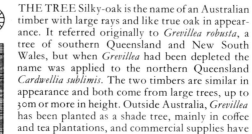
(x10)

THE TREE Silky-oak is the name of an Australian timber with large rays and like true oak in appearance. It referred originally to *Grevillea robusta*, a tree of southern Queensland and New South Wales, but when *Grevillea* had been depleted the name was applied to the northern Queensland *Cardwellia sublimis*. The two timbers are similar in appearance and both come from large trees, up to 30m or more in height. Outside Australia, *Grevillea* has been planted as a shade tree, mainly in coffee and tea plantations, and commercial supplies have been obtained from East Africa.

THE WOOD Silky-oak resembles true oak in its silver grain, visible when quarter-cut, and its pinkish-brown colour has some likeness to that of red oak. Plantation timber is pale in colour. Like oak, it has a rather coarse texture, but it is only about two-thirds the weight of oak.

TECHNICAL PROPERTIES Silky-oak must be carefully dried, but once dry can be readily sawn and it works well, taking a good finish. The wood nails without splitting and holds nails and screws well. It can also be peeled or sliced for veneer. For its weight, it is strong and tough, but it cannot equal the strength and outdoor durability of true oak. It is a good timber for steam-bent work.

USES Silky-oak is a decorative wood, used in Australia for furniture and panelling and for general building purposes, where it is common. It is also good for flooring, for coachwork and for the staves of casks. Outside Australia it is seen mainly as veneer.

HORNBEAM
Carpinus betulus

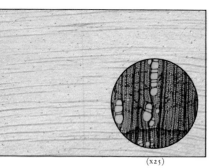
(x25)

THE TREE The only hornbeam of commercial interest is that found in Europe, Turkey and Iran. It is a medium-sized tree, like beech in appearance though not usually as large. Its stem is often rather poorly formed.

THE WOOD Hornbeam is a cold, white wood with a fine texture and an almost featureless appearance. Its grain may be irregular and it may show undulating growth rings, especially if the stem is misshapen. It is one of the denser temperate hardwoods and somewhat heavier than beech or oak.

TECHNICAL PROPERTIES Hornbeam is in many respects like a dense beech. It dries readily and well, but, like beech, it is not very stable under conditions of changing humidity. In strength it compares well with beech, being particularly resistant to splitting. Because of its greater weight, it is rather more difficult to work than beech, but it turns well and takes a very smooth finish. It is a good bending wood. Hornbeam is not resistant to fungal attack, but can be treated with preservatives for outdoor use.

USES Because supplies are limited, it is now a special-purpose wood. It was traditionally used for turned and machined parts in windmills and watermills and, because of its high resistance to splitting, for tool handles, jack-planes, etc. It was sometimes stained black and used in place of ebony. Nowadays it is used in billiard cues, for drumsticks, and in piano actions, where it compares favourably with maple. It makes a hard-wearing, splinter-free floor.

HICKORY
Carya spp.

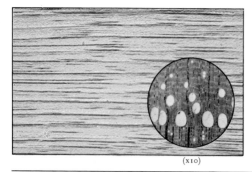
(x10)

THE TREE Hickory is an American timber, growing in southern Canada and in the eastern and southern United States. Most true hickory is produced by four species of *Carya*, but other species produce pecan hickory, which is considered somewhat inferior to true hickory. Hickory is a medium-sized to large tree and an important source of timber, used both in America and elsewhere.

THE WOOD Hickory has a white sapwood and a red-brown heartwood. It is ring-porous and therefore coarse-textured, but straight-grained when well grown. It is a dense wood, some 15 per cent heavier than ash, but, as with all ring-porous woods, its density is influenced by the width of the rings. As a rule, the toughest hickory has fewer than sixteen growth rings in 25mm of radial growth.

TECHNICAL PROPERTIES Quite the most important feature of hickory is its exceptional combination of high-strength properties. It is stiff and hard, and is especially able to withstand suddenly applied loads, being much superior to ash in this respect. It dries slowly, with a high shrinkage, and because of its weight it is fairly hard on saws and cutting tools.

USES Hickory is used where toughness is needed, for hammer, axe and pick handles, railway shunting poles, etc. White wood is generally preferred but the red heartwood is as serviceable; it is more important to avoid wood with a wavy or cross grain. Lower-grade timber is used for less exacting structural purposes. Hickory is much favoured for smoking food.

CHESTNUT
Castanea spp.

(x10)

THE TREE More familiar for its fruit than its wood, the European edible or sweet chestnut, *C.sativa*, occurs naturally in the Mediterranean countries, but has long been grown elsewhere. Chestnuts also occur in America and Japan, but the American chestnut, *C.dentata*, has been badly ravaged by blight. In Europe, the chestnut is a large tree often characterized by a spiral fissuring of the bark. Often the wood, too, has a spiral grain, which may contribute to the tendency of large stems to split badly when felled. Chestnut is also grown on short rotations of about twenty years as coppice for poles.

THE WOOD Chestnut is a pale-brown wood with prominent growth rings. It is like oak in appearance when flat-sawn, but because it has no large rays it lacks the silver grain figure of oak when quarter-sawn. It is about 20 per cent lighter in weight than oak.

TECHNICAL PROPERTIES Chestnut is softer and weaker than oak, but is generally easier to work and more stable in use. It dries only slowly and has a tendency to collapse. Like oak, its heartwood has high natural durability and tends to corrode iron fastenings in moist conditions.

USES Naturally durable, it is an excellent timber for outdoor use; even coppice poles have only a narrow band of sapwood and when treated make good stakes and poles. The timber is often cleft for fencing; it is only occasionally used for furniture. Chestnut veneer makes attractive, if somewhat plain, panelling.

Hickory

Chestnut

Black Bean

Cedar

Ceiba

Katsura

BLACK BEAN
Castanospermum australe

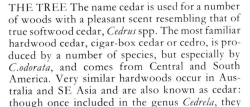

(x25)

THE TREE Black bean is an Australian wood from eastern Queensland. It comes from a small to medium-sized tree, up to 35m high and with a bole 1m in diameter. Logs commonly have a high proportion of sapwood and, as only the heartwood is of commercial interest, yields are often low.

THE WOOD The timber is medium-brown with fine, pale streaks giving a decorative effect on both flat- and quarter-sawn surfaces. Its decorative appearance is sometimes enhanced by an interlocked grain, which gives a striped figure, but more usually the grain is straight. It is fairly coarse in texture and moderately heavy, about the weight of oak or beech.

TECHNICAL PROPERTIES Black bean must be dried slowly and with some care; it should be allowed some time to season naturally before it is put in the kiln. Once dry, it is moderately stable in use. The wood can be sawn and worked by hand or machine to a good finish. It produces a good veneer, but care is needed in gluing because it has a somewhat greasy surface. It is stiff and strong, though inclined to be brittle. Black bean is reputed to be resistant to both termites and fungi.

USES It is one of Australia's more attractive woods, used for furniture, cabinetwork and high-class joinery. Because of its decorative appearance and its superficial resemblance to walnut, it is popular for panelling and inlay and is also used for fancy turned items. However, the quantity is limited and local demand restricts its availability elsewhere.

CEDAR
Cedrela spp.

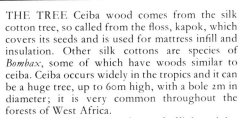

(x10)

THE TREE The name cedar is used for a number of woods with a pleasant scent resembling that of true softwood cedar, *Cedrus* spp. The most familiar hardwood cedar, cigar-box cedar or cedro, is produced by a number of species, but especially by *C.odorata*, and comes from Central and South America. Very similar hardwoods occur in Australia and SE Asia and are also known as cedar: though once included in the genus *Cedrela*, they are now designated *Toona* spp.

THE WOOD It is similar to mahogany in colour but coarser in texture, lighter in weight and occasionally resinous. The weight and colour of the wood vary considerably with different conditions of growth.

TECHNICAL PROPERTIES Cedar has a useful combination of technical properties. It dries readily and when dry is very stable in use; its strength, though not high, is good for its weight; it works easily and well and is very durable, resisting both fungus and termite attack.

USES Combining an attractive appearance with its stability, durability and ease of working, cedar is much sought after, both for local use and in overseas markets. In tropical America it is a preferred timber for lightweight construction and for joinery, furniture and all kinds of domestic use, and because of local demand the quantities available for overseas use are limited. As a result, it is used abroad for special purposes: its traditional use is for the packaging of cigars, and it is also used for the building of racing boats.

CEIBA
Ceiba pentandra

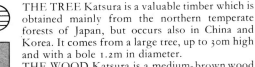

(x10)

THE TREE Ceiba wood comes from the silk cotton tree, so called from the floss, kapok, which covers its seeds and is used for mattress infill and insulation. Other silk cottons are species of *Bombax*, some of which have woods similar to ceiba. Ceiba occurs widely in the tropics and it can be a huge tree, up to 60m high, with a bole 2m in diameter; it is very common throughout the forests of West Africa.

THE WOOD Ceiba is a pale wood of light weight, lighter than obeche but heavier than balsa. However, it lacks the bright appearance of these two woods, partly because it is coarser in texture and also because it is difficult to keep its surface clean and free from the fungal discoloration to which the timber is prone.

TECHNICAL PROPERTIES Despite its light weight, ceiba can be difficult to handle. When first cut it contains a large amount of water and it must be dried quickly, for it is readily attacked by fungi; however, it dries easily and well. Because it is a soft, lightweight wood, saw-teeth and machine cutters must be kept very sharp or a smooth finish cannot be obtained. It can be peeled to give a good-quality veneer. The wood has low strength properties, even allowing for its weight, and, though easily nailed, its nail-holding properties are poor.

USES Ceiba is used mainly for veneer, especially the core veneers of plywood faced with other woods, and in blockboard: its light weight makes it a good insulating material. It is also used for packaging where strength is not important.

KATSURA
Cercidiphyllum japonicum

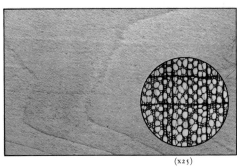

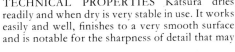

(x25)

THE TREE Katsura is a valuable timber which is obtained mainly from the northern temperate forests of Japan, but occurs also in China and Korea. It comes from a large tree, up to 30m high and with a bole 1.2m in diameter.

THE WOOD Katsura is a medium-brown wood of fine texture and generally straight grain. It has a high lustre but a plain appearance, relieved only by a growth-ring figure visible on flat-sawn surfaces. Though somewhat darker in colour, it is rather like poplar in appearance and of similar light weight.

TECHNICAL PROPERTIES Katsura dries readily and when dry is very stable in use. It works easily and well, finishes to a very smooth surface and is notable for the sharpness of detail that may be obtained by hand or machine tools. It can be peeled to give a good veneer. It is not very strong, and because it is light in weight does not hold nails or fastenings well.

USES Katsura is used mainly in Japan, but is sometimes exported in small quantities as square-edged boards. Its fine, even texture and small movement make it particularly suitable wherever detail in finish and stability in use are important, e.g. in foundry patterns, for mouldings and engravings, in lacquer ware and for drawing boards. It is used in cabinetwork, in the manufacture of pencils and for the Japanese shoes called "geta". Made into plywood or veneer, it presents an attractive, if somewhat plain, surface and is used for panelling.

Commercial Hardwoods

Iroko   Ceylon Satinwood

IROKO
Chlorophora excelsa

(x10)

THE TREE Iroko is an important African wood, used throughout its wide distribution from the Ivory Coast to Angola, from the Sudan to Moçambique – and in Europe. It is a very large tree, up to 50m high, and yields fine, cylindrical logs, which for export, mainly from West Africa, are usually squared or have the sapwood removed.

THE WOOD Yellow-brown to deep-brown in colour, iroko has a distinctive appearance due to the pale, soft tissue associated with its vessels. The grain is typically interlocked and sometimes irregular. Iroko is sometimes likened to teak, but it is coarser in texture, a little lighter, and lacks the distinctive odour and greasiness to the touch characteristic of teak. White stony deposits sometimes occur in cavities in its wood.

TECHNICAL PROPERTIES Though it is not quite so strong, iroko compares favourably with teak in its other properties. It has the same outstanding durability and in its stability in use is slightly superior, if the direction of the grain is straight. It saws and machines well, though stony deposits can damage saw teeth and cutters.

USES Iroko has a useful combination of properties, and is comparatively inexpensive. It is suitable for many of the purposes for which teak is used, although it lacks the decorative effect of teak and it is therefore not an important furniture wood. It is used in ship- and boat-building, for high-class joinery, counter and bench tops, for park and garden seats and for parquet flooring, even where underfloor heating systems are installed.

CEYLON SATINWOOD
Chloroxylon swietenia

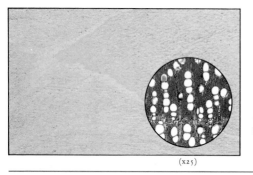

(x25)

THE TREE Ceylon or East Indian satinwood, although not the original satinwood (which came from the West Indies), is botanically related and has been accepted commercially as a "true" satinwood for more than 100 years. Today, it is the more common commercial satinwood and comes from Sri Lanka, though it also grows in central and southern India. The tree is of modest size, giving often somewhat misshapen logs 30 to 50cm in diameter.

THE WOOD Ceylon satinwood is a handsome and distinctive wood, pale-yellow to golden-yellow in colour, and with a fine, even texture. It has a variable grain, interlocked and sometimes wavy, giving a highly decorative stripe or mottle figure, but its appearance may be marred by gum veins.

TECHNICAL PROPERTIES Satinwood is both strong and durable, but its importance lies in its decorative appearance. With care, it can be dried satisfactorily, but it tends to surface-check and distort. It is hard on both hand and machine tools, but can be worked to a very fine finish. It may be sliced to give a fine veneer.

USES Typically more highly figured than West Indian satinwood, though not quite the same clear yellow, Ceylon satinwood is a cabinet-maker's wood. It was popular for furniture-making in the 19th century; today it is seen mainly in the form of quarter-cut veneer, used as inlay in fine cabinetwork. However, it is also used in the solid for small turned items, for the backs and handles of hair-brushes, and for recorders.

CAMPHORWOOD
Cinnamomum camphora

THE TREE True camphorwood comes from a tree found in China, Taiwan and Japan, and at one time was an important source of crystalline camphor. Nowadays camphor is made synthetically and camphorwood is no longer of great commercial interest. The name today is often applied to other woods with a similar smell, even though their appearance and properties may differ considerably. True camphorwood comes from a tree of medium size, up to 30m high, with a bole 1m in diameter.

THE WOOD Camphorwood is distinctive both because of its smell and its appearance: it is yellowish, with red or reddish-brown streaks, and as the grain is often rather irregular, its appearance is, perhaps, more striking then decorative. Its weight is variable but typically fairly light.

(x25)

TECHNICAL PROPERTIES Camphorwood is said to dry readily, if with some tendency to distort, but once dry, to be stable in use. It is a lightweight wood and not particularly strong, but it has a reputation for durability.

USES Camphorwood is only occasionally seen outside the Far East, and then mainly as veneer, used for its decorative effect in marquetry and inlay. In the Far East it is highly prized, and used especially for clothes-chests, trunks, wardrobes and book-cases, for it has a reputation for repelling moths and other insects. It is also the traditional and preferred wood for coffins. Seamen's chests were formerly made from, or lined with, camphorwood, because it was widely supposed to have preservative properties

CORDIA
Cordia spp.

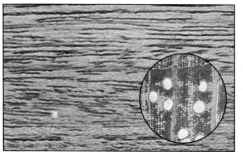

(x25)

THE TREE *Cordia* species occur in many tropical countries, but commercial timber comes mainly from West Africa, as African cordia or omo, and from South America, especially Brazil, as freijo and louro. Cordia trees are mostly of moderate size, up to 30m high and 60cm to 1m in diameter.

THE WOOD There are several types of cordia timber. Occasionally, it is hard, heavy and dark-brown, but that of commercial interest is typically medium-brown and light or only moderately heavy. Freijo and louro resemble teak in appearance, but are 10 to 15 per cent lighter in weight. West African cordia varies in colour from pale-yellow brown to chocolate, sometimes with a pinkish tint; it is coarse-textured and normally light in weight.

TECHNICAL PROPERTIES The various cordia woods dry readily; once dry, they have a reputation for stability in use. African cordia is not strong, but the heavier American timbers are only 15 to 20 per cent weaker than teak. All are easily sawn, can be finished well, nail easily and hold fastenings well. Though cordia is generally durable, the pale, lightweight wood should not be exposed to conditions favouring decay.

USES Cordias are attractive woods, used in tropical America for a variety of decorative, joinery and construction purposes. The small quantities shipped from South America are used for furniture both as veneer and in the solid. African cordia is used for exterior cladding and for the lipping of flush-panelled doors.

Camphorwood

Cordia

Dogwood

Rosewood

Tulipwood

African Blackwood

DOGWOOD
Cornus spp.

(x25)

THE TREE Dogwood, or cornel, is found in many temperate countries, but the only source of commercial timber of more than local use is eastern North America. Well known for its showy flowers, it is there a small tree, 5 to 10m in height, yielding logs no more than 15cm in diameter. Of similar size, but less important as a source of timber, is European dogwood or cornelian cherry, *Cornus mas*, found through much of Europe and into western Asia.

THE WOOD Dogwood is hard, heavy, very fine in texture and usually straight-grained. It is typically yellow to pinkish-brown, sometimes with a small dark-brown core of heartwood, which is usually excluded from commercial supplies. Dogwood is about 15 per cent heavier than beech.

TECHNICAL PROPERTIES Dogwood dries well but slowly. It is noted for its strength, particularly for its hardness and toughness. Although because of its weight it is hard to machine, its working properties are good and it finishes very smoothly. It is not resistant to decay.

USES A special-purpose wood, dogwood is traditionally used in the textile industry for shuttles and spindles, because it is both hard-wearing and has the ability to remain smooth despite continuous use; commercial supplies are prepared in the form of blanks of a length and section suitable for their manufacture. European cornel is occasionally used for hammer-handles, as it is tough, and fine and even in texture. In the Middle East it has also served as a dye-wood.

ROSEWOOD
Dalbergia spp.

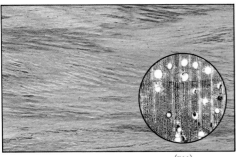
(x25)

THE TREE The highly decorative rosewoods have long been sought after. They are typically small to medium-sized trees, found in a variety of forms in many parts of the world. Commercial supplies today come mainly from India and Brazil, but also from Madagascar and Honduras and other Central American countries.

THE WOOD Often a wide band of pale sapwood is cut away leaving a core of highly figured heartwood, which is extracted in short lengths. Typically the heartwood is purple-brown, with darker, almost black, markings, and varying in weight from heavy to very heavy. Indian or Bombay rosewood has a purplish tint, but can be very dark. Brazilian or Rio rosewood is brown and often highly figured; kingwood, another Brazilian

variety, is finely striped and very heavy. Honduras, and some Malagasy, rosewood is finely figured.

TECHNICAL PROPERTIES Rosewood is not unduly difficult to dry or machine, though very heavy wood demands greater care. The fresh heartwood often has a pleasant rose-like odour when it is machined. Rosewood slices well for veneer.

USES For more than two centuries rosewood has been highly prized for fine cabinetwork, and it retains its popularity in furniture of quality today. It is also used for turned items, for example the handles of cutlery and other stainless steel items. Honduras rosewood is the preferred wood for the keys of xylophones and marimbas; figured wood is used for inlay work.

TULIPWOOD
Dalbergia spp.

(x25)

THE TREE Tulipwood is quite distinct in its appearance from the other rosewoods. Brazil is the long-established source of tulipwood, *D.frutescens* var. *tomentosa*, but a similar wood, *D.oliveri*, has been obtained in the past from Burma. The Brazilian tree is small, and its bole often misshapen; it yields heartwood billets 60 to 120cm in length and 10 to 20cm in diameter. The Burmese tree is larger and often of good form, producing large logs of very good-quality timber.

THE WOOD Tulipwood is very remarkable for its colour, a creamy yellow with pink to reddish-purple stripes; the bright colours fade somewhat on exposure, but the wood retains a decorative appearance. It has a straight to interlocked grain and a fine texture; it is very heavy, comparable to

the denser types of rosewood.

TECHNICAL PROPERTIES Tulipwood is like the denser rosewoods in its technical behaviour. It is prone to splitting on drying and is difficult to machine because it tends to splinter easily, but once smoothly finished, it will take a high polish. It slices satisfactorily to give a decorative veneer. Like many rosewoods, it gives off a faint fragrance when it is worked.

USES Though it has never been available in quantity, tulipwood has often been popular for inlay and marquetry, notably in the ornate furniture of 18th-century France. It is used for similar purposes today, e.g. as quartering in the surrounds of panels, and as inlay in musical boxes and jewellery cases. It is seldom seen in the solid.

AFRICAN BLACKWOOD
Dalbergia melanoxylon

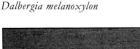

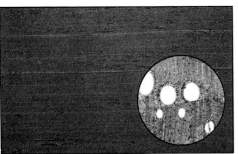
(x25)

THE TREE African blackwood is unusual among species of *Dalbergia*, or rosewoods, for its black colour. Though it occurs widely in tropical Africa, commercial supplies are obtained mainly from Tanzania and Moçambique. The wood comes from small, often misshapen trees, and is shipped in the form of short billets.

THE WOOD The wood is dark-brown to black, with a fine, even texture and a straight to irregular grain. It is extremely heavy, only marginally lighter than lignum vitae and, like lignum vitae, slightly oily to the touch.

TECHNICAL PROPERTIES African blackwood must be dried with great care if excessive splitting is to be avoided. Once dry, it is very slow to pick up moisture and so alter its dimensions. Its

strength properties are of little significance, excepting its tendency to be brittle. It is hard to saw, causing rapid blunting, and, because it is hard and brittle, care is needed in machining.

USES The primary use of African blackwood, now that the traditional cocuswood is no longer available, is for woodwind instruments – for oboes, flutes and clarinets, and for the chanters of bagpipes. Besides giving a good tone, blackwood is particularly suitable for use in musical instruments because of its working properties: it can be turned and bored to give a smooth finish, it is hard enough to be tapped for the screw-threads of the metal pillars of the keys, and its movement is very small. Blackwood is also used for turned items, and especially in African carvings.

Commercial Hardwoods

Cocobolo

Basralocus

COCOBOLO
Dalbergia retusa

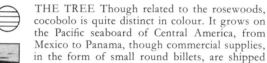

(x25)

THE TREE Though related to the rosewoods, cocobolo is quite distinct in colour. It grows on the Pacific seaboard of Central America, from Mexico to Panama, though commercial supplies, in the form of small round billets, are shipped mainly from Costa Rica and Nicaragua.
THE WOOD Freshly cut cocobolo has a rainbow hue, but the bright colours fade on exposure and the wood darkens to a deep orange-red, sometimes with darker stripes or mottling. Cocobolo has a medium texture and a grain that may be straight to irregular, according to the shape of the log. It is very heavy, comparable to the denser rosewoods.
TECHNICAL PROPERTIES Cocobolo is usually dried in small sizes; once dry, it has a reputation for stability. It is not unduly difficult to work,

considering its weight, and is noted for its good turning properties. When machined, it gives off a mild fragrance from a natural oil which contributes to its good finishing qualities but makes it difficult to glue. When worked, it produces a fine dust, which may be irritant – causing a form of dermatitis and staining the skin orange.
USES Cocobolo was formerly important, especially in America, for knife-handles: it is not only attractive in appearance and capable of an excellent finish, but it can withstand repeated washing without deterioration. It is also used for the backs of hair-brushes, for tool-handles, for chessmen, and for decorative turned items in general; it is sometimes seen as inlay on jewellery boxes and other small decorative items.

BASRALOCUS
Dicorynia guianensis

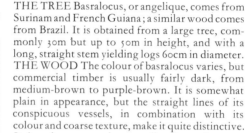
(x25)

THE TREE Basralocus, or angelique, comes from Surinam and French Guiana; a similar wood comes from Brazil. It is obtained from a large tree, commonly 30m but up to 50m in height, and with a long, straight stem yielding logs 60cm in diameter.
THE WOOD The colour of basralocus varies, but commercial timber is usually fairly dark, from medium-brown to purple-brown. It is somewhat plain in appearance, but the straight lines of its conspicuous vessels, in combination with its colour and coarse texture, make it quite distinctive. It is a heavy wood, generally about the same weight as oak.
TECHNICAL PROPERTIES Basralocus is a difficult wood to dry, and though when green it can be sawn fairly easily, it is very abrasive when

it is dry. It is strong and very durable, and withstands the attack of fungi, of marine borers and, reputedly, of termites.
USES Basralocus is one of a comparatively small number of timbers available in large sizes for heavy structural use. For this purpose the tree is of excellent shape, yielding poles and piles from 12 to 18m in length and 40 × 40cm in cross-section, which combine strength and great durability. The main uses are for docks, harbours and marine structures, but the timber is also used for decking, in boat-building and for flooring. It has found favour both in Europe and in America; in Europe, it has been used especially in Holland, but up to now has been seen only in small quantities in Britain.

EBONY
Diospyros spp.

THE TREE The use of ebony is as old as Ancient Egypt: though scarce today, its intense black colour is still familiar. However, not all ebonies are black: Macassar ebony and coromandel are striped or mottled, and many others, like persimmon, are pale – though few of these "white" ebonies are of commercial interest. Ebony occurs in many parts of the world, but black ebony, once chiefly obtained from India and Sri Lanka, today comes mainly from tropical Africa. It is shipped in the form of short heartwood billets.
THE WOOD Though more familiar as a jet-black wood, ebony may be medium-brown to dark-brown with black stripes, as is Macassar ebony, or with grey or brown mottling, like coromandel. It is fine and even in texture and extremely heavy.

TECHNICAL PROPERTIES Ebony is not easy to handle. It must be dried with great care, and it must be skilfully processed, for it is hard and brittle. However, it can, with care, be machined to an excellent finish.
USES Ebony has always been used in Europe and Asia and especially for furniture and carving in the ancient courts of Egypt, Persia and India. Though it is available today only in small sizes, it is still used wherever it can be shown to decorative advantage – for handles, door knobs; the backs of brushes, the butts of billiard cues, and for a variety of turned ware. It is used in a wide range of musical instruments, for the fittings of violins, for organ stops, castanets, and, traditionally, for the black keys of keyboard instruments.

PERSIMMON
Diospyros virginiana

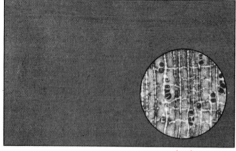

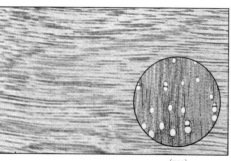
(x10)

THE TREE Persimmon is a "white" ebony: though it belongs to the genus that produces the familiar black ebony, any coloured wood persimmon may have is confined to a very small core of heartwood, and it is the pale sapwood that is of commercial interest. Persimmon grows in the middle and southern regions of the United States, reaching a modest size – up to 30m in height and with a bole 50 to 75cm in diameter. There are, elsewhere, other "white" ebonies, but few are important for their timber, although the Asian kaki, *D.kaki*, like persimmon, has an edible fruit.
THE WOOD Commercial persimmon is off-white in colour, with a greyish tint. It has a fine, even texture and a straight grain. It is a dense wood, about 15 per cent heavier than beech.

PROPERTIES Persimmon dries fairly easily, but shrinks very considerably. When dry, it moves appreciably under changing conditions of humidity. For its traditional uses, its instability is of minor importance; more significant are its strength properties, exceeding those of beech, and its excellent finishing and wearing properties.
USES Persimmon is used only for special purposes. For the textile industry it is prepared and dried in the form of blanks to be made into shuttles, for which purpose persimmon is very suitable, as it can be machined to the intricate detail and very smooth finish required. It is also traditionally used for golf-club heads as it is exceptionally hard, has a very high resistance to impact, and takes a very smooth finish.

Ebony

Persimmon

Keruing/Gurjun/Yang

Paldao

Kapur

Jelutong

KERUING/GURJUN/YANG
Dipterocarpus spp.

(×10)

THE TREE One of the most important trees of SE Asia, *Dipterocarpus* is known by a distinctive name according to its country of origin – gurjun in India and Burma, yang in Thailand, keruing in Malaysia and Indonesia, and apitong in the Philippines. Commercial shipments include timber of a number of species which, though similar, differ somewhat in character. Where only a limited number of species occurs, as in Burma, Thailand and the Philippines, supplies are more uniform in character than shipments from areas having a wide range of species.

THE WOOD The different timbers are a medium- to dark-brown, generally straight-grained, coarse in texture and of plain appearance. Some exude a sticky resin. They vary in weight, but are on average a little heavier than oak.

TECHNICAL PROPERTIES The wood dries slowly, but once dry, it is strong and fairly resistant to fungal attack. Because it is abrasive and often resinous, it can be difficult to process. It is not very stable in use.

USES Because of their commercial availability, their quality and their low price, these woods have been used in many countries. They lack the exceptional durability of oak, but are suitable for many outdoor uses in temperate countries, in building, for cladding and for sills and thresholds, in lorry and truck construction, in boats, and for telegraph cross-arms. They are apt to exude resin, which can be troublesome if they are used for flooring. They are used for plywood in Asia.

PALDAO
Dracontomelum spp.

(×10)

THE TREE Paldao, or dao, and the closely related New Guinea walnut are decorative woods obtained from the Philippines and from Papua–New Guinea. Paldao is a tall tree, 30m or more in height, with a cylindrical bole above large buttresses.

THE WOOD Paldao is like walnut, though not botanically a true walnut. It resembles Queensland walnut more closely than European walnut as it is grey to brown with fairly regular dark stripes. Only the heartwood is figured and the sapwood, which often forms a high proportion of the log, is pale and featureless. The grain is usually interlocked, but occasionally logs have a wavy grain, which produces a particularly decorative effect. It compares closely in texture with true walnut, but is about 10 per cent heavier.

TECHNICAL PROPERTIES Because paldao, with its wide sapwood of no commercial interest, is seldom converted for use in the solid, little technical information is available about it. The timber is said to dry with a tendency to distort, but to machine well. For the production of veneer, it both peels and slices well and can be finished to give a fine surface.

USES Paldao has been exploited commercially for more than 50 years; New Guinea walnut is of more recent introduction. Both are among the most attractive of the walnut-like woods, though lacking the outstanding decorative appearance of true walnut. They are seen mainly in the form of sliced veneer, used for figured surfaces of furniture and cabinetwork and for panelling.

KAPUR
Dryobalanops spp.

(×25)

THE TREE Kapur is a SE Asian timber, obtained commercially from western Malaysia but important also in Sumatra and Borneo. Typically, it is a very tall tree, up to 60m high, with a long, straight, cylindrical bole. Kapur is sometimes known as Borneo camphorwood, because of its pleasant smell, which is particularly noticeable when the wood is being worked.

THE WOOD A fairly plain yellow-brown to reddish-brown in colour, kapur is somewhat like keruing in appearance, but it is more uniform in character; it has a finer texture and it is not resinous. Its appearance is often marred by numerous, very small worm-holes which are barely visible until the wood is machined. It is about the same weight as keruing and a little heavier than oak.

TECHNICAL PROPERTIES Kapur is a strong wood; though it is prone to attack by worms, any attack ceases once the wood is dry and does not detract from its technical performance. It is more stable and durable in use than keruing, but like keruing it is abrasive to cutting tools because it contains silica.

USES Though plain, kapur is strong, stable and durable when used in temperate conditions. Though not so popular as keruing, its good technical properties make it suitable for many of the purposes keruing and oak serve. It is very suitable for outside joinery, e.g. sills and thresholds, for cladding, for estate work as gates and posts, and for farm buildings. In large sizes it is suitable for piles, piers and other marine works.

JELUTONG
Dyera costulata

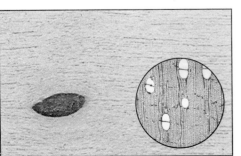

(×10)

THE TREE Jelutong is a SE Asian timber, found in Malaysia and in the Indonesian islands of Sumatra and Kalimantan. Though important for its wood, the tree is also tapped for a milky latex, which is used in chewing gum. It is a very tall tree, up to 60m high, with a straight, cylindrical stem. It is exported in the form of sawn wood.

THE WOOD Jelutong is a plain, straw-coloured wood, typically with occasional, and sometimes with abundant, horizontal cavities – these are the latex traces, lens-shaped on flat-sawn surfaces, and may be 10mm or more in height. The wood is usually straight-grained, with a fine texture, and lacking any distinctive figure. It is light in weight, about the equal of poplar, and has an unusually soft but firm texture.

TECHNICAL PROPERTIES Jelutong dries rapidly and well; once dry, it is stable in use. It is soft and easily indented; it is a weak wood, similar to obeche in its strength properties. It saws easily, machines to give a very smooth finish, and can be carved to give fine detail with sharp tools. Its resistance to fungal attack is low, and it is readily discoloured if drying is delayed.

USES Jelutong is noted for its fine, even texture and good working properties. These, allied with its stability in use, have made it popular for engineers' patterns as an alternative to yellow pine. Wood free from latex-trace cavities is rarely obtained and it is therefore used mainly in small sizes, for handicraft work, toys and models, or where appearance does not matter.

Commercial Hardwoods

Queensland walnut   Sapele

QUEENSLAND WALNUT
Endiandra palmerstonii

(x25)

THE TREE Queensland walnut is so called because it is similar to walnut in appearance, although, botanically, it is unrelated; it is sold as oriental wood in America. It comes from a large tree, about 40m in height, with a long, straight bole yielding logs 1.5m in diameter. The tree occurs in a limited area of northern Queensland, but there it is quite common.

THE WOOD Among the so-called walnuts, the Queensland wood comes closest to European walnut in appearance. It is typically grey to medium-brown in colour, with darker markings: these tend to be straight, and when combined with an interlocked grain produce a more regularly striped figure than true walnut normally has. In texture and in weight it compares closely with true walnut.

TECHNICAL PROPERTIES Two technical aspects, drying and machine processing, are of special significance with Queensland walnut. It dries well in thin sizes, but care is needed with thicker boards. Once dry, it is an abrasive wood to saw and work with hand and machine tools because it contains silica, which blunts cutting edges. Also, when first sawn it has an unpleasant smell, though this disappears when the wood is dried. It can be sliced to produce a good veneer.

USES Queensland walnut is available commercially both as veneer and as solid timber. As solid timber, it is used for high-class joinery and fittings in shops, banks and public buildings, and it makes attractive flooring. Selected veneers are used in decorative panelling and for furniture.

SAPELE
Entandrophragma cylindricum

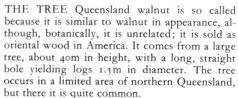

(x10)

THE TREE Sapele, though it has the name of a Nigerian river port, occurs widely in tropical Africa, from Sierra Leone to Uganda and Zaire. It is a well-known and important commercial timber, shipped mainly from the West African countries between the Ivory Coast and Cameroon. It comes from a very large tree, yielding cylindrical logs 1m or more in diameter.

THE WOOD Sapele is a mahogany-like wood. Noted for its stripe figure when quarter-cut or sliced, occasionally it has a fiddle-back or mottle figure. It is darker in colour, with a finer texture, and heavier than African mahogany.

TECHNICAL PROPERTIES The drying and machining properties of sapele are particularly influenced by the presence of interlocked grain, which causes flat-sawn boards to distort when they are dried, and makes it necessary to plane quarter-sawn surfaces with care if tearing is to be avoided. Harder and heavier than African mahogany, it is also stronger, more difficult to work, and more durable, but not quite so stable in use.

USES Commonly used as a mahogany, sapele combines an attractive appearance with strength and durability. It is used for high-class joinery both indoors and outdoors, for window frames, staircases, shop fittings and flooring. Quarter-cut to produce a decorative veneer, it is used on doors, on pianos, and on the surfaces of furniture when mahogany is in fashion. It is rotary-peeled for plywood, which, if suitably bonded, is acceptable for marine use.

UTILE
Entandrophragma utile

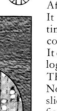

(x10)

THE TREE Utile, or sipo, is one of the oustanding African woods; it occurs from Sierra Leone to Uganda and Angola, but commercial supplies come mainly from the Ivory Coast and Ghana. It comes from a very large tree, 60m or more in height, with a long, straight bole yielding logs 2m or more in diameter and giving very wide boards.

THE WOOD Utile is like sapele in appearance, but coarser in texture, and though typically it has an interlocked grain, it has a wider and not so decorative stripe figure. In colour it is mahogany-red; it is appreciably heavier than African mahogany and marginally heavier than sapele.

TECHNICAL PROPERTIES Utile is not generally quite so troublesome in drying as sapele, unless it has a pronounced cross grain; once dry, it is more stable in use. It saws easily and machines well, though, as with any wood with interlocked grain, care is needed if a good finish is to be obtained on quarter-sawn surfaces. It is rated more durable than sapele and mahogany.

USES Utile is one of the mahogany-like woods that is most sought after for high-class joinery; it is used for window frames and sills, for exterior doors and door surrounds, for cladding and for shop and office fittings. Its strength and durability make it a useful constructional timber, for road vehicles and in ship- and carriage-building. It is used for furniture, but it is not as decorative as sapele. It is peeled for plywood, in which form, combining durability with strength, it is suitable for the most exacting uses.

TASMANIAN OAK
Eucalyptus spp.

(x25)

THE TREE Tasmanian oak is the export name for three eucalypt timbers known and sold by distinctive names in Australia and there generally called ash; botanically, it is unrelated to either true oak or true ash. The trees occur in southeastern Australia from New South Wales to Tasmania, and may reach a very great height, up to 90m, with a clear, straight stem 1m or more in diameter.

THE WOOD Tasmanian oak has some resemblance to flat-sawn oak, but lacks any silver grain when quarter-sawn. It is typically coarse in texture and straight-grained, with a fairly plain appearance occasionally marred by dark gum-lines. It is a wood of medium and somewhat variable weight, but not as heavy as true oak, and readily distinguished, as it is not ring-porous.

TECHNICAL PROPERTIES Tasmanian oak dries readily, but has a tendency to collapse on the surface, although this can be remedied by a conditioning treatment after drying; it is moderately stable in use. Its strength properties are good and it can be sawn and machined without difficulty. It is moderately durable, but not the equal of true oak in conditions favouring decay.

USES Tasmanian oak is popular in Australia for a wide range of purposes in construction, joinery and furniture. It is used in coach and truck construction, and, because it is tough, for sports goods. It is popular also for packing-cases and cooperage, and provides good paper pulp. Outside Australia, it is seen mainly in the form of strips, which make good flooring.

Utile

Tasmanian Oak

Karri

Spotted Gum

Jarrah

Blackbutt

KARRI
Eucalyptus diversicolor

(x25)

THE TREE Karri is an Australian timber confined to a limited area of southwestern Western Australia, but is there very common and available in quantity. It comes from one of the huge trees of Australia, reaching a height of 80 to 90m, with a long, straight stem up to 3m in diameter, capable of yielding very large sizes of timber.

THE WOOD Karri is reddish-brown, coarse-textured, and very like jarrah in appearance, though marginally paler and a little heavier. Like jarrah, it is occasionally figured.

TECHNICAL PROPERTIES Karri is more difficult to dry than jarrah and has a tendency to split on the surface. In large sizes it is air-dried; in smaller dimensions it may be kilned, but this requires care. Karri is a very strong wood, even compared to jarrah, and is especially hard, stiff and tough. Being heavier than jarrah, it is more difficult to saw and to machine, but it can be finished to a good surface. It is durable in conditions favouring fungal attack, but in this respect it is not the equal of jarrah.

USES Available in very large sizes, karri is used both in Australia and in many other parts of the world for a range of structural purposes, but especially for the superstructure of bridges, piers and wharves. For these purposes it is superior to jarrah in strength, but inferior if there is a high risk of fungus, marine borer or termite attack. It is used also for truck and wagon construction, for the cross-arms of telegraph poles and for the manufacture of plywood.

SPOTTED GUM
Eucalyptus maculata

(x25)

THE TREE Spotted gum is an Australian timber, obtained in Queensland and New South Wales, and important in South Africa, where it has been introduced. It comes from a tree of moderate size for a eucalypt, about 40m in height, with a straight bole yielding logs 1m in diameter.

THE WOOD Spotted gum is a pale wood, grey-brown to brown in colour, coarse in texture, and with a straight or an interlocked grain, but without a distinctive figure. It is very heavy; Australian timber approaches the weight of greenheart, but South African may be somewhat lighter.

TECHNICAL PROPERTIES Spotted gum is difficult to dry, tending to split on the surface and to distort if cross-grained. It is a strong wood and noted for its toughness; though hard, it is not particularly difficult to saw and can be machined to a good finish. It is rated moderately resistant to fungal attack.

USES Spotted gum is an important wood in Australia and of increasing interest in South Africa, but outside these countries it is not often seen. Its strength and especially its toughness make it a valuable structural wood – for bridge-building, truck and wagon construction and for boatwork. It is one of the best Australian woods for hammer, axe and pick handles and it is used for these in South Africa. It makes a hard-wearing floor, in both strip and block form. Young stems are given preservative treatment and used in South Africa for transmission poles, and it has been exported to Europe for telegraph cross-arms.

JARRAH
Eucalyptus marginata

(x25)

THE TREE Jarrah is an Australian species of limited distribution: it is confined to the coastal region south of Perth in Western Australia, but there it is so common that more jarrah is cut than any other Australian timber. It comes from a tree which reaches a modest height for a eucalypt, up to 45m, with a stem of 1.5m diameter.

THE WOOD Jarrah is a coarse-textured, medium to dark reddish-brown wood, sometimes with darker markings. It has a distinctive but not particularly decorative appearance, although the occasional combination of interlocked and wavy grain gives a figured wood; its appearance can be marred by the presence of gum-veins and pockets. It is a dense wood, somewhat variable in weight but averaging about 15 per cent heavier than oak.

TECHNICAL PROPERTIES Jarrah is air-dried when used in large sizes, and even in small dimensions may be air-dried before kilning. It is stronger and more durable than oak and resistant to both termite and marine borer attack. Though hard, it is not unduly difficult to saw and machine.

USES Jarrah is strong, very durable, and available in large sizes: it is used in Australia, and exported to many parts of the world, for dock, harbour and bridge construction, as piles and as framing and decking timber. It is used in ship-building and in many countries for railway sleepers, e.g. on the underground lines in London. It is suitable as flooring for all but the most exacting industrial conditions, and in Australia it is used for high-class exterior and interior joinery.

BLACKBUTT
Eucalyptus pilularis

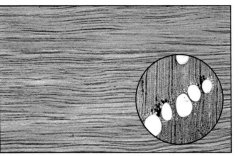

(x25)

THE TREE Blackbutt, one of the most important eucalypt timbers of eastern Australia, is especially common in parts of New South Wales and Queensland. The tree is large, reaching 50m in height and yielding logs about 1m in diameter. It is said to derive its name from the charring by bush fires of the fibrous bark at the base of the tree.

THE WOOD Blackbutt is pale-brown, with a moderately coarse texture, a straight to interlocked grain, and little decorative appearance; it is often characterized by gum-veins or streaks. It is heavy – about 20 per cent heavier than oak – but not unduly so compared to many eucalypts.

TECHNICAL PROPERTIES Blackbutt is difficult to dry without degrading, chiefly in the form of surface splitting, but the wood is also prone to surface collapse, though this defect can be remedied after drying. It is a strong, tough and stiff wood; though hard, it can be sawn and it works readily with both hand and machine tools to give a good finish, but it cannot be nailed unless holes are bored first. It is rated resistant to fungal attack.

USES Blackbutt is available in quantity and used extensively in Australia, but it is not often seen outside that country. It is important in Australia for structural purposes, especially for carcassing in house construction, but also for weather-boarding, and flooring and other building uses. It is used for telegraph and electricity transmission poles and for railway sleepers, and it is a useful estate wood, especially for fence posts and rails.

Commercial Hardwoods

Saligna Gum

West Indian Satinwood

SALIGNA GUM
Eucalyptus spp.

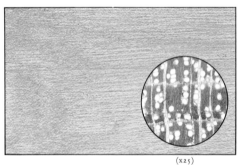

(x25)

THE TREE Saligna gum, properly *E.saligna*, is an Australian tree, planted in many parts of the world and very extensively and successfully in South Africa. However, many of these plantings, though called saligna, are believed to be *E.grandis* or a hybrid between *saligna* and *grandis*. In favourable conditions the tree grows very rapidly, producing commercial timber within six to eight years, and reaching a height of 30m or more in twenty.
THE WOOD In Australia, saligna (or Sydney blue gum as it is there usually known) is a heavy, medium-red to dark-red wood; plantation timber, especially from vigorously grown or young trees, is light in weight and pale-pink, though in older trees it is heavier and darker. Saligna is typically straight-grained, with a fairly fine and even texture.

TECHNICAL PROPERTIES Saligna logs, particularly of plantation timber, are liable to split, especially if left after felling; further splitting may occur on conversion. Sawn timber dries fairly readily; once dry it can be machined to a good finish, although both its working and its strength properties vary appreciably. Heavy wood is moderately durable, but lighter wood is perishable.
USES In Australia, saligna is a general-purpose building and flooring timber. In South Africa, its most common use is for mine props, but it is also used after preservative treatment for telegraph and electricity transmission poles, and sawn for building timber and for boxes and flooring. It is pulped for paper and fibreboard and it is a source of cellulose for rayon.

WEST INDIAN SATINWOOD
Fagara flava

(x25)

THE TREE West Indian satinwood, the original satinwood of commerce, is a timber of southern Florida and the West Indian islands. Though other, mainly African, species of *Fagara* have attracted commercial interest, their yellow woods lack the outstanding decorative appearance of West Indian satinwood. This comes from a small tree, 10 to 12m high; its logs may be occasionally 50cm in diameter but are usually much smaller.
THE WOOD Satinwood is a creamy to golden-yellow, with a fine, even texture; even when straight-grained, it has a distinctive appearance due to its bright, satiny sheen; occasionally its appearance is further enhanced by a wavy grain giving a mottle figure. It is a heavy wood, though not quite so dense as East Indian satinwood.

TECHNICAL PROPERTIES Because of its specialized use, the technical information available is limited. Satinwood can be dried satisfactorily; once dry, it can be machined to an excellent finish, though it is rather hard on tools. It turns well. When worked, it has a characteristic scent resembling that of coconut, but its fine dust is reported to irritate the skin.
USES Satinwood is a very beautiful wood used extensively in the 18th century in the fine cabinetwork of Adam, Sheraton and Hepplewhite. Today, it is seen only in small quantities; as veneer, it is used for inlay and for marquetry, and in the solid for small turned items, such as bobbins, and occasionally for the backs of hair-brushes and for hand mirrors.

BEECH
Fagus spp.

(x25)

THE TREE Beech is a northern temperate wood, in Europe rivalling oak as the most commonly used hardwood, and of commercial interest also in Japan and the central and Atlantic states of North America. European beech, *F.sylvatica*, or *F.orientalis* in parts of the Balkans, Turkey and Iran, is a medium-sized to large tree, 30m or more in height, often forming dense stands in which its cylindrical bole is best developed.
THE WOOD Beech is a white or pale-brown wood, sometimes steamed to a pinkish colour. It has a conspicuous growth-ring figure on flat or rotary-cut surfaces and on quartered surfaces a characteristic fleck due to its medium-sized rays. It is typically straight-grained, with a fine, even texture, and is of medium, though variable, weight.

TECHNICAL PROPERTIES Beech dries readily, though tending to distort; when dry, it moves appreciably in changing conditions of humidity. The wood, particularly that grown in northern Europe, is strong, but works easily, takes a good finish, and turns especially well. It can be steam-bent and is rotary-peeled to give an excellent veneer. Beech is perishable and unsuitable for outdoor use unless it has been subjected to preservative treatment.
USES Beech is an outstanding furniture wood, and used especially for the turned and bent members of chairs. It is also used for many domestic purposes, for wooden spoons and other kitchen utensils, for the handles of tools and brushes, and for toys; it makes a hard-wearing domestic floor.

QUEENSLAND MAPLE
Flindersia brayleyana

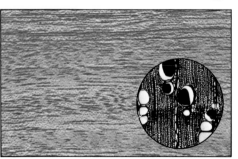

(x25)

THE TREE Queensland maple is a decorative wood, popular and highly valued in Australia. It is quite unlike maple in appearance and is not botanically related. It comes from a tree of medium height, about 30m, but it has a massive trunk, 1m or more in diameter. It occurs in the tropical forests of northeastern Queensland.
THE WOOD Queensland maple is a pale to medium pinkish-brown to brown in colour, and noted for its lustrous appearance. It has a moderately coarse but even texture, and a grain which is straight, interlocked or wavy, and may give a stripe or a mottle figure. It is sometimes called silkwood as a combination of mottle figure and high lustre can give an appearance like watered silk. In weight it is comparable to mahogany.

TECHNICAL PROPERTIES Queensland maple dries well, but has a tendency to distort if the grain is interlocked or irregular. It is light in weight, but its strength properties are good, and it works well by hand or machine, though the finishing of quartered surfaces requires care. Its resistance to fungal attack is uncertain, but probably low.
USES In Australia, Queensland maple is considered the equal of mahogany and classical timbers for fine cabinetwork, panelling and high-class joinery. Since it is both strong and light, it is used for framing and interior fittings in boats and railway carriages. It is one of the few woods used successfully for rifle-stocks. It is also peeled and sliced for veneer and it makes an excellent plywood. Outside Australia it is seen mainly as veneer.

Beech

Queensland Maple

Ash

Gmelina

Ramin

Agba

ASH
Fraxinus spp.

THE TREE Ash occurs widely in the northern temperate regions and is important commercially in Europe, North America and Japan. Typically, it comes from medium- to large-sized trees 20 to 35 m in height, with boles 60cm to 1m in diameter.
THE WOOD Ash is white, with a pale-pinkish tint when first cut. It is ring-porous with a conspicuous growth-ring figure, especially on flat-sawn surfaces. Typically straight-grained, its decorative appearance is sometimes enhanced by a wavy grain, particularly in Japanese ash. Its weight can vary appreciably: on average, it is almost the weight of beech, but slow-grown wood is lighter.
TECHNICAL PROPERTIES Ash dries readily and is moderately stable in use. It is a strong wood and is especially noted for its toughness. It saws

(x10)

and machines easily, taking a good finish, and responds very well to steam-bending. It is perishable and unsuitable for use out of doors unless treated.
USES Ash is variable in character: some is exceptionally tough and is selected for use in sports goods – especially tennis racquets, hockey sticks, baseball bats, gymnasium equipment and cricket stumps – and for ladder-rungs and the handles of striking tools, such as axes, picks and hammers. For these purposes, straight-grained timber of vigorous growth is required, and European, and especially English, ash is preferred. Other uses of ash are for the handles of garden tools, for the framing of lorries and buses, for the wooden parts of agricultural machinery, and for bent work in furniture and boat-building.

GMELINA
Gmelina arborea

THE TREE Although gmelina has been up to now of little commercial interest, it must become increasingly important as extensive plantations yield appreciable quantities of wood. It occurs naturally in India, Bangladesh and Burma as a medium-sized tree, 30m high, with a bole 60cm in diameter. Seed, mainly from Burmese trees, has been planted in many parts of the world, in Malaysia, West Africa, the West Indies and, on a large scale in recent years, in Brazil. Plantation gmelina grows quickly, producing sawmill logs 4 to 5m long and 20 to 40cm in diameter, in ten to fifteen years.
THE WOOD Gmelina is a plain, pale, straw-coloured wood with a high natural lustre; it has a medium to coarse texture and slightly interlocked

(x25)

grain, giving a figure to quartered surfaces. It is about the same weight as European redwood.
TECHNICAL PROPERTIES Gmelina dries well, though rather slowly, especially in thicker sizes; once dry it is very stable in use. Its strength properties are modest: it is harder but otherwise appreciably weaker than European redwood, but it saws easily and machines well to give a good finish. It is not resistant to fungi.
USES Even when vigorously grown, gmelina is firm and of sufficient density to be suitable for light construction, general carpentry, packaging and parts of furniture. The extensive plantations, especially those in Brazil, are likely to provide fibre for paper-making and board production, and plywood timber.

RAMIN
Gonystylus spp.

THE TREE Ramin is one of the important timbers introduced in about 1950. It was obtained from Sarawak, from a tree with a small bole, rarely exceeding 60cm, but abundant in extensive tracts of coastal swamp-forest. First shipped as logs, and discolouring badly in shipment, it almost failed to gain commercial acceptance, and did not become successfully established until it was sawn and dried in Sarawak. Obtained now from Indonesia, Sarawak and also from western Malaysia, it has been so extensively cut and is in such demand that future supplies are in jeopardy.
THE WOOD Ramin is almost white, moderately fine and even in texture and generally straight-grained. Its unpleasant smell when green disappears on drying. It is of medium weight.

(x10)

TECHNICAL PROPERTIES Ramin dries easily, but has a tendency to split and to discolour, especially in thicker sizes. A little lighter than beech, it compares well with it in many strength properties, though it is not so tough and splits more readily. It works easily and finishes well. Ramin is not suitable for use out of doors.
USES Ramin attracted interest because of its plain appearance. Being pale in colour, it could be stained and finished to match more decorative woods, and, having a straight grain, it was popular for machine processing. It was used especially in the furniture industry as an alternative to beech. Ramin is also used for mouldings, dowelling, handles and toys, and wherever a clean appearance and a smooth surface are required.

AGBA
Gossweilerodendron balsamiferum

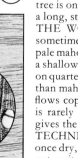

THE TREE Agba, or tola, is a West African wood obtained mainly from Nigeria and Angola. The tree is one of Africa's largest, up to 60m high with a long, straight bole yielding logs 2m in diameter.
THE WOOD Agba is straw-brown in colour, sometimes with a pinkish tint, and resembles a pale mahogany. It has a moderately fine texture and a shallowly interlocked grain giving a broad stripe on quarter-cut surfaces. It is a little lighter in weight than mahogany. Agba contains a dark resin which flows copiously when some logs are cross-cut; it is rarely troublesome once the wood is dry but gives the wood a slightly resinous odour.
TECHNICAL PROPERTIES Agba dries easily; once dry, it is stable in use. It is of modest strength, not quite the equal of African mahogany, and

(x25)

occasional pieces are particularly brittle. Agba saws easily and machines to a good finish. Though light in weight, its heartwood is durable and resistant to fungi.
USES Agba is an attractive wood combining light weight with natural durability. It is therefore used for exterior and interior joinery, for cladding, and for the planking and the laminated frames of boats. It is often available in wide boards and it is popular for chair-seats and the tops of tables and desks, particularly for use in schools. It makes an attractive domestic floor and can be used over underfloor heating. It is rotary-peeled for plywood, which is suitable for outdoor and marine use. Because of its slight odour it cannot be used in contact with food.

Commercial Hardwoods

Maracaibo Boxwood   Lignum Vitae

MARACAIBO BOXWOOD
Gossypiospermum praecox

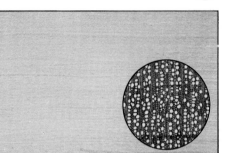

(x25)

THE TREE Maracaibo boxwood, or zapatero, has long been accepted as a boxwood, although it is botanically unrelated to true box. It occurs in Cuba and in the Dominican Republic, but commercial supplies come from Colombia and Venezuela – it was first obtained from the region around Lake Maracaibo. Like many prized woods, it comes from a comparatively small tree; formerly it was shipped as logs or billets, 15 to 25 cm in diameter, but nowadays is more often exported as sawn timber.

THE WOOD Closely resembling true boxwood, Maracaibo boxwood has a very fine, even texture and a yellow colour, slightly paler than true boxwood and sometimes almost creamy-white. It has a straight grain and featureless appearance. It is

about 20 per cent lighter than true boxwood.

TECHNICAL PROPERTIES Maracaibo boxwood, like true boxwood, is noted for its excellent working properties: it is not quite so hard, and so marginally easier to work than true boxwood, but it tends to split more readily. However, it can be finished very smoothly and engraved with very fine detail. Its resistance to fungi is low, and the wood readily becomes discoloured if it is kept in conditions favouring attack.

USES Maracaibo has the advantage over true boxwood in that it is available in larger sizes. It is particularly suitable for rules and drawing instruments. It is used for turned and carved items and for engravers' blocks. It is occasionally sliced to give a plain but evenly coloured veneer for inlay.

LIGNUM VITAE
Guaiacum officinale

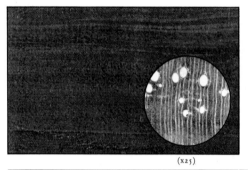

(x25)

THE TREE Lignum vitae (wood of life) is so called because when it was first discovered in the 16th century its resin was believed to cure many diseases. Nowadays it is of interest for its remarkable timber, the heaviest in commercial use. It comes from a small tree and is marketed in the form of short billets up to 3m long and usually 75 to 300mm in diameter. The best lignum vitae comes from the West Indian islands and the coastal region of Colombia and Venezuela, but it is also shipped from parts of Central America.

THE WOOD Lignum vitae is a very distinctive greenish-black, with a high resin content and of exceptional weight, some 70 to 80 per cent heavier than oak. It has a fine texture and a closely interlocked grain.

TECHNICAL PROPERTIES Lignum vitae is very strong and exceptionally hard, with a high resistance to abrasion. It is difficult to dry and hard to work, but it is an excellent wood for turning. It is exceptionally durable.

USES The outstanding industrial use of lignum vitae is for the bearing surfaces for ships' propeller shafts. End-grain blocks, machined to fit the shaft, are packed into the stern-tube. Its combination of density, fine texture and the property of self-lubrication (due to its high resin content) makes lignum vitae uniquely suited to this purpose. A long-established use is for flat-green and crown-green bowls, but difficulties in obtaining lignum vitae of adequate size have caused other materials to be used today.

GUAREA
Guarea spp.

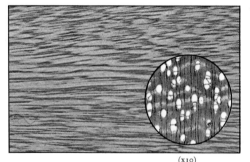

(x10)

THE TREE Guarea is a tropical timber occurring widely in America and Africa but shipped in commercial quantities mainly from the Ivory Coast and Nigeria. It is produced by two species, distinguished by the colour of their bark as white or scented guarea, *G.cedrata*, and black guarea, *G. thompsonii*. They are of somewhat different character, but often mixed. Both are large trees, up to 50m in height, with boles 1m in diameter.

THE WOOD Guarea is a mahogany-like wood, but rather firmer, of a finer texture and a little heavier than mahogany. Scented guarea, so-called because it smells like cedar, can be somewhat gummy. It often has the more attractive appearance, and a stripe or mottle figure; black guarea is plainer and slightly heavier.

TECHNICAL PROPERTIES Both timbers dry well, but black guarea requires greater care; once dry, they are stable in use. They have good strength for their weight, with black guarea marginally the stronger. They can be sawn and machined to a smooth finish, though scented guarea is more abrasive because it contains silica. The fine dust produced in processing often irritates the skin, eyes and throat. Guarea is rated very resistant to fungi and moderately resistant to termites.

USES Guarea is an attractive wood, used in furniture, especially where a wood similar to but stronger than mahogany is required, and for joinery, for the framing of trucks and lorries and for flooring. It makes an excellent plywood which is suitable for use out of doors.

BUBINGA
Guibourtia spp.

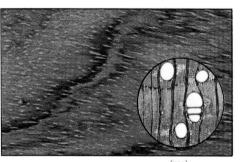

(x25)

THE TREE Bubinga is an African timber obtained mainly from Cameroon and Gabon; because it is often more highly figured, timber from Gabon is sometimes distinguished as kevazingo. Three species produce bubinga; all are large trees, up to 30m in height, yielding fine, cylindrical logs. A similar wood occurring in south central Africa is known as Rhodesian copalwood.

THE WOOD Bubinga is red to red-brown, with purple veining, particularly conspicuous in freshly cut wood, but fading on prolonged exposure. Its texture is medium and the grain often somewhat interlocked; but when it is wavy or irregular, the wood has a very decorative appearance on both flat- and quarter-sawn surfaces. Bubinga is a heavy wood, comparable in weight to rosewood.

TECHNICAL PROPERTIES Bubinga dries slowly but well; once dry it is stable in use. It is strong but not particularly resilient. It saws easily, considering its weight, and when dry can be machined to a fine finish, though care is needed if the grain is irregular. It has a reputation for durability, including resistance to termite attack.

USES Somewhat resembling rosewood, bubinga has a distinctive and often highly decorative appearance – especially where an irregular or wavy grain combines with its natural variation of colour. It is seen mainly as veneer and is used for inlay and similar decorative purposes. In the solid, it is used for knife handles and the backs of brushes and it is, though heavy, potentially an attractive furniture wood.

Guarea

Bubinga

Ovangkol

Mengkulang

Holly

Merbau

OVANGKOL
Guibourtia ehie

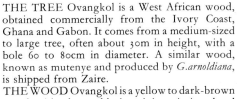

(×25)

THE TREE Ovangkol is a West African wood, obtained commercially from the Ivory Coast, Ghana and Gabon. It comes from a medium-sized to large tree, often about 30m in height, with a bole 60 to 80cm in diameter. A similar wood, known as mutenye and produced by *G.arnoldiana*, is shipped from Zaire.

THE WOOD Ovangkol is a yellow to dark-brown wood with almost black veining; it is related botanically to bubinga, but the two timbers are of quite different appearance. Ovangkol resembles walnut, but as it has a fairly regular stripe, it is more like Queensland walnut than European walnut. It is heavy, with a medium texture and an interlocked grain, and its appearance is sometimes marred by numerous small white flecks. Mutenye is similar to ovangkol in appearance, but is heavier, has a finer texture, and is sometimes more decorative, because of its wavy grain.

TECHNICAL PROPERTIES Ovangkol dries slowly but without distorting and, once dry, it is stable in use. It saws fairly easily but is somewhat difficult to machine, and cutters must be kept sharp to achieve a good finish. It is believed to be moderately resistant to fungi and highly resistant to termites.

USES Ovangkol is a comparatively recent introduction to the market and has been seen mainly as veneer. With an attractive, walnut-like appearance it has been used for furniture, cabinetwork and panelling. It makes a hard-wearing floor. Mutenye is used for similar purposes.

MENGKULANG
Heritiera spp.

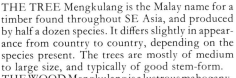

(×10)

THE TREE Mengkulang is the Malay name for a timber found throughout SE Asia, and produced by half a dozen species. It differs slightly in appearance from country to country, depending on the species present. The trees are mostly of medium to large size, and typically of good stem-form.

THE WOOD Mengkulang is a lustrous mahogany-red, variable from pink to red-brown in colour. It has a moderately coarse texture and typically an interlocked grain, which gives a stripe figure on quarter-sawn surfaces; it has a distinctive fleck, seen on accurately cut quartered surfaces, due to its fairly large rays. It is moderately dense, heavier than mahogany and about the weight of utile and West African niangon, which it closely resembles in physical appearance.

TECHNICAL PROPERTIES Mengkulang dries quickly and well; once dry, it is stable in use. It is a fairly strong wood and compares favourably with the West African red woods, such as utile and niangon. It is, however, more difficult to saw and machine, especially when dry, as it contains silica; it can be peeled to give an attractive veneer. It is rated moderately resistant to fungi.

USES Mengkulang has been exploited on an increasing scale in Malaysia, at first as sawn timber, but because of its abrasive character, it is more often peeled, and is nowadays one of the most familiar Malayan plywoods. In the solid, it has been used for carpentry, joinery, the framing of lorries and the floors of containers; as plywood, it is put to structural and general use.

HOLLY
Ilex spp.

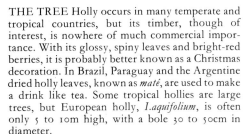

(×25)

THE TREE Holly occurs in many temperate and tropical countries, but its timber, though of interest, is nowhere of much commercial importance. With its glossy, spiny leaves and bright-red berries, it is probably better known as a Christmas decoration. In Brazil, Paraguay and the Argentine dried holly leaves, known as *maté*, are used to make a drink like tea. Some tropical hollies are large trees, but European holly, *I.aquifolium*, is often only 5 to 10m high, with a bole 30 to 50cm in diameter.

THE WOOD Holly is white or greyish-white and featureless; it has a fine, even texture and a straight or somewhat irregular grain, depending on the shape of the log. It is a heavy wood, about the weight of hornbeam.

TECHNICAL PROPERTIES Holly is difficult to dry, and, in order to prevent distortion, it is best cut to small sizes; once dry, it is prone to movement under changing conditions of humidity. It is a hard wood, and somewhat difficult to machine, especially if the grain is irregular, but with care it can be finished to a very smooth surface. It can be stained very effectively. It is not resistant to fungal attack.

USES Holly is available only in limited quantities, and, because it is difficult to dry and the grain is often irregular, it is usually converted to small sizes. It is used for fancy turned items, for engraving, and, perhaps surprisingly in view of its white colour, it is dyed black to substitute for ebony. It is an excellent wood for burning.

MERBAU
Intsia spp.

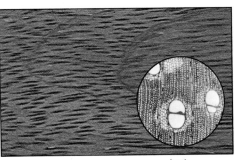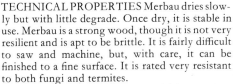

(×25)

THE TREE Merbau is the Malay name for a timber which occurs in mainland SE Asia, through the East Indies and the islands of the southwestern Pacific. It comes from two species of *Intsia*, trees of somewhat variable stem-form but often productive of large logs, 1 to 1.5m in diameter. Merbau is the Asian equivalent of African afzelia and both were for a long time included in the same genus.

THE WOOD Merbau is a dark, mahogany-red wood, like afzelia in appearance; it is handsome, if plain. It is course in texture and has a straight or, more usually, an interlocked grain, giving a stripe figure to quarter-sawn surfaces. It is a little heavier than afzelia and some 15 to 20 per cent heavier than oak. Some pores contain a yellow deposit used for making dyes.

TECHNICAL PROPERTIES Merbau dries slowly but with little degrade. Once dry, it is stable in use. Merbau is a strong wood, though it is not very resilient and is apt to be brittle. It is fairly difficult to saw and machine, but, with care, it can be finished to a fine surface. It is rated very resistant to both fungi and termites.

USES Merbau attracted commercial interest as a supplement to supplies of afzelia; it is especially popular in Holland. It is used for high-class joinery both indoors and outdoors, although exposure to rain may cause its red colour to leach out and stain stonework unless the wood is effectively protected. Merbau is a structural timber and makes a fine floor for all but the most exacting purposes. Selected timber makes attractive panelling.

Commercial Hardwoods

Walnut   African Mahogany

WALNUT
Juglans spp.

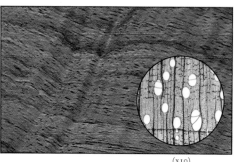

(×10)

THE TREE The name walnut is applied to a number of figured brown woods, but only those produced by species of *Juglans* are true walnuts. Walnut occurs in the warmer temperate regions of the northern hemisphere, extending through Central America to the Andean region of South America. Two timbers are of particular importance, European walnut, *J.regia*, which has been widely grown in Europe for a very long time but originated in the mountainous regions of SW Asia, and American black walnut, *J.nigra*, of the United States and Canada. Walnut is a medium-sized tree, 20 to 30m high; it is as important for its nuts as for its timber.

THE WOOD European walnut is grey-brown with almost black streaks and generally more variable in colour than American walnut, which is typically a uniform dark purple-brown. The grain is straight or occasionally wavy; texture medium. Walnut is a little lighter than beech in weight.

TECHNICAL PROPERTIES Walnut dries slowly, but once dry, it is moderately stable in use. It works easily and is noted for its excellent finish. It is moderately resistant to fungi.

USES Walnut is one of the world's outstanding decorative woods, long used for cabinetwork, and particularly associated with furniture of the Queen Anne period. Today, walnut is used in furniture and for decorative panelling mainly as veneer; it is used in the solid for fine joinery and for bowls and other turned items, and it is the preferred timber for the butts and stocks of rifles and guns.

AFRICAN MAHOGANY
Khaya spp.

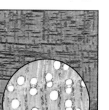

(×10)

THE TREE African mahogany first came into general use towards the end of the 19th century to supplement the limited supplies of true mahogany from tropical America. Though distinct, it is closely related botanically to the American wood, and is today universally accepted as mahogany. It is produced by five species, widely distributed from Portuguese Guinea to Angola and from the Sudan to Moçambique, but commercial timber is mainly *K.ivorensis* and *K.anthotheca*, shipped from West Africa.

THE WOOD Mahogany is pale-pink to red-brown, with a medium texture and typically an interlocked grain which gives a stripe figure. It is mostly light in weight, only a little heavier than European redwood.

TECHNICAL PROPERTIES Easy to dry and stable in use, Africa mahogany is noted for its good working and finishing properties, although care is needed with quarter-cut surfaces. It has moderate strength, but some timber from the centre of large logs is very brittle. It is sliced for veneer. It is moderately resistant to fungi.

USES Although demand for mahogany furniture is to some extent dictated by fashion, mahogany is in continuous use in reproduction furniture, office desks and in contract work, both in the solid for framing the sides of drawers, etc., and as decorative veneer. It is used for joinery, especially doors, and for shop and bank fittings. It is a popular boat timber, being light and moderately durable. It makes a useful general-purpose plywood.

LABURNUM
Laburnum anagyroides

(×25)

THE TREE Laburnum occurs naturally in parts of central and southern Europe, but has for long been cultivated elsewhere as a parkland and garden tree, notable for its pendulous racemes of brilliant yellow flowers. It is at best a small to medium-sized tree, rarely more than 10m high, often yielding only short lengths of small-diameter timber because it branches low down on the stem.

THE WOOD Freshly cut laburnum is bright-yellow with a greenish tint, but it darkens on exposure to golden-brown and finally to deep-brown. The coloured heartwood contrasts with the narrow, almost white sapwood. It is generally straight-grained with a fairly fine texture and a lustrous surface. Flat-sawn timber has a growth-ring figure and accurately sawn quarter-cut timber has a decorative fleck due to the ray tissue. It is dense, somewhat heavier than oak.

TECHNICAL PROPERTIES Only limited technical information is available for laburnum. It is said to dry readily and, though heavy, to work well with both hand and machine tools. The wood also takes a good finish and is reputed to be durable.

USES Laburnum is an attractive wood, but rarely seen because supply is difficult and because it generally comes in small sizes. It is an excellent wood for turning, and has been used for parts of musical instruments, especially bagpipes, and for knife handles and other handles. It is sliced to produce an attractive veneer, used for inlay; when cut across the end-grain it produces the "oyster" seen in period furniture.

AMERICAN RED GUM
Liquidambar styraciflua

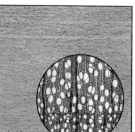

(×25)

THE TREE American red gum, sap gum or sweet gum is a timber of the eastern and southern parts of the United States, extending through Mexico to Central America. It comes from a medium-sized to large tree, 30 to 45m high, at its best in the southern states on low-lying land liable to flood. The tree produces a vanilla-scented gum or resin, storax, once used in medicine and perfumery.

THE WOOD The name red gum refers to the lustrous, reddish-brown heartwood, which is sometimes figured with darker markings, and well defined from the pale and typically wide sapwood, known as sap gum. Fine and even in texture, red gum often has an irregular and interlocked grain, giving a stripe figure to quartered surfaces. It is of medium weight, comparable to soft maple.

TECHNICAL PROPERTIES Red gum is a difficult wood to dry; it is liable to distort because of its interlocked grain. It is moderately hard and has good strength properties, especially shock-resistance and stiffness. Red gum works fairly easily, but because of its interlocked grain requires care in finishing. Its resistance to fungi is low.

USES Red gum is used extensively and almost exclusively in the United States, though once it was an important timber in overseas markets. Before World War II it was popular in Britain, where the heartwood was known as satin walnut, and the sapwood as hazel pine. Today, in America, it is used in furniture, interior joinery, dry cooperage and for boxes, crates and pallets. It is made into plywood and it is used for pulp.

Laburnum

American Red Gum

American Whitewood

Ekki

African Walnut

Magnolia

AMERICAN WHITEWOOD
Liriodendron tulipifera

THE TREE American whitewood, or yellow-poplar as it is known in America, is produced by the tulip tree, familiar as a parkland tree in many countries besides America. It is important over much of the eastern United States, and particularly common in the southern Appalachian regions of North Carolina and Tennessee. It comes from a medium-sized to large tree, up to 60m in height and typically of good stem-form.

THE WOOD Whitewood is variable in colour; much is white, especially the wide sapwood from young trees of vigorous growth, but the heartwood, especially from older trees, is pale-brown and often streaked with green, black, blue or red. Whitewood is fine and even in texture, straight-grained, and with little figure other than that due to colour variation. It is moderately light in weight, especially if the timber is from old trees, when it is about the weight of cottonwood and aspen.

TECHNICAL PROPERTIES Whitewood dries readily, and when dry is stable in use. Soft and low in strength, especially in its resistance to bending and to shock, it saws and machines well to give a smooth finish. It should be treated for outdoor use.

USES Shipped in large quantities in the 1920s and 1930s, selected pale wood, given the name whitewood, was used in Britain for interior joinery, for the sides of drawers, and for linings in furniture. Nowadays, it is rarely seen outside America, but there it is popular for furniture, joinery, packaging and pallets, and for dry cooperage. It is also used for pulp and wood-flour.

(x25)

EKKI
Lophira alata

THE TREE Ekki, often more familiar as azobé, is a West African timber occurring from Sierra Leone to Gabon. It is particularly important in Cameroon and is shipped mainly from that country. The wood comes from one of the largest African trees, 50 to 60m high, with a clear, straight stem suitable for the production of long lengths of timber in large sizes.

THE WOOD Ekki has a distinctive appearance, being a dark red-brown or purple-brown with many very fine, almost white flecks associated with its vessel-lines. It has a coarse texture and a typically interlocked and sometimes irregular grain; it is an exceptionally dense wood, of the same order of weight as greenheart and about 50 per cent heavier than oak.

TECHNICAL PROPERTIES Ekki is difficult to dry as it is liable to split and to distort. Being so heavy, it is extremely strong and very difficult to work. It can be sawn when green, but dry wood blunts saws rapidly. It cannot be nailed or fastened unless it has been bored. It is extremely durable, being resistant to fungi, termites and marine borers.

USES Ekki is an outstanding timber for structural use, very popular in France and elsewhere in Europe, but not so common in Britain, where greenheart is preferred. Available in large sizes, it is used for harbour work, piers, jetties, piling and railway sleepers. An unusual use is for the running track of the rubber-wheeled trains on the Paris Métro.

(x25)

AFRICAN WALNUT
Lovoa trichilioides

THE TREE African walnut is so called for its resemblance to true walnut, though botanically it is related to mahogany, not to walnut. It is a West African timber, obtained from a large tree up to 40m high, and shipped from the timber-producing countries from the Ivory Coast to Gabon. Other species of *Lovoa* occur in East Africa, but they produce a somewhat plainer wood.

THE WOOD In its grain, texture and weight, African walnut is like African mahogany, but it has a warm, golden-brown colour. It is commonly marked with long, straight, black streaks or veins, which give the wood some resemblance to walnut, more especially on flat-sawn surfaces, as a typically interlocked grain produces a stripe figure when exposed on quarter-cut surfaces.

TECHNICAL PROPERTIES African walnut can be dried quickly and well, and once dry is stable in use. It is comparable to African mahogany in strength and appreciably easier to cut and machine than true walnut, though care is needed to avoid tearing because of the interlocked grain. It is moderately resistant to fungi but not to termites.

USES African walnut is the most readily available of the so-called walnuts and comparatively cheap. Though not particularly decorative, it sufficiently resembles walnut to be used in the solid in furniture, joinery and shop fittings, often in conjunction with a more highly figured walnut, used in veneer to cover the larger surface areas. It is popular for turned items, such as bowls, platters and lamp standards.

(x25)

MAGNOLIA
Magnolia spp.

THE TREE Though more than 70 species of *Magnolia* occur in eastern Asia, the United States and Central America, only three American species are important commercially for their timber; many others, often shrubs or small trees, are of horticultural interest for their decorative flowers. The three American species are known by distinctive names, cucumber-tree, southern magnolia and sweetbay, or collectively as magnolia. They are medium-sized to large trees of the eastern and southern United States.

THE WOOD Magnolia is a pale wood, yellow to brown in colour according to the species, and typically with a very fine, even texture. The grain is straight and the only figure is a growth-ring pattern on flat-sawn surfaces.

TECHNICAL PROPERTIES Magnolia dries satisfactorily but with moderate shrinkage. When dry, it is moderately hard and stiff, has good shock-resistance, but is below average in its compression and bending strengths. It can be machined to a good finish and turns well; it nails without splitting. It is not resistant to fungi and should not be used out of doors unless it has been treated.

USES Magnolia is a plain, featureless wood which works well and is used for the framing and lining of furniture, for packaging and pallets, for interior joinery, including doors, and for venetian blinds. It is sliced for veneer. In those parts of its range where it is available in only modest quantities, it is often marketed with similar woods, especially American whitewood, or yellow-poplar.

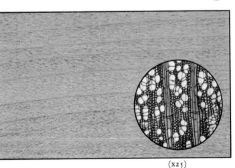

(x25)

Commercial Hardwoods

Apple   Mansonia

APPLE
Malus sylvestris

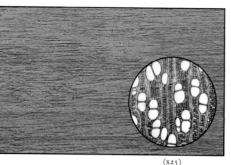

(x25)

THE TREE Better known and commercially far more important for its fruit than for its timber, apple is a tree of Europe and western Asia nowadays cultivated in many parts of the world. At best it is a small tree, 8 to 10m high, often with a misshapen stem. There is little difference between the wood of wild and crab apple trees and that produced by the many cultivated varieties.
THE WOOD Apple is a pale to medium pinkish-brown wood, with a very fine and even texture, though not quite so fine as that of pear; in misshapen trees the grain is often spiral and distorted. Apple is a moderately heavy wood, about the same weight as beech.
TECHNICAL PROPERTIES Apple is slow to dry and in drying tends to warp badly and to split;

once dry, it is said to be stable in use. It is a hard and strong wood with a reputation for toughness, and it is difficult to split. It saws well and can be machined to give a good finish, though it is somewhat hard to work; care is needed to avoid tearing the surface if the grain is irregular. It takes stains and polishes well. It is perishable in conditions favouring decay.
USES Its limited supply and small size combine to make apple a wood for craft rather than industrial use. It turns extremely well and, with a fine texture, can be carved to show intricate detail. It has been used for cog wheels, wooden screws, shuttles, golf-club heads, and for carpenters' tools, such as planes, mallets and the handles of saws. It makes an attractive decorative inlay.

MANSONIA
Mansonia altissima

(x25)

THE TREE Mansonia is a West African wood obtained mainly from the Ivory Coast, Ghana and Nigeria. It comes from a tree of modest size up to 30m in height, giving logs 60 to 80cm in diameter.
THE WOOD Mansonia is a medium-brown to dark-brown wood, often with a grey or purple tint, and sometimes faintly striped; on exposure it can lose its dark colour and may after a long time become a pale yellow-brown. It has a fine texture and a straight or interlocked grain; it is intermediate in weight between mahogany and walnut.
TECHNICAL PROPERTIES Mansonia dries quickly with little degradation apart from a tendency to split; once dry, it is moderately stable in use. For its weight it is a strong wood, comparing well with beech, except in stiffness. It saws easily

and machines to a good finish, but is an unpleasant wood to process, especially when dry, as its fine dust commonly causes irritation to the skin, eyes, nose and throat. In the factory special attention must be given to the facilities for dust extraction, and the use of face masks and barrier creams is recommended. It is a very durable wood: it is resistant to fungi and reported to be highly resistant to termites.
USES Mansonia is an attractive if fairly plain walnut-like wood which would find wider use but for the discomfort experienced by many operatives in its processing. With care, these problems can be minimized and then the wood is used for high-class joinery and furniture, for radio and television cabinets and, in cars, for fascias and other fittings.

ZEBRANO
Microberlinia brazzavillensis and *M. bisulcata*

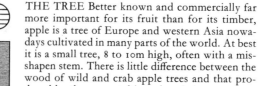

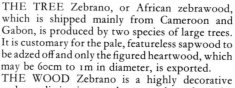

(x10)

THE TREE Zebrano, or African zebrawood, which is shipped mainly from Cameroon and Gabon, is produced by two species of large trees. It is customary for the pale, featureless sapwood to be adzed off and only the figured heartwood, which may be 60cm to 1m in diameter, is exported.
THE WOOD Zebrano is a highly decorative and very distinctive wood, straw-coloured or pale-brown with many dark-brown or almost black stripes. Its appearance varies with the log and according to the method of conversion: on quartered surfaces it has typically a regular and parallel stripe and on flat or rotary-cut surfaces a far less regular pattern. Normally the grain is interlocked and the texture is coarse. It is a moderately dense wood, somewhat heavier than oak.

TECHNICAL PROPERTIES Zebrano is difficult to dry as it is liable to twist. When used in the solid, it should be quarter-sawn to keep distortion to a minimum. It has good strength properties and is noted for its shock resistance. It saws readily and can be worked with hand and machine tools to give a good finish; it slices and peels well to give a good but sometimes fragile veneer. It is believed to be resistant to fungi.
USES Zebrano is mainly seen in the form of sliced veneer, which is used for decorative inlay in furniture and cabinetwork and sometimes for panelling. Solid wood is suitable for brush-backs, handles and small turned items and, because it has good shock resistance, it has been suggested it might be used for skis and tool handles.

PANGA PANGA/WENGÉ
Millettia spp.

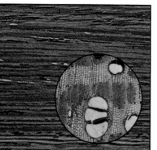

(x25)

THE TREE Species of *Millettia* occur in many tropical countries, but the only timbers of commercial interest are panga panga, *M.stuhlmanii*, from Moçambique, and wengé, *M.laurentii*, from Zaire, Cameroon and Gabon. Both timbers come from trees of only modest size, reaching 18m in height and giving logs 50 to 60cm in diameter.
THE WOOD Panga panga and wengé have similar properties; both are brown to almost black with a distinctive appearance due to fine lines of paler tissue. They are coarse and somewhat variable in texture, but the grain is usually straight. Both are dense woods, panga panga being about 10 per cent and wengé some 20 per cent heavier than oak.
TECHNICAL PROPERTIES Though drying slowly both timbers dry well with little degrada-

tion; once dry, they are very stable in use. They are strong woods, noted for their shock resistance, which is said to compare favourably with that of hickory. Dense and coarse in texture, they cause fairly rapid blunting of cutting tools, which must be kept sharp to obtain a good finish; both turn well. They can be sliced, but are not usually rotary-peeled. Both are believed to be resistant to fungi.
USES These are highly decorative woods which have attracted some interest, particularly in Europe, for cabinet-making and high-class joinery and, in veneer form, for decorative panelling. They make attractive and particularly stable flooring, but are better suited to normal pedestrian than to exacting use because of their coarse texture. Wengé has been used for small turned items.

Zebrano

Panga Panga/Wengé

Abura

Opepe

Southern Hemisphere Beech

Tupelo

ABURA
Mitragyna ciliata

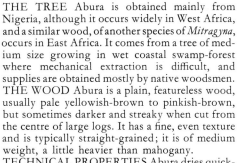

(x25)

THE TREE Abura is obtained mainly from Nigeria, although it occurs widely in West Africa, and a similar wood, of another species of *Mitragyna*, occurs in East Africa. It comes from a tree of medium size growing in wet coastal swamp-forest where mechanical extraction is difficult, and supplies are obtained mostly by native woodsmen.
THE WOOD Abura is a plain, featureless wood, usually pale yellowish-brown to pinkish-brown, but sometimes darker and streaky when cut from the centre of large logs. It has a fine, even texture and is typically straight-grained; it is of medium weight, a little heavier than mahogany.
TECHNICAL PROPERTIES Abura dries quickly and with little degradation; once dry, it is stable in use. It has modest strength properties, generally comparable to those of mahogany, but because of its straight grain it tends to be stronger than mahogany in small dimensions. Its sawing and machining properties vary; it works fairly easily, but some timber is siliceous and is abrasive to cutting tools. It has a low resistance to fungi and insects and is unsuitable for outdoor use.
USES Abura is a plain but nevertheless popular wood, used as much for its uniformity of character as for any technical merit, although its stability in use makes it attractive for parts of furniture such as framings, lippings, drawer sides, etc., and it can be stained to match more decorative woods. It is a useful wood for interior joinery and, because of its fine texture and straight grain, it is excellent for mouldings.

OPEPE
Nauclea diderrichii

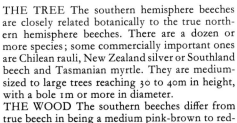

(x10)

THE TREE Opepe is obtained from the principal West African timber-producing countries from the Ivory Coast to Zaire. It comes from a large tree, which often reaches 40m in height with a bole 1.5m in diameter.
THE WOOD Opepe is a very distinctive bright orange-yellow colour, sometimes with pinkish streaks. It has a coarse texture and a grain which may be straight, interlocked or irregular in direction. It is about the same weight as oak.
TECHNICAL PROPERTIES Opepe dries slowly and often develops surface splitting, especially if it is dried in large sizes or if irregular grain is present. Once dry, it is stable in use. It is a strong wood, though when used in small sizes its mechanical performance depends very much on the incidence of cross-grain. It saws and machines well, considering its weight, but care is needed in finishing quartered surfaces with interlocked grain. It is resistant to fungi and marine borers and moderately resistant to termites.
USES A heavy, strong, durable wood, opepe is available in large sizes and long lengths and is used for structural purposes in shore and riverbank defence, in lock, harbour and dock construction and for piling. It is an excellent timber for wharf and pier decking and it has been used for boat decking, but is prone to surface checking. It is used in place of oak for the structural parts of fishing boats and, in many countries, for railway sleepers. It is suitable for exterior joinery and makes an attractive domestic floor.

SOUTHERN HEMISPHERE BEECH
Nothofagus spp.

THE TREE The southern hemisphere beeches are closely related botanically to the true northern hemisphere beeches. There are a dozen or more species; some commercially important ones are Chilean rauli, New Zealand silver or Southland beech and Tasmanian myrtle. They are medium-sized to large trees reaching 30 to 40m in height, with a bole 1m or more in diameter.
THE WOOD The southern beeches differ from true beech in being a medium pink-brown to red-brown, lacking the distinctive ray-figure of true beech. They have a fine, even texture and generally a straight grain. They vary in weight from moderately light to moderately heavy: rauli and silver beech are about the weight of mahogany, Tasmanian myrtle the same weight as true beech.
TECHNICAL PROPERTIES The less heavy woods dry readily and well, but Tasmanian myrtle requires some care to prevent checking and collapse. Strength properties vary with weight, Tasmanian myrtle comparing well with beech, but rauli and silver beech are appreciably weaker. All are readily sawn. Silver beech and Tasmanian myrtle are rated not resistant to fungi, but rauli has a reputation for durability.
USES Except where high strength is required the southern beeches are suitable for many of the purposes for which true beech is used. They are attractive furniture woods and are used for joinery, flooring, mouldings and turnery; the Australian and New Zealand timbers are used for boxes and crates in the dairy products industry.

TUPELO
Nyssa aquatica & *N.sylvatica*

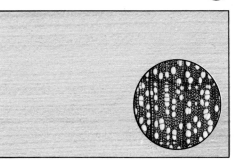

(x25)

THE TREE Tupelo is an American wood from the southern and eastern parts of the United States. It comes mainly from two species, sometimes marketed separately as tupelo gum, *N.aquatica*, and black gum, *N.sylvatica*. Both are tall trees, up to 30m high, with a straight bole about 1m in diameter, but buttressed at the base, especially tupelo gum, which is a swamp-forest tree.
THE WOOD Tupelo is yellow to pale-brown, occasionally with darker patches but generally somewhat featureless. It has a fine, even texture but an irregular and often interlocked grain. It is moderately light in weight, especially tupelo gum, which tends to be softer and more open in texture than black gum; wood from the swollen base of the tree can be exceptionally light.
TECHNICAL PROPERTIES Tupelo dries readily but has a marked tendency to distort. It is not a strong wood, but black gum especially is noted for its resistance to splitting. Though soft and light, it can be somewhat difficult to work and care is needed to produce a good finish because of its irregular grain; it nails well. It is not durable in conditions favouring decay.
USES Tupelo is a general-purpose wood used in the solid, but is often also rotary-peeled for veneer and plywood. It is used in furniture and for interior joinery, mouldings and packaging, and it makes a hard-wearing floor. It is particularly resistant to splitting and has been used in the United States, after preservative treatment, for railway sleepers. It is used as a plywood core veneer.

Commercial Hardwoods

Balsa   Greenheart

BALSA
Ochroma pyramidale

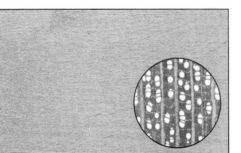

(x10)

THE TREE Balsa occurs widely in tropical America and has been planted elsewhere in the world, but the main supply and best commercial wood is shipped from Ecuador. The tree grows very rapidly, reaching 20m in height, with a bole 60cm in diameter, in five to six years.

THE WOOD Balsa is the lightest wood in general use; nevertheless, it varies appreciably in weight and commercial timber is usually selected to be in the density range 100–250 kg/m³ (6–15 lb/cu. ft) when dry, or about 1/7th to 1/3rd the weight of a piece of beech or oak of the same size. The best wood is white, sometimes with a slight pinkish tint, and straight-grained and highly lustrous with a soft, velvety feel. It has a moderately coarse but even texture.

TECHNICAL PROPERTIES Balsa is easily dried and, when dry, is a stable wood. It has low strength but is firm; it is easy to work, but tools must be kept sharp to give a good finish. It is noted for its low thermal conductivity and its high sound-absorption. It is perishable.

USES Balsa is an important industrial wood though probably better known for its use in model-making and for theatrical props. It is used in very large quantities as insulation in ships carrying liquid gas; for cold stores and similar refrigeration purposes; for sound absorption; and for buoyancy articles such as life-saving equipment. The Kon-Tiki raft was made of balsa logs. As core stock in metal-faced sandwich construction it has been used for aircraft floors and partitions.

GREENHEART
Ocotea rodiaei

(x25)

THE TREE Greenheart is the outstanding commercial wood of Guyana. It comes from a medium-sized to large tree, up to 40m high, with a long, straight, cylindrical bole up to 1m in diameter.

THE WOOD Greenheart is pale yellow-green to dark olive-brown in colour, sometimes with darker, almost black markings. It has a fine, even texture and a straight or interlocked grain. It is a very dense wood, about 50 per cent heavier than oak, and will not float in water, even when dry.

TECHNICAL PROPERTIES Greenheart dries very slowly and is liable to split and check, particularly in thicker sizes. It is exceptionally strong, even allowing for its weight, with almost twice the strength of oak as regards bending, compression and stiffness. It is hard to saw and difficult

to machine; greenheart splinters have a reputation for causing septic wounds. It is exceptionally resistant to both fungus and marine borer attack.

USES With its straight, cylindrical stem, greenheart is available in long lengths of up to 20m and in large sizes up to 600 × 600mm in cross section. It is commonly imported as hewn baulks and sawn timber. Combining great strength with durability and resistance to abrasion, it is used in large sizes for littoral defence, for piles and for lock gates. Sawn timber is used for structural parts – struts, bearers, walings and decking – of wharves, piers and jetties. Greenheart is also used as bearers for engines, for heavy-duty floors and for vats and filter presses in the chemical industry. It makes a very tough fishing rod.

OLIVEWOOD
Olea spp.

(x25)

THE TREE Olivewood is obtained from the olive tree, *O.europaea*, grown in the Mediterranean countries mainly for its edible fruit and as a source of olive oil. Other species occur in parts of Africa, and East African olive, *O.hochstetteri*, and loliondo, *O.welwitschii*, are important timbers in Kenya and Tanzania. European olive is usually a small and misshapen tree; African olive can reach 25m in height but it, too, is often of poor stem-form.

THE WOOD Olivewood has a handsome appearance: it is pale-brown to medium-brown in colour with darker, grey or black markings. The texture is fine and the grain sometimes straight but often irregular, or, in East African olive, shallowly interlocked. Loliondo is plainer in appearance and a

little lighter in weight than both European and East African olive, which are on average about 20 per cent heavier than beech.

TECHNICAL PROPERTIES Olivewood dries slowly with some tendency to check and split; it has good strength and, combining high density and fine texture, is noted for its resistance to abrasion. It is fairly hard to saw, but works well with hand and machine tools; it can be finished to a very smooth surface and stains and polishes well. It is believed to be moderately resistant to fungi.

USES European olive is seen mainly as small turned and carved items for sale to tourists in southern Europe. African olive is also used for carved items, but is a fine flooring wood, having good appearance and excellent wearing properties.

LANCEWOOD
Oxandra lanceolata

(x25)

THE TREE Lancewood, also known as asta in the United States, is a West Indian wood obtainable in small quantities on Jamaica, Cuba and Hispaniola. It comes from a small tree and is traditionally marketed in short lengths, called spars, from 3 to 5m long and 12 to 18cm in diameter. Another type of lancewood known as degamé, or as lemonwood in the United States, is obtained from parts of Central and South America.

THE WOOD Lancewood has a dark heartwood, but commercial preference is for the pale, creamy-yellow sapwood, which has some resemblance to a dull satinwood. The wood is straight-grained with a very fine and even texture. It is very dense and almost as heavy as greenheart, making it one of the world's heaviest timbers.

TECHNICAL PROPERTIES Only limited information is available on the technical behaviour of lancewood. It is noted for its good strength properties, especially its resilience and its ability to absorb energy. It is hard to saw and difficult to machine because of its weight, but it can be finished to a smooth surface and is noted for its good turning properties. It is not resistant to fungi.

USES Long used as a special-purpose wood, lancewood, when available nowadays, is only obtainable in small quantities. It was once used for wheel spokes and carriage shafts because of its resilience, and also for fishing rods, billiard cues, archery bows and drumsticks. Degamé, which is similar in character but not quite so dense, is more readily available and used for the same purposes.

Olivewood

Lancewood

Tchitola

White Seraya

Purpleheart

Afrormosia

TCHITOLA
Oxystigma oxyphyllum

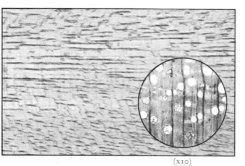

(×25)

THE TREE Tchitola, shipped mainly from Nigeria and Angola, comes from a large tree, commonly 40m in height, with a straight, cylindrical bole 1m or more in diameter. In the forest it is liable to be confused with agba, and in Angola the two timbers have a common name, tola. However, tchitola is usually distinguished as tola mafuta and agba as tola branca or white tola.

THE WOOD Tchitola is straw-yellow to medium-brown with darker veins or streaks giving some resemblance to walnut. Like agba, it is often somewhat gummy. The grain is straight or shallowly interlocked and occasionally wavy, when a fiddle-back figure enhances its decorative colour. It has a moderately coarse texture and medium weight, a little lighter than walnut.

TECHNICAL PROPERTIES Tchitola dries well with little degradation and once dry is stable in use. Its strength properties compare favourably with those of walnut, and it saws and machines well, apart from some tendency for gum to build up on cutting edges; it peels and slices to give a good veneer. It is normally moderately resistant to fungi, but occasional pieces have a less satisfactory performance.

USES Tchitola has been used in Europe as a general-purpose brown wood, but more especially for the decorative surfaces of furniture and television cabinets in veneer form, when it is selected for its figure. In the past, it has been a popular general-purpose plywood in South Africa, made from logs shipped from Angola.

WHITE SERAYA
Parashorea malaanonan

(×10)

THE TREE White seraya is a Sabah timber which also occurs elsewhere in Borneo and in the Philippines; in the Philippines it is known as bagtikan, but is sometimes shipped, mixed with similar woods, as white lauan. White seraya comes from a very large tree, up to 60m high, with a straight cylindrical bole 1.5m or more in diameter, above large buttresses. Sabah seraya is in general equivalent to Malayan meranti, but white seraya is quite different from white meranti. Like white seraya, but somewhat heavier, is Malayan gerutu.

THE WOOD White seraya is straw-yellow or a pale pinkish-white, with a moderately coarse texture and a shallowly interlocked grain, giving a wide stripe on quartered surfaces. In weight it is comparable to light-red meranti and mahogany.

TECHNICAL PROPERTIES White seraya is easily and quickly dried and, when dry, is stable in use. It compares closely in strength with light-red meranti and mahogany, and it presents little difficulty in sawing and machining. It can be rotary-peeled to give a good veneer. It has low resistance to fungal attack.

USES White seraya is an attractive, if fairly plain and coarse-textured wood. It first attracted attention as a timber for ships' decking, due to its pale colour, light weight and availability in fairly long lengths. However, it lacks the durability of teak and is best used under cover. It makes an attractive domestic floor, and is suitable for interior joinery, furniture framing and light structural uses. It is used as general-purpose plywood.

PURPLEHEART
Peltogyne spp.

THE TREE Purpleheart, also known as amaranth, is a wood of northern South America, especially the Guianas and Brazil. It is produced by about 20 species, many of which are large trees 30 to 45m high, yielding logs 1m or more in diameter.

THE WOOD When first cut, Purpleheart is a dull brown, but on exposure rapidly becomes a bright purple and is then among the most distinctive and vividly coloured of commercial woods. The purple colour fades on prolonged exposure and the wood tones to a rich brownish-red. The texture varies according to species from fine to moderately coarse; the grain is sometimes straight or may be interlocked and is occasionally wavy. Purpleheart is a heavy wood, varying in weight according to species, but not quite so dense as greenheart.

TECHNICAL PROPERTIES Purpleheart dries well with little degradation, but rather slowly in thick sizes; once dry, it is stable in use. It is a very strong wood and, because of its density, it is hard to saw. It causes fairly rapid blunting of tools when it is machined, though it can be finished to a good surface. It is very durable.

USES Purpleheart, though brightly coloured, is only rarely used for its decorative effect and then mainly for small turned items, such as the butts of billiard cues, or as veneer for inlay. Its more usual use is in heavy construction work, for dock, bridge and pier work and for similar purposes demanding good strength and durability. It is used for chemical vats and for filter press frames and makes a hard-wearing floor.

AFRORMOSIA
Pericopsis elata

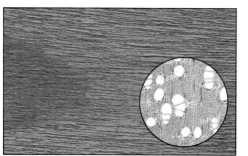

(×25)

THE TREE Afrormosia is obtained mainly from the Ivory Coast and Ghana, but it also occurs in Zaire. It comes from a tall tree, up to 40m high, with a long but often irregular bole 1m or more in diameter.

THE WOOD Afrormosia is yellow-brown to brown and has some resemblance to teak; it is about the same weight but finer in texture, it commonly has an interlocked grain giving a stripe figure, and it lacks the leathery smell and greasy feel of teak. In moist conditions, it reacts with iron fixings, which cause the wood to blacken. TECHNICAL PROPERTIES Afrormosia dries slowly but well, with little degradation, and once dry is very stable in use. It is a strong wood and rather better than teak in most strength properties.

It saws and works well, taking a fine finish, although care is needed with quarter-sawn surfaces because of its interlocked grain. It is a very durable wood, resistant to both fungi and termites. USES Afrormosia is a strong, stable, durable timber. When it was first marketed in the years following World War II, it found immediate acceptance for many of the purposes for which teak is used, especially in ship- and boat-building and high-class joinery. While teak furniture was fashionable, afrormosia was often used for solid framing, but it soon became accepted as a decorative wood in its own right. It is widely used nowadays for cabinetwork and chairs, both in the solid for structural parts and as veneer for large decorative surfaces. It makes an attractive and stable floor.

Commercial Hardwoods

Snakewood

Plane

SNAKEWOOD
Piratinera guianensis

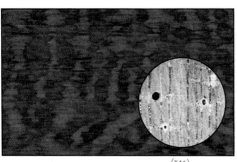

(x25)

THE TREE Snakewood, or letterwood, is a tropical American wood, which has always been exported only in small quantities, mainly from the Guianas. It is nowadays of mainly historical interest, as one of the heaviest woods to have been used commercially, exceeding even lignum vitae in weight. It came from a tree of medium size, up to 25m high, with a bole sometimes 80cm in diameter, but as only the figured heartwood was of commercial interest, the wide band of pale sapwood was cut off and the timber shipped in small billets, 1 to 2m long and 7 to 18cm in diameter. THE WOOD Snakewood is a distinctive wood, deep mahogany-red with irregularly shaped dark markings, resembling the markings on certain snakes or hieroglyphic characters – hence its names,

snakewood and letterwood. The grain is typically straight and the texture fine and even; it is very dense, about 80 per cent heavier than oak.
TECHNICAL PROPERTIES Only limited information is available. Snakewood is strong but brittle and tends to splinter when worked; it splits readily along the grain. It is hard to cut, but can be finished to a smooth surface and takes a high natural polish. It is highly resistant to fungi.
USES At one time snakewood was in demand for walking sticks, and highly figured pieces commanded a very high price; it was also used for umbrella handles and, as sawn veneer, for inlay and other decorative work. It was used for bows for classical stringed instruments such as viols, but for violin bows it is not the equal of brazilwood.

PLANE
Platanus spp.

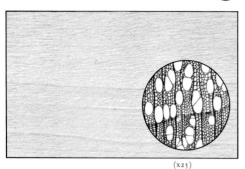

(x25)

THE TREE The eastern plane, *P.orientalis*, of SE Europe, Turkey and Iran, and the western plane, *P.occidentalis*, of the eastern United States, called by the Americans buttonwood or sycamore, have hybridized to produce the European plane, *P.hybrida*, the familiar urban landscape tree. The American and European planes are tall trees giving large logs; the eastern plane is smaller.
THE WOOD Plane is a pale-brown wood with large rays clearly visible on flat-sawn surfaces; when accurately quarter-cut, the rays give the wood a very distinctive fleck-figure and it is then known as lacewood. Plane has a fine, even texture and usually a straight grain. European plane is about 15 per cent lighter in weight than beech, and American wood lighter again.

TECHNICAL PROPERTIES Plane dries readily but has a tendency to distort. It has moderate strength properties generally comparable to those of mahogany, but it is not quite as stiff. It saws well, apart from a slight tendency to bind, and it machines easily to give a good finish, if cutters are kept sharp to avoid tearing ray tissue on quartered stock. It is perishable.
USES European plane is usually seen as quartered veneer used for panelling, e.g., in railway carriages, or as inlay or in marquetry. Sawn wood is sometimes used for such general purposes as handles for brushes and non-striking tools. American plane is more readily available and is used in furniture and for pallets, packaging, butchers' blocks and, as veneer, for panelling and door skins.

POPLAR
Populus spp.

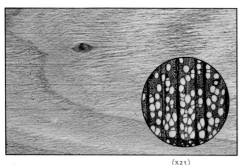

(x25)

THE TREE A large number of species and cultivated varieties occurring in many temperate countries, and often extensively planted, produce poplar. Their woods are similar though sometimes known by a distinctive name, e.g., aspen. In North America the timber is known as cottonwood. Poplars are medium-sized to large trees, some reaching 30m in height; often of vigorous growth they are among the most productive temperate hardwoods.
THE WOOD Poplar is white, sometimes with a pink or brown tint; it is generally straight-grained, with a fine, even texture. It is light in weight, comparable to spruce, but aspen may be firmer and heavier.
TECHNICAL PROPERTIES Poplar dries readily and is moderately stable in use. Though not strong,

it is tough for its weight and does not easily split or splinter. It saws and works easily, but with a tendency to a woolly surface, and tools must be kept sharp. It peels to give a good veneer. It is rapidly attacked in conditions favouring decay.
USES Poplar is an important commercial timber. It is used as sawn wood for joinery and light structural work, for flooring, for boxes and crates and for many small domestic articles, such as wooden shoes and kitchen utensils. Poplar veneer is used for fruit baskets and to make a general purpose plywood; and for matches, which are cut from veneer, aspen is the timber most commonly used. Poplar is an important pulp wood, especially in America, and is used for wood wool and a board product made of fairly large, thin flakes.

CHERRY
Prunus spp.

(x25)

THE TREE Cherry is a northern temperate genus with two timbers of special interest, European cherry, obtained through much of Europe and into Asia Minor, and American black cherry of the eastern United States. European cherry, though fairly common, is a tree of only modest size. American cherry is somewhat larger, reaching 30m in height with logs commonly 35 to 50cm in diameter.
THE WOOD Cherry is a fine-textured, generally straight-grained wood, which is pale pinkish-brown when first cut but darkens on exposure, and after prolonged use can become almost mahogany-red. Its appearance is occasionally marred by gum-streaks and flecks. European cherry is marginally heavier than American and intermediate in weight between mahogany and walnut.

TECHNICAL PROPERTIES Cherry needs care in drying as it tends to distort, but, once dry, is moderately stable in use. It has good strength for its weight though it is not particularly stiff. It saws easily and can be worked with both hand and machine tools to give an excellent finish. It is of uncertain durability, but is probably best not used in conditions favouring decay.
USES Cherry is a decorative wood used for furniture and cabinetwork. European cherry is available in only limited quantities and tends to be used for special items of furniture. American cherry is more readily available and of commercial importance in the middle-Atlantic states. It is used for panelling, furniture, high-class joinery and as the backing blocks for electrotype plates.

Poplar

Cherry

Muninga

Padauk

Pterygota

Pear

MUNINGA
Pterocarpus angolensis

(x10)

THE TREE Muninga is an East African wood occurring southwards from Tanzania to the northern parts of the Republic of South Africa. It comes from an "open-forest" tree of small to medium size and variable stem-form. Botanically it is a padauk, i.e., a species of *Pterocarpus*, but it has long been known by its distinctive name.

THE WOOD Muninga is variable in colour from golden-brown to deep-brown, sometimes with a reddish tinge and with irregular darker markings. Its attractive appearance is sometimes marred by white spots or blotches. It has a medium texture and a grain which varies from straight to irregular depending on the form of the log. Its weight varies with its growth; at its lightest it is comparable to mahogany, but is usually comparable to walnut.

TECHNICAL PROPERTIES Muninga dries slowly but well, with a very small shrinkage and little distortion; when dry, it is quite exceptionally stable in use, and superior in this respect to most other commercial timbers. It is of variable strength according to its density, but generally low in shock resistance and especially stiffness. It saws easily, works well and takes a fine finish. It is very resistant to both fungi and termites.

USES Muninga is a fine wood with outstanding technical properties. It is used for high-class joinery and, in both veneer and solid form, for furniture and panelling. It makes an attractive floor for domestic use and is especially suitable over under-floor heating systems. It is a highly valued wood in South Africa.

PADAUK
Pterocarpus spp.

(x10)

THE TREE Timbers known as padauk come from West Africa, mainly Cameroon and Nigeria, and from the Andaman Islands; they were formerly of commercial importance from Burma. A highly figured East Indian padauk is known as amboyna, and East African wood as muninga. Typically the padauks are medium-sized to large trees.

THE WOOD Padauks are brightly coloured, especially African padauk, which is purple-red when freshly sawn but darkens to a deep purple-brown; Andaman padauk is crimson-brown to red-brown with darker markings and not quite so vivid. Typically they are coarse in texture with an interlocked grain and are heavy or moderately heavy.

TECHNICAL PROPERTIES Padauks dry slowly but well and, when dry, are noted for their exceptional stability in use. They are strong; Andaman padauk is superior in its strength properties to oak. Though heavy, they are not unduly difficult to saw and can be machined to give a fine finish. Padauks are noted for their high resistance to decay.

USES The padauks combine a fine appearance with excellent technical properties. Andaman padauk in particular was once popular for high-class joinery, especially shop and bank counters and fittings. It is an excellent boat-building wood: *Lively Lady*, the boat used by Sir Alex Rose for his single-handed journey round the world, was made of padauk. African padauk, today more readily available than the Andaman wood, is used in joinery, for tool handles and spirit levels, and as hard-wearing and exceptionally stable flooring.

PTERYGOTA
Pterygota spp.

(x10)

THE TREE Pterygota occurs in Africa and Asia, but commercial supplies come from the West African timber-producing countries. Two species occur in Nigeria with woods which are similar and are not distinguished commercially; they are trees of medium size, 30 to 35 m in height, with boles 1 m in diameter above the large buttresses.

THE WOOD Pterygota is creamy-yellow to almost white, with a coarse texture and straight or shallowly interlocked grain. It is characterized by an attractive fleck on accurately quarter-cut surfaces due to its large rays. It varies in weight according to species, from a little heavier than mahogany to about the weight of birch.

TECHNICAL PROPERTIES Pterygota dries fairly rapidly and with little distortion; once dry it is moderately stable in use. It is variable in strength depending on its density, with the heavier wood not quite as strong as beech and the lighter wood comparing more nearly to mahogany in strength. It saws and machines satisfactorily, if with some tendency to a woolly finish, and can be sliced and peeled to give an attractive veneer. It has a low resistance to fungi.

USES Pterygota is a white wood very prone to stain blue or grey if conversion and drying are delayed, and for this reason it was for some years difficult to market successfully. Nowadays, with improved handling giving clean wood, pterygota is attracting increasing interest, both as a sliced veneer and for use in the solid in furniture and for interior joinery.

PEAR
Pyrus communis

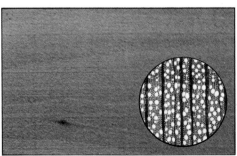

(x25)

THE TREE Pear is a northern temperate tree widely cultivated for its fruit, and most of the timber available commercially comes from old orchard trees. The tree is at best of only modest size, 10m or exceptionally 15m high, and often of poor stem-form. Similar woods are produced by the mountain ash, or rowan, by wild service and by whitebeam, although these tend to be a little paler than pear.

THE WOOD Pearwood is pale pinkish-brown with a uniform and very fine texture, finer than that of apple, and a straight or irregular grain according to the shape of the stem. It is moderately heavy, about the same weight as beech.

TECHNICAL PROPERTIES Pear dries slowly with a tendency to distort, especially if the grain is irregular. It is strong, with a reputation for toughness, and is difficult to split. It is fairly hard to saw and, though it machines well and takes a fine finish, it has a moderately blunting effect on cutting tools and care is necessary with wood of irregular grain. It is not resistant to decay.

USES Because supply is irregular and availability mostly in small sizes, the use of pearwood is limited to small turned and carved items and to inlay and marquetry. It turns extremely well and is used for bowls and the backs and handles of brushes; it is sought after as a carving wood due to its very fine texture. It has been used for drawing instruments, T-squares, etc., for recorders and as veneer in cabinetwork; a special use is for laps for polishing the jewels used in clocks and watches.

Commercial Hardwoods

White Oak

Red Oak

WHITE OAK
Quercus spp.

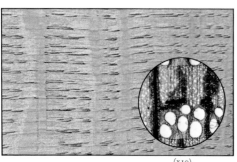

(×10)

THE TREE European oak, Japanese oak and American white oak, though distinguished commercially, are similar in character and can be considered together as white oaks. The best oak timber comes from forest-grown trees, which can reach 30m in height and have a long, straight bole up to 1m in diameter; old parkland trees sometimes have a wider bole, which often branches low down.

THE WOOD White oak is a pale yellow-brown, typically with a straight grain and a coarse texture due to its ring-porous structure. It has a conspicuous growth-ring figure on flat or rotary-cut surfaces and a characteristic silver-grain figure when quarter-cut or split. It is variable in weight but typically moderately heavy; American oak is somewhat heavier than European oak.

TECHNICAL PROPERTIES The properties of oak are particularly affected by its growth. Quickly grown, it is often dense and is a strong, tough, durable wood that can be fairly hard to work; slowly grown timber is lighter in weight, milder, not as strong and easier to work.

USES Many fine examples of the structural, joinery and decorative uses of oak are to be seen in historic buildings. Today it is still an important structural wood where traditional appearance or strength and durability are required. It is used for boat-building, specially fishing boats, and in estate work for gates and posts. It is an important furniture wood and is used for joinery, panelling and domestic floors; a special use is for the staves of whisky and sherry casks.

RED OAK
Quercus spp.

(×10)

THE TREE Red oak is a northern temperate wood, important in parts of North America and Iran, but without the widespread commercial significance of the white oaks. In Europe, red oak is sometimes cultivated: Turkey oak, for instance, though its wood has a limited usefulness, is widely grown, though usually as a parkland and not as a forest tree. Many red oaks reach a large size, 35m or more in height, with a bole commonly 1m and up to 1.8m in diameter in old trees.

THE WOOD Red oak has a distinctly pinkish tint compared with white oak. It is often of vigorous growth with a coarse texture and, when well grown, a straight grain. Like white oak, it has a silver-grain figure on quartered surfaces, but it is not generally so well marked, as the rays tend to

be shorter. Red oak is a dense wood, marginally heavier than American white oak and 10 to 15 per cent heavier than European oak.

TECHNICAL PROPERTIES Though heavier, red oak is typically more permeable and less durable than white oak. Its greater weight makes it more difficult to saw and machine and it is often more troublesome to dry.

USES Red oak lacks the outstanding durability of white oak and is not suitable for exterior structural, joinery and building purposes unless it is effectively treated. It is acceptable for interior work, for panelling, flooring and in furniture, though generally regarded as inferior to white oak. Persian oak is unusual in being impermeable and is used for barrel staves.

EVERGREEN OAK
Quercus spp.

(×10)

THE TREE Many species of oak growing in warm temperate and some tropical countries of Central America and SE Asia are evergreen and have a distinctive type of timber, called evergreen oak. In Europe, the familiar evergreen oaks are holm oak, or holly oak, and cork oak, which is more important for its bark – the source of commercial cork – than for its timber. Cork oak produces a thick layer of light-weight bark, which can be stripped at intervals of eight to ten years without damaging the tree.

THE WOOD Evergreen oak is harder and heavier than the more familiar white oak. Typically, it is a pale to medium brown; it has very large rays giving a silver-grain figure that is conspicuous though not usually as decorative as that of the best

white oak. It differs, too, from the common commercial oaks in not having a ring-porous structure, so that its texture is finer and more even, and a growth-ring figure, if present, is poorly marked.

TECHNICAL PROPERTIES Evergreen oak is difficult to dry as it distorts and is very prone to split. It is reputed to be strong and durable, but its weight makes it hard to saw and machine, and it is difficult to finish to a smooth surface, especially when the grain is irregular.

USES A timber of minor commercial importance compared with white oak, evergreen oak is far more difficult to use and is rarely exported. It is used locally for rough structural work, including posts and fencing, and sometimes for agricultural implements; in many countries it is mainly used for fuel.

MANGROVE
Rhizophora spp.

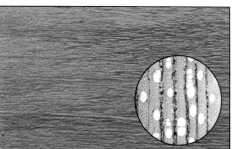

(×25)

THE TREE Mangroves occur along the littoral of many tropical countries. They are produced by a number of genera, but one of the most important is *Rhizophora*, with three species, which occurs throughout the tropical regions. Though species of *Rhizophora* can reach a height of 30m, with a bole up to 60cm in diameter, they are frequently much smaller; the trees are often much branched and supported on stilt-like roots.

THE WOOD Mangrove is a very heavy, hard, moderately fine-textured but generally featureless wood. It is a pale to dark reddish-brown, with a straight, irregular or sometimes spiral grain. It is of the same order of weight as greenheart.

TECHNICAL PROPERTIES Mangrove is a difficult wood to dry as it has considerable

shrinkage and tends to split on the ends and check on the surface. It is a strong wood, at least half as strong again as oak, except in shock resistance. It is hard to saw and difficult to work, but takes a fine finish. It is moderately resistant to fungi.

USES Mangrove is a particularly valuable wood for local use in the countries where it occurs. Its bark is a source of tannin; the wood is used for building poles, and especially for fuel and the production of charcoal. It has been shipped from parts of SE Asia as wood chips for pulp production in Japan. In West Africa, it has been used for pit props and for the staves of barrels for palm oil. It is resistant to abrasion and is suitable for flooring. With its exceptional strength, it has been suggested for tool handles.

Evergreen Oak

Mangrove

Robinia

Willow

Sandalwood

Light-red Meranti

ROBINIA
Robinia pseudoacacia

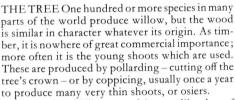

(×10)

THE TREE Robinia, known in the United States as black locust, is an American wood now grown in many European countries and in parts of Africa and Asia. It comes from a small to medium-sized tree, varying in size and form according to its conditions of growth. Though exceptionally 20 to 25 m in height, it more commonly branches low down, making for a smaller tree. It is valued as a windbreak tree protecting against soil erosion.
THE WOOD When first cut, robinia is yellow with a greenish tint, but it darkens, on exposure, to a medium golden-brown. It is a coarse-textured wood, with a conspicuous growth-ring figure and a grain that is straight or somewhat irregular, depending on the stem-form. It is moderately heavy, about the same weight as oak.

TECHNICAL PROPERTIES Robinia dries slowly, with a tendency to distort. Once dry, it is strong, comparing well with ash in toughness; it saws well, but is moderately hard to machine, though it can be finished to give a good surface. It resists fungi and termites.
USES Though not much used in Britain, robinia is popular in some European countries, especially Hungary, where it is extensively planted. It grows quickly and, even as a small stem, has only a narrow band of sapwood, making it very suitable for poles and estate work. Because it is tough, it is used for agricultural implements, wheel spokes and ladder rungs; like ash, it is particularly suitable for bending. It is occasionally used for furniture.

WILLOW
Salix spp.

(×25)

THE TREE One hundred or more species in many parts of the world produce willow, but the wood is similar in character whatever its origin. As timber, it is nowhere of great commercial importance; more often it is the young shoots which are used. These are produced by pollarding – cutting off the tree's crown – or by coppicing, usually once a year to produce many very thin shoots, or osiers.
THE WOOD Willow wood is very like that of poplar and often the two timbers are not easily distinguished except under a microscope. Willow is a pale, very fine-textured, mainly featureless wood; it is about the same weight as poplar, but when vigorously grown is somewhat lighter.
TECHNICAL PROPERTIES Willow dries quickly and well; once dry, it is stable in use.

Though not particularly strong, it can absorb energy without splintering and it is this property that makes its best wood suitable for cricket bats. It works easily and finishes well, but is perishable in conditions favouring decay.
USES Willow is a traditional wood for artificial limbs as it combines light weight and toughness; like poplar, it is used for fruit baskets and clogs, and, because it does not readily ignite by friction, for the brake blocks in colliery winding gear. It is used for flooring where resistance to splintering is required. Cricket bats are made from a special type of willow cultivated to give fast and uniform growth. Osiers are used for wickerwork and basketwork and thicker, pollarded shoots for stakes and wattle hurdles.

SANDALWOOD
Santalum album

(×25)

THE TREE Sandalwood has been prized since classical times both for its oil and its wood. True sandalwood from India is considered to produce the best-quality wood and best yield of oil, but closely related species with aromatic woods occur in Australia and on many Pacific Islands. In many parts of the world, quite unrelated but scented woods are sometimes known as sandalwood. Indian sandalwood, from southern India, mainly Mysore, comes from a small tree parasitic on the roots of other trees; the wood is so valuable that the tree is not felled but uprooted, and every piece containing heartwood is kept.
THE WOOD Sandalwood has a very fine and even texture and a straight or irregular grain; it is pale yellow-brown when freshly cut, but darkens

on exposure to a medium brown; it has a characteristic and persistent aromatic scent and a slightly oily feel. It is heavy, about the weight of rosewood.
TECHNICAL PROPERTIES Sandalwood dries slowly but without splitting; it saws readily and is noted for its excellent working properties; it carves especially well. It is very durable.
USES Sandalwood has two main uses. Heartwood chips and shavings are distilled to give sandalwood oil, used throughout the world for perfumery; solid wood is used for carving, especially for fancy boxes, picture-frames, combs, paper-knives and similar small objects. It is traditionally burnt as incense at Hindu funeral services. Oil from East African muhuhu may be used as a substitute for sandalwood oil.

LIGHT-RED MERANTI
Shorea spp.

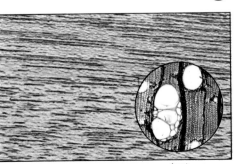

(×25)

THE TREE The meranti timbers, produced by species of *Shorea*, are woods of light to medium weight, distinguished commercially according to their colour as light-red, dark-red, yellow and white meranti. The paler, lightweight red woods from peninsular Malaysia are known as light-red meranti, from Sabah as light-red seraya and from the Philippines they are included in shipments of white lauan. Typically, the many species of *Shorea* producing light-red meranti are large trees, 60m in height with a bole 1m or more in diameter.
THE WOOD Light-red meranti is a pale pink to red, medium-textured wood with a shallowly interlocked grain giving a broad stripe on quartered surfaces. It is variable in weight but typically light, about the same weight as African mahogany.

TECHNICAL PROPERTIES Light-red meranti dries easily with some tendency to cup, but otherwise little degradation; it is stable in use. Light-red meranti varies in strength according to its weight, but is generally comparable to African mahogany; as with mahogany, occasional pieces from the centre of some logs can be very brittle. It works easily, taking a good finish, and rotary-peels to give an excellent veneer.
USES Light-red meranti is of outstanding commercial importance. It is shipped from peninsular Malaysia to Europe and Australia as lumber, and from the SE Asian islands to Japan, Korea and Taiwan for plywood as well as sawn timber. It is widely used in joinery and furniture and for light structural purposes, both solid and as plywood.

Commercial Hardwoods

Dark-Red Meranti   Yellow Meranti

DARK-RED MERANTI
Shorea spp.

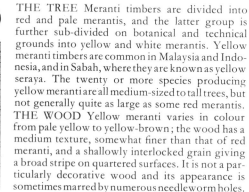

(x25)

THE TREE Medium-weight meranti of a red to dark-red colour is known as dark-red meranti in contrast to the paler light-red meranti; dark-red seraya and red lauan are similar. Like other merantis, dark-red meranti is a mixture of species, but *S.pauciflora* is a common source in Malaysia; it is a very large tree, reaching 70m in height with a bole up to 1.5m in diameter.

THE WOOD Dark-red meranti is distinguished by its medium-red to dark red-brown wood, often with conspicuous, white dammar, or resin, streaks. It has a medium to coarse texture and a shallowly interlocked grain. It is heavier than light-red meranti and is used for more exacting purposes; though variable in weight, it is on average about the same weight as West African utile.

TECHNICAL PROPERTIES Dark-red meranti dries more slowly than light-red meranti and has some tendency to split and check in thick sizes; once dry it is a stable wood. It has good strength, comparing well with utile, but is more easily split because the grain is straighter. It saws easily and works well, taking a good finish; it is more resistant to fungi than the paler, light-red woods.

USES Dark-red meranti is an important commercial timber, shipped from Malaysia to Europe and elsewhere and from the Philippines to the United States, where it is known as Philippine mahogany. It is popular for joinery, including outdoor use in temperate countries, for vehicle construction, cladding, panelling, floors and, especially in the United States, for boat-building.

YELLOW MERANTI
Shorea spp.

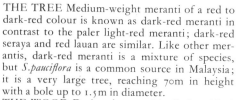

THE TREE Meranti timbers are divided into red and pale merantis, and the latter group is further sub-divided on botanical and technical grounds into yellow and white merantis. Yellow meranti timbers are common in Malaysia and Indonesia, and in Sabah, where they are known as yellow seraya. The twenty or more species producing yellow meranti are all medium-sized to tall trees, but not generally quite as large as some red merantis.

THE WOOD Yellow meranti varies in colour from pale yellow to yellow-brown; the wood has a medium texture, somewhat finer than that of red meranti, and a shallowly interlocked grain giving a broad stripe on quartered surfaces. It is not a particularly decorative wood and its appearance is sometimes marred by numerous needleworm holes.

Yellow meranti is variable in weight but, on average, it is about 10 per cent lighter than oak.

TECHNICAL PROPERTIES Yellow meranti dries slowly but with little degradation; once dry, it is moderately stable in use. Its strength properties vary with the weight of the wood, but are on average somewhat better than those of light-red meranti. It saws easily, works well and rotary-peels to give a good veneer. It is moderately resistant to fungi but not to termites.

USES Yellow meranti is a useful general-purpose wood, but the fact that it is often marred by many small worm holes limits its use where appearance is important. It is suitable for interior joinery, for truck and lorry framing and for domestic floors; it is commonly used for plywood.

WHITE MERANTI
Shorea spp.

(x25)

THE TREE White meranti is the palest though not the lightest in weight of the meranti timbers. It is produced by more than twenty species, which are widely distributed from Burma, through SE Asia, to the Philippines, and particularly common in Malaysia and Indonesia. Many of the trees producing white meranti reach a large size and often have straight cylindrical boles. Other groups of meranti and seraya are generally similar, but white meranti is quite distinct botanically and technically from white seraya.

THE WOOD Like other meranti timbers, white meranti varies in colour, but much of the wood is almost white when freshly cut, though it darkens to yellow or pale-brown on exposure. It has a moderately coarse texture and a slightly interlocked

grain producing a broad stripe on quartered surfaces. It is of medium though variable weight, on average about 10 per cent lighter than oak. Unlike other merantis, it has minute aggregates of silica in its wood.

TECHNICAL PROPERTIES Though white meranti dries well and with little degradation, its silica content makes it so difficult to machine that it is rarely used as sawn wood. However, like other medium-weight siliceous woods, it can be rotary-peeled to give a good veneer.

USES White meranti is used almost entirely for the production of plywood. Today, with mersawa, it is important as face-veneer for Malaysian white plywood. As ply, it is used for structural and general purposes.

SAL/BALAU
Shorea spp.

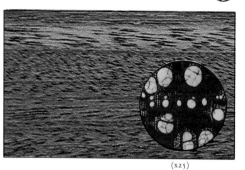

(x25)

THE TREE Though many species of *Shorea* have timbers of the meranti type, others have a much denser wood; these vary in colour and the heavy red woods are sometimes distinguished commercially from those that are shades of brown. The brown woods are of particular importance throughout their wide distribution from India to the Philippine Islands, though they are not often exported. They are known by distinctive names according to their country of origin: as sal in India, chan in Thailand, balau and selangan batu in parts of Malaysia, and yakal in the Philippines.

THE WOOD The heavy *Shorea* woods are mostly yellow-brown to dark-brown, with a fine texture and an interlocked grain; they vary in weight, but are typically 20 to 40 per cent heavier than oak.

TECHNICAL PROPERTIES Sal and balau are slow-drying and tend to split on the surface; in large sizes they cannot be effectively dried, though for many purposes this may not matter. They are noted for their strength, especially toughness, but their weight makes them hard to saw and machine. They are durable, even under tropical conditions.

USES The heavy *Shoreas* are among the most important structural woods in their countries of origin. Sal is widely used in India for all kinds of building and for railway sleepers; balau is used for similar purposes in Malaysia and for transmission poles, as a mining timber and for ship-building. In Europe and elsewhere they are usually used for heavy-duty purposes in large sizes, such as dock and harbour work, wharf decking and piling.

White Meranti

Sal/Balau

Whitebeam

American Mahogany

Teak

Indian Laurel

WHITEBEAM
Sorbus aria

(x25)

THE TREE The many species of *Sorbus*, which occur in the temperate countries of the northern hemisphere, have similar woods. Whitebeam is described as it is typical, though not of special commercial interest. It occurs widely through Europe and, when well grown, the tree reaches a height of 10 to 12m and has a stem up to 50cm in diameter. Other European species are the service, wild service and rowan, or mountain ash; all are small to medium-sized trees.

THE WOOD Whitebeam varies from almost white to pale pinkish-brown, closely resembling pear. It has a fine, even texture, and the grain is straight. It is hardly figured apart from a faint growth-ring. It is about the same weight as pear and beech. Rowan is similar, but not quite so heavy.

TECHNICAL PROPERTIES Whitebeam must be dried slowly and carefully to avoid distortion. When dry, it is a strong wood, specially noted for its toughness. It is somewhat difficult to saw and, being hard, blunts cutting tools fairly quickly, but it can be machined to give an excellent finish; it turns and carves well. It is not durable in conditions favouring decay.

USES Whitebeam, like other closely related timbers, including the fruit woods, is available only sporadically and then usually in small sizes. It is used for small turned items, such as bobbins, spools, rollers and tool-handles; it is a fine wood for carving and is sliced to give a veneer used for inlay. Other species of *Sorbus* are used for similar purposes.

AMERICAN MAHOGANY
Swietenia spp.

(x25)

THE TREE American mahogany is the original commercial mahogany. Cuban or Spanish mahogany was brought to Europe by the Spaniards from their West Indian colonies in the late 16th century, but it was in the 18th century that mahogany came to prominence in furniture designed by Chippendale, Hepplewhite and Sheraton. Since then it has been in continuous use for fine cabinetwork in a variety of styles. Today, Cuban mahogany, *Swietenia mahogani*, is a rare wood, and most American mahogany, mainly *Swietenia macrophylla*, comes from the mainland and is sold as Honduras, Brazilian, Peruvian mahogany, etc., according to its origin.

THE WOOD Mahogany is variably a medium to deep red-brown: Cuban mahogany is dark and heavy, and mainland wood mostly lighter in weight

and paler. It has a medium texture and is often of plain appearance, but occasional logs with an irregular grain give a highly figured wood.

TECHNICAL PROPERTIES American mahogany is easily dried and stable in use. It saws easily and can be machined to an excellent finish.

USES Though American mahogany available today is a milder wood than that used in older furniture, it is still regarded as the finest of the commercial mahoganies. It is used for high-class furniture, especially reproduction work, for fine joinery and for panelling. It is used for planking and fittings in ships and boats as it combines stability and durability with light weight, and for engravers' blocks and engineers' patterns as it is stable and can be carved to give fine detail.

TEAK
Tectona grandis

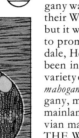

(x25)

THE TREE Teak is one of the world's outstanding woods and one against which others are often judged. Though many woods are called teak, only that produced by *T.grandis* is true teak. It occurs naturally in Burma and Thailand, the main sources of commercial timber, and in India, Indo-China and Java, but it has been planted in many parts of the world. It comes from a tree of variable size and form, which, when well grown, can reach 40m in height and have a bole 1 to 1.5m in diameter, though the bole in large trees is often fluted.

THE WOOD Teak is usually a uniform golden-brown, but may be a medium to deep brown and have darker, almost black markings. It is ring-porous, with a conspicuous growth-ring figure on flat-sawn surfaces and a coarse and often uneven

texture. It has a greasy feel and a characteristic leathery smell. Teak is a medium-weight wood, heavier than mahogany but lighter than oak.

TECHNICAL PROPERTIES Teak dries slowly but well, and once dry it is noted for its stability in use. It is strong for its weight, but is hard to saw and machine because of its abrasive nature. It is extremely durable.

USES Teak is foremost a ship-building wood, used where strength, stability and durability are demanded, and preferred to all other woods for decking. It is used for fine joinery and for laboratory fittings, especially bench tops, and in chemical works on account of its acid resistance. It has been fashionable in recent years in domestic furniture, often as veneer.

INDIAN LAUREL
Terminalia spp.

(x10)

THE TREE Indian laurel is a common and popular timber in many parts of India and Burma, but it is rarely seen nowadays outside these countries. The tree is of variable size, but when well grown can reach 30m in height and have a clean, straight bole 1m in diameter.

THE WOOD Indian laurel is a decorative wood, sometimes pale-brown with few markings but usually selected for its figure, when it varies from pale-brown to dark-brown, sometimes with a greyish tint, and has dark-brown to almost black markings. The texture is fairly coarse and the grain is straight or irregular. It is a dense wood, about 20 per cent heavier than oak.

TECHNICAL PROPERTIES Indian laurel is a difficult wood to dry, especially in thick sizes,

as it distorts and is liable to split and check. It is a strong wood, comparing favourably with oak, but its weight makes it difficult to saw and to work with hand or machine tools, especially if irregular or interlocked grain is present. It is rated moderately resistant to both fungi and termites.

USES Indian laurel is a fine wood, especially when selected for its decorative appearance. It is valued in India and has never been available elsewhere in quantity. It has some resemblance to walnut and has been used, both in the solid and as veneer, in furniture and cabinetwork, for high-class joinery, such as staircases and doors, and for panelling. In India, it is used for building, for vehicles and railway carriages, for tool-handles and many general purposes.

Commercial Hardwoods

Indian Silver-Grey Wood   Idigbo

INDIAN SILVER-GREY WOOD
Terminalia bialata

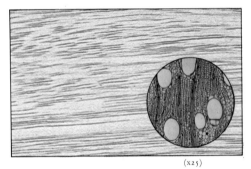

(x10)

THE TREE Indian silver-grey wood is obtained from the Andaman Islands, from a tree which reaches 50m in height and has a bole 1m or more in diameter. In some trees the wood is a uniform pale-yellow throughout and the timber is then known as white chuglam; more rarely, there is a coloured heartwood, known as silver-grey wood.

THE WOOD Silver-grey wood is a pale to medium brown with dark-grey to almost black streaks resembling some types of European walnut; it has a medium texture, is generally straight-grained, and is almost the same weight as walnut. White chuglam, though different in colour, is similar in grain, texture and weight.

TECHNICAL PROPERTIES Silver-grey wood dries readily and well and has a reputation for stability in use. It has good strength properties, approaching those of beech, and is said to saw readily, to work well, and to take an excellent finish. It is rated moderately resistant to fungi but not to termites. White chuglam is similar in its drying, strength and working properties, but is less durable in conditions favouring decay.

USES Though once available in modest quantities, silver-grey wood is seen only occasionally today and then usually as veneer. It was formerly used for high-class joinery, for furniture and cabinet-work, and for decorative panelling in public buildings, railway coaches and ocean liners. White chuglam is used in India for furniture, joinery, flooring, general construction and boat-building; it is also used for tea chests.

IDIGBO
Terminalia iverensis

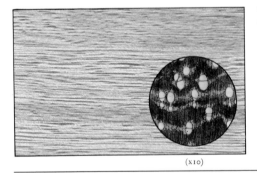

(x25)

THE TREE Idigbo is a West African wood exported from Nigeria, Ghana and the Ivory Coast. It comes from a large tree, often 40m in height with a bole 1m or more in diameter, though large logs often contain a brittle core. Medium-sized logs yield better timber.

THE WOOD Idigbo has some resemblance to flat-sawn oak. Though not ring-porous, it has a conspicuous growth-ring figure and resembles oak in its yellow-brown colour. When quarter-sawn it lacks the conspicuous ray-figure of oak and may, if the grain is interlocked, show a striped figure. On average it is about the same weight as mahogany and some 25 per cent lighter than oak, but it is often variable, and some wood, which commonly has a pinkish tint, is very light in weight.

TECHNICAL PROPERTIES Idigbo dries rapidly and well and, once dry, is noted for its quite exceptional stability in use. It has modest strength, generally comparable to that of mahogany, but some wood is very brittle. It saws easily and machines to a good finish. It is resistant to fungi and moderately resistant to termites.

USES Idigbo is attractive, durable and stable though not strong, and it is sometimes necessary to exclude brittle timber. It is very suitable for interior and exterior joinery, though in damp conditions non-ferrous fixings should be used as it corrodes iron. It is used for mouldings, cladding and as domestic flooring, and in furniture for its oak-like appearance; it should not be used in moist conditions, where it may stain fabrics.

AFARA/LIMBA
Terminalia superba

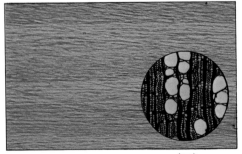

(x10)

THE TREE Afara is the name used in English-speaking countries for the timber known elsewhere as limba. It comes from a tree which occurs widely and often abundantly in West Africa, and which reaches a large size, 40m in height with a bole 1m or more in diameter. Commercial timber comes mainly from Zaire and western central Africa.

THE WOOD Afara varies in colour; often it is a pale straw colour, when it is known as light afara, or limba clair, but sometimes the wood is darker with irregular, grey markings and is then called dark afara, or limba noir. It has a medium texture and a straight or slightly interlocked grain. Though, on average, about the weight of mahogany, it can vary from very light to moderately heavy.

TECHNICAL PROPERTIES Afara dries rapidly and well and is stable in use. It has moderate strength, though this varies with density and some timber is particularly brittle. It saws easily and works well, taking a good finish; it peels to give a good veneer. It is readily attacked by fungi in conditions favouring decay.

USES Pale afara is a general-purpose wood, and popular in continental European countries. It is used for the solid parts of furniture and for interior joinery, though it is necessary to exclude brittle timber where good strength is required, and is often made into plywood, which is suitable for indoor use. Darker, figured wood is sliced to give a decorative veneer with a walnut-like appearance used for panelling and inlay.

MAKORÉ
Tieghemella heckelii

THE TREE Makoré is a West African wood obtained mainly from Ghana and the Ivory Coast. It comes from a large tree of excellent stem-form, reaching 40m or more in height with a bole 1 to 2m in diameter.

THE WOOD Makoré varies from pale-pink to deep-red. It has a fine texture and often a straight grain, when it is a plain though fairly lustrous wood; occasionally the grain is interlocked and wavy giving the wood a moiré, or watered-silk, figure and makoré is then one of the most handsome of African woods. It is of medium weight.

TECHNICAL PROPERTIES Though drying slowly, makoré dries well and is stable in use. It has good strength properties, appreciably better than those of mahogany, but not quite the equal of utile.

It is troublesome to process, especially when dry, because its silica content makes it abrasive to cutters, and because the fine dust produced in machining and sanding commonly causes irritation of the eyes, nose and throat; it can, however, be peeled and sliced without trouble to give excellent veneer. Makoré is a very durable wood, resistant to both fungus and termite attack.

USES An attractive wood, makoré is difficult to handle because of its abrasive and irritant properties, but with care these can be overcome and it is then used for high-class joinery, in furniture, for vehicle and boat construction, and for coach panelling. Makoré is an important plywood timber as its outstanding durability makes it suitable for marine and other exacting uses.

Afara/Limba

Makoré

Lime

Obeche

Elm

Virola

LIME
Tilia spp.

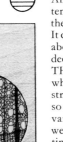
(x25)

THE TREE Lime, or basswood as it is known in America, is widely distributed in the northern temperate regions and is of commercial interest in the United States, Canada, Europe and eastern Asia. It comes from a medium-sized to large tree, usually about 20m in height, which is often grown for its decorative effect in avenues and parkland.

THE WOOD Lime is a pale, almost white wood which darkens to pale-brown on exposure. It has a straight grain and a fine and very uniform texture so that it is generally a featureless wood. It is of variable weight; European lime is about the same weight as mahogany, but American and Japanese timber is appreciably lighter, comparable to poplar.

TECHNICAL PROPERTIES Lime dries rapidly and well, apart from some tendency to distort, and once dry is moderately stable in use. It is not a strong wood; European lime has strength properties similar to those of sycamore, but the American and Japanese timbers are appreciably weaker. Lime is noted for its excellent working properties. It is readily attacked in conditions favouring decay.

USES Lime is one of the best woods for carving; it has long been used for this purpose and the remarkable detail that can be obtained is to be seen in the flower and fruit carvings of Grinling Gibbons, made in the late 17th century. Other traditional uses are for hat blocks and piano parts; it is used for dairy and domestic utensils as it is free from taint, and basswood is the preferred wood for beehive frames. Other uses are for small turned items, toys, bobbins and sometimes plywood.

OBECHE
Triplochiton scleroxylon

(x25)

THE TREE Obeche is one of Africa's leading export timbers. Though it occurs widely, it is particularly common in the Ivory Coast, Ghana, Nigeria and Cameroon. It comes from a very large tree, 50m or more in height with a long, straight bole up to 1.5m in diameter.

THE WOOD Obeche is of interest as it is one of the lightest hardwoods in common use, generally about the same weight as the lightest softwoods, but still about twice the weight of balsa. It is also noted for its attractive appearance; it is a pale straw-yellow wood with a high natural lustre giving it a silky sheen. It has a medium texture and an interlocked grain, so that quartered surfaces have a pronounced stripe, which further enhances its decorative effect.

TECHNICAL PROPERTIES Obeche tolerates very rapid drying with little degradation and, once dry, is very stable in use. Being light in weight, it is not strong; it saws easily and works well, taking a good finish provided that cutters are kept sharp. Special care is needed in cutting end-grain surfaces as they tend to crumble. It peels to give an excellent veneer. It is perishable and rapidly attacked by fungi in conditions favouring decay.

USES Obeche is very popular where attractive appearance, ease of working but little mechanical strength are demanded. It is used for the framing of cabinetwork and for drawer sides, especially in whitewood furniture, and for mouldings and interior joinery. It is used for the solid core of blockboard and very commonly for plywood.

ELM
Ulmus spp.

(x10)

THE TREE The elms occur widely in the northern hemisphere; their woods differ somewhat in character though are generally similar. Specially important are the rock and white elms of America, the English, Dutch and wych elm of Europe, and Japanese nire.

THE WOOD Elm has a characteristic appearance, with a prominent growth-ring figure, coarse texture and often irregular grain. The wood is pale-brown, sometimes with a reddish tint or, as in wych elm, a greenish cast. Rock elm is a pale and heavy wood; wych elm, too, is fairly dense, but most other elms are moderately light in weight.

TECHNICAL PROPERTIES Elm dries readily though with a tendency to distort if irregular grain is present. Rock elm is noted for its strength, but generally elm is not a strong wood; it can, however, be bent very successfully. It works well and, though not particularly resistant to fungi, has a long life in waterlogged conditions – short lengths, bored out and used as drain pipes, have been dug up in sound condition after centuries of use.

USES Elm is used for structural purposes where the wood is permanently wet, in fishing boats and barges, for dock work and piling – the Rialto in Venice is said to stand on elm piles. It is the traditional wood for coffins, weather-boards and the seats of Windsor chairs. Today it is widely used in furniture for its decorativeness, and as flooring. It is turned for bowls and used for butchers' chopping blocks as it resists splitting. Low-grade timber is used in mining and estate work.

VIROLA
Virola spp.

(x10)

THE TREE Virola is an American wood shipped mainly from Brazil, where it is known as ucuuba, from Surinam, as baboen, and from Colombia. It is produced by many species, but commercial timber comes from those trees which reach a large size and have a long, straight bole, though this is often of quite modest diameter.

THE WOOD Virola is a pale pinkish-brown, with a medium texture and generally straight grain. It is featureless with little distinctive appearance. It varies in weight according to its origin, but is typically light, about the weight of obeche to approaching that of mahogany.

TECHNICAL PROPERTIES Although it is light in weight, virola needs care in drying as it shrinks appreciably, tends to distort and may check and split. It is low in strength, but saws easily and works well, taking a smooth finish if cutters are kept sharp. It rotary-peels to give a good veneer. Virola is readily attacked in conditions favouring decay: because it is so perishable, logs have to be extracted from the forest quickly to prevent the timber deteriorating.

USES Virola is one of the leading export timbers of tropical America, especially Brazil, whence it is shipped mainly to the United States. It is a light-weight, general-purpose wood, used extensively for mouldings, for drawer sides and other non-structural parts of furniture, and for interior joinery, and the lippings and cores of doors. Virola makes excellent plywood, though it is suitable only for interior use.

Commercial Softwoods

Fir

Kauri

FIR
Abies spp.

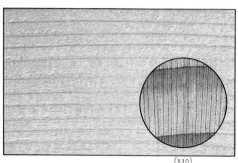

(×10)

THE TREE Only species of *Abies* produce true fir, though the name is also used for other woods, such as Douglas fir and Scots fir, or Scots pine. The true firs occur through central and eastern Asia, in central and southern Europe, and they are particularly important in North America. They are mostly 20 to 40m in height, though American noble fir can reach 60 to 70m.

THE WOOD The many species of fir have similar woods, which are creamy-white to pale-brown in colour; they are like spruce but not so lustrous. The timber is non-resinous, odourless and non-tainting when dry, and is usually straight-grained. The firs are lightweight woods; European fir is about the weight of European spruce, but other firs are somewhat lighter.

TECHNICAL PROPERTIES The firs dry easily and well. They are mostly low in strength, tending to be brittle, but are easy to work, though because they are soft tools must be kept sharp to give a good finish. They are not resistant to fungi and are difficult to preserve effectively.

USES Though often locally important, the firs are not traded internationally on the scale of some other softwoods. When sold they are often mixed with other woods, for example with hemlock from western Canada and with spruce from central Europe. As sawn wood they are used in building, for joinery and for packaging. Fir is an important pulpwood in North America, with much of the vast production in eastern Canada obtained from small logs of balsam fir and spruce.

KAURI
Agathis spp.

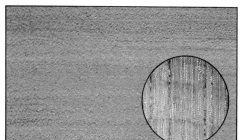

(×10)

THE TREE Kauri is obtained in peninsular Malaysia, the East Indies, some Pacific Islands, and in Queensland and New Zealand. It is shipped in small quantities from Malaysia and Fiji, but elsewhere almost all the timber is used locally. The kauris are medium-sized to large trees; New Zealand kauri reaches 30m, the Queensland trees up to 45m and the East Indian kauri as much as 60m.

THE WOOD Kauri varies in colour from a pale yellow-brown to pinkish-brown and is noted for its lustrous surface. Growth rings are barely present so that the wood has a plain appearance and a fine and even texture; it is typically straight-grained and non-resinous. Woods from the East Indies and Queensland are light, but those from Fiji and New Zealand are appreciably heavier.

TECHNICAL PROPERTIES The timbers dry easily and well though with a tendency to distort. They vary in strength according to their weight, but all work readily, taking a fine finish. New Zealand kauri is rated moderately resistant to decay, but other types are less durable.

USES New Zealand kauri was once an important commercial wood, available in large sizes and free from knots, but its popularity led to serious over-cutting and today it is rarely seen outside New Zealand. Its special uses were for chemical vats and tanks and in boat-building. Queensland kauri is used for building and joinery, food boxes and engineers' patterns. Fiji and East Indian kauri are used in joinery and furniture and are peeled to make a general-purpose plywood.

PARANA PINE
Araucaria angustifolia

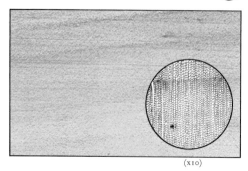

(×10)

THE TREE Parana pine is obtained mainly from the state of Parana in Brazil, but its range extends into parts of Paraguay and northern Argentina. It comes from a medium-sized to large tree, up to 40m in height, with a long straight bole and a flat-topped crown. The Chile pine, or Monkey Puzzle tree, is closely related and similar, often planted but of little importance for its timber.

THE WOOD Parana pine is straw-coloured to pale-brown with occasional bright-red streaks. It has a very fine and even texture due to the almost complete absence of growth rings, and is typically straight-grained. It is about the same weight as European redwood.

TECHNICAL PROPERTIES Parana Pine is one of the most difficult softwoods to dry, needing care to prevent distortion and splitting. It is moderately strong, comparing closely with European redwood except in toughness; it saws easily and works well with both hand and machine tools to give a smooth finish. It is not durable.

USES Parana pine is the most important export timber from Brazil, valued for its fine, even texture and virtually knot-free condition in long lengths. It is outstanding for indoor joinery, especially staircases, but is also used for cabinet framing and drawer sides, especially in whitewood furniture, for mouldings, in shop-fitting, and for vehicle-building. It is specially popular for do-it-yourself work because of its ease of working and absence of knots. It is peeled in Brazil for a general-purpose plywood.

INCENSE CEDAR
Calocedrus decurrens

(×10)

THE TREE Incense cedar occurs in a limited area of the western United States, with production centred mainly in northern California and southern Oregon. It is a tree of modest size, usually about 30m but sometimes up to 50m in height, with a straight but irregularly shaped bole, 1m or more in diameter.

THE WOOD Incense cedar has some likeness to western red cedar, but is paler, typically a pinkish-brown colour. It has a faintly spicy odour, and timber from old trees has narrow rings with very little dense latewood, so that it has a very fine and even texture. Its appearance is sometimes marred by pockets of rot, which develop in the tree, when the wood is described as pecky. It is light, about the same weight as western red cedar.

TECHNICAL PROPERTIES Incense cedar dries quickly and well, with a very small shrinkage. It is low in strength, especially shock resistance and stiffness, but is noted for its excellent working properties, giving a smooth finish provided that tools are kept sharp. It is very durable.

USES Outside America, incense cedar is only seen as pencils, for which it supplements the limited supply of pencil cedar. It is somewhat softer than pencil cedar, but, because of its fine and even texture, it whittles easily and cleanly. In the United States it is used for the laths of venetian blinds, for storage chests and roofing shingles. Lower-grade timber is used out of doors, where its natural durability makes it particularly suitable for poles, estate work and for railway sleepers.

Parana Pine

Incense Cedar

Cedar

Port Orford Cedar

Yellow Cedar

Sugi

CEDAR
Cedrus spp.

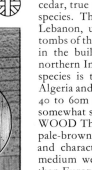

(×10)

THE TREE Though many woods are known as cedar, true cedar is a softwood produced by three species. The cedar of antiquity is the cedar of Lebanon, used in the construction of the royal tombs of the early kings of Egypt and by Solomon in the building of the Temple; the deodar of northern India is almost as famous, and the third species is the Atlas cedar of the mountains of Algeria and Morocco. All are large trees, reaching 40 to 60m in height; as parkland trees they are somewhat shorter, with wide, spreading crowns.

WOOD The wood of the three species is similar, pale-brown, with a fairly well-defined growth ring, and characterized by a fragrant smell. It is of medium weight for a softwood, a little heavier than European redwood.

TECHNICAL PROPERTIES Cedar dries readily though with a tendency to distort. It is inclined to be brittle and, generally, is not a strong wood; it works easily and well and takes a fine finish. It is noted for its resistance to both fungi and termites.

USES Cedar has long been prized for its exceptional durability and there are examples in the Middle East and India of its having lasted for many centuries. Today it is rarely available outside its countries of origin other than from occasional parkland trees. It is a fine joinery and furniture wood, and though used for building is somewhat weak for this purpose; it is an excellent estate wood. It is sliced and used as a veneer for decorative surfaces.

PORT ORFORD CEDAR
Chamaecyparis lawsoniana

(×10)

THE TREE Port Orford cedar is an American wood obtained from a limited area of southern Oregon and California. It comes from a tree of large size, reaching 40 to 50m in height and up to 1.5m in diameter. It has been extensively planted, often as an ornamental tree, and is commonly known as Lawson's cypress.

THE WOOD Like many so-called cedars, Port Orford cedar has a fragrant, spicy smell. The wood is pale-yellow to pale-brown, non-resinous, with a straight grain and a fine, even texture, especially in slowly grown American wood. It is light in weight, marginally heavier than European spruce. Timber from plantation trees is generally similar, though of more vigorous growth and with a somewhat coarser texture.

TECHNICAL PROPERTIES Port Orford cedar dries well with little degradation and, once dry, is very stable in use. It is a strong wood for its weight, stiff and with good shock resistance. It works easily and well with hand and machine tools, taking an excellent finish. It is noted for its resistance to decay.

USES An attractive wood, though rarely seen outside the United States, Port Orford cedar is popular for ship- and boat-building as it combines light weight, strength and durability, and for indoor and outdoor joinery. It is used for the framing of furniture and for clothes chests as it is thought to deter moths. A traditional use was for acid battery separators, made from veneer produced by rotary-cutting, slicing or sawing.

YELLOW CEDAR
Chamaecyparis nootkatensis

THE TREE Yellow cedar is a timber of the Pacific coast region of North America from Alaska to southern Oregon. It comes from a tree of modest size in comparison with others in these forests, reaching 30m in height, with a bole up to 1m in diameter.

THE WOOD Yellow cedar is pale-yellow and a little darker than Port Orford cedar. It has a strong odour when freshly cut, but this largely disappears on drying, leaving only a faint, raw, potato-like smell. It is typically a very slowly grown wood with a fine and even texture and a straight grain. It is about the same weight as Port Orford cedar and a little heavier than spruce.

TECHNICAL PROPERTIES Yellow cedar dries well with little shrinkage, and once dry is noted for its stability in use. It is a moderately strong wood, stiff for its weight and with good shock resistance. It works easily and well, taking a good finish with both hand and machine tools. Like Port Orford cedar it is noted for its resistance to decay.

USES Yellow cedar is an attractive, valuable but not very abundant wood, shipped in small quantities, mainly from British Columbia. It has many of the characteristics of Port Orford cedar and is used for similar purposes, in boat-building and for joinery and furniture. It is an excellent wood for use out of doors; among its more specialized uses are engineers' patterns, surveyors' poles, and it has always been considered one of the best woods for battery separators.

SUGI
Cryptomeria japonica

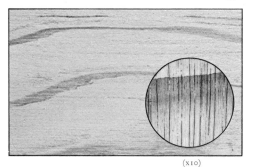

(×10)

THE TREE Sugi is Japan's most important commercial softwood, obtained from a fast-growing tree which reaches 40m or more in height and has a bole 1m in diameter. It occurs in parts of China as well as Japan and has been planted extensively elsewhere in the world.

WOOD Sugi is pale yellow-brown to pinkish-brown or, in some plantation timber, almost pink. Natural-growth timber, though it has conspicuous growth rings, has a fine, even texture and is faintly scented; plantation timber is typically of vigorous growth and, cut from young trees, it has a somewhat coarse texture. It has a straight grain and is one of the lightest-weight commercial softwoods, only marginally heavier than western red cedar.

TECHNICAL PROPERTIES Sugi dries quickly and well and, once dry, is noted for its stability in use. It is low in strength, appreciably softer and marginally weaker in bending, compression and impact than western red cedar, but somewhat more resilient. It works easily and takes an excellent finish provided that cutters are kept sharp. It is reported moderately durable, though young plantation timber is likely to be less so.

USES Though light in weight, sugi is used in Japan in large sizes for construction work, and in smaller sizes for joists, rafters, studding and domestic flooring. It is an attractive joinery wood, and is used for panelling, in furniture, and for boat planking. Other uses include cooperage, packaging, lacquerware and telegraph poles.

Commercial Softwoods

Cypress

Alerce

CYPRESS
Cupressus spp.

(×10)

THE TREE The true cypress is a tree of the warm temperate parts of the world; it is nowhere very common and is usually of only minor commercial interest where it grows naturally. It has, however, been introduced as a plantation tree, especially in East Africa, but also in South Africa, New Zealand, Australia and elsewhere. The trees grow quickly on favourable sites, reaching 20m in height with a bole 60cm in diameter in twenty-five to thirty years. The European cypress is of only local importance for its timber. Of more commercial interest is the Leyland cypress, a hybrid between true cypress and yellow cedar.

THE WOOD Cypress is a pale yellow-brown with a fine texture and generally straight grain. Plantation timber is often vigorously grown, tends to be knotty, and is fairly light, about the same weight as European whitewood. It is non-resinous, but has a faint cedar-like odour.

TECHNICAL PROPERTIES Cypress dries rapidly and well. It is moderately strong, comparing closely with whitewood, and works easily to give a good finish, though knots and tearing of associated distorted grain can be troublesome in some plantation timber. It is durable.

USES Plantation timber is used mostly in the countries where it is grown, but occasional shipments of East African wood are available. Cut from fairly young trees, it tends to be knotty and is used for general construction and for boxes and crates; selected timber is suitable for joinery, including outdoor use.

ALERCE
Fitzroya cupressoides

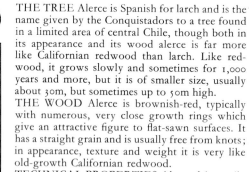

(×10)

THE TREE Alerce is Spanish for larch and is the name given by the Conquistadors to a tree found in a limited area of central Chile, though both in its appearance and its wood alerce is far more like Californian redwood than larch. Like redwood, it grows slowly and sometimes for 1,000 years and more, but it is of smaller size, usually about 30m, but sometimes up to 50m high.

THE WOOD Alerce is brownish-red, typically with numerous, very close growth rings which give an attractive figure to flat-sawn surfaces. It has a straight grain and is usually free from knots; in appearance, texture and weight it is very like old-growth Californian redwood.

TECHNICAL PROPERTIES Alerce dries easily and well. It is not a strong wood, but, with a fine and even texture, it works well with both hand and machine tools, taking a fine finish in spite of its light weight. It is durable.

USES Alerce is an attractive wood, though not often seen. It is available in only limited quantity because it is difficult to extract from swamp forests, which are often remote and can be worked for only a limited period each year. It is suitable for many of the purposes for which sequoia, or Californian redwood, is used, such as joinery and light structural use where durability is required. In Chile it is an important building wood as its exceptionally straight grain enables it to be split to give boards and roofing shingles. It is used for barrels and musical instruments and, as small round timber, for posts and masts.

PENCIL CEDAR
Juniperus spp.

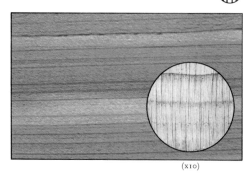

(×10)

THE TREE The first cedar, or more correctly juniper, used for pencils was the Virginian pencil cedar of the eastern half of the United States, but as supplies became scarce another pencil cedar, growing in the mountains of East Africa, was introduced. The East African wood comes from a tree which reaches 30m in height and often has a fluted and fissured stem; the American tree is usually somewhat smaller, 12 to 15m in height.

THE WOOD Pencil cedar is a fine-textured, reddish-brown wood, non-resinous but with a pleasant cedar-like smell. It is of medium weight for a softwood, with the African wood marginally heavier than the American.

TECHNICAL PROPERTIES Pencil cedar is slow to dry and care is needed to avoid distortion and splitting; once dry it is stable in use. The wood is not particularly strong, though the African wood is somewhat better in this respect than the American, but it is noted for its high resistance to decay and its excellent working qualities – it cuts easily and cleanly and can be finished to give a very smooth surface.

USES This is the traditional wood for pencils, used for its excellent whittling properties, but only limited quantities of acceptable quality are available today. Much lower-grade wood is used where durability is important and in America it is popular for fence posts. Other uses are in joinery and furniture, especially the linings of wardrobes and linen and blanket chests, as it is said to repel insects.

LARCH
Larix spp.

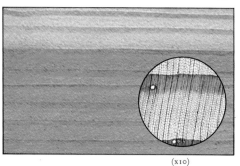

(×10)

THE TREE Larch occurs widely in the more northern parts of Asia and North America, and sporadically in mid-Europe. Both European larch and Japanese larch have a limited natural distribution but are, with their hybrid, widely planted. Larches are unusual among softwoods in casting their leaves in winter. The different species vary in size; some reach 40m or more in height, with a bole 1m in diameter, but others are smaller. American larch is known as tamarack.

THE WOOD Larch is a pale to medium red-brown, with conspicuous growth rings having a marked contrast between early wood and latewood. It is resinous and faintly scented and can have many, usually small, dead knots. It is one of the denser softwoods, heavier than European redwood.

TECHNICAL PROPERTIES Larch dries fairly rapidly but with some tendency to distort; once dry it is stable in use. It is one of the hardest and toughest commercial softwoods, with only pitch pine generally stronger. It works easily and finishes well, though knots can be troublesome. It is moderately resistant to decay.

USES Its strength and moderate durability make larch a useful timber for outdoor use. It is a traditional timber for boat planking, especially of fishing boats, and is an excellent estate timber. It is used both as sawn wood and in the round, though small stems are best treated with preservative. American larch is used for poles, railway sleepers, mining timber, building work, boxes and as a source of pulp.

Pencil Cedar

Larch

Spruce/Whitewood

Pitch Pine

Radiata Pine

Yellow Pine

SPRUCE/WHITEWOOD
Picea spp.

(×10)

THE TREE Spruce is, after pine, the most commonly used softwood. It occurs throughout the northern coniferous forest and is particularly important in North America, as white spruce and Sitka spruce, and in Europe, where it is known as whitewood. The trees vary in size; those of eastern America are mostly small, 12 to 20m high, but Sitka spruce, in the west, reaches 50 to 60m with a bole 1 to 2m in diameter; European spruce reaches 30m or more in height.

THE WOOD Spruce is almost white, sometimes with a pinkish tint in Sitka spruce, and noted for its high natural lustre. Compared with pine of the redwood type, spruce is less resinous, has less conspicuous growth rings, and is about 10 per cent lighter in weight.

TECHNICAL PROPERTIES Spruce dries quickly and well. It has good strength, especially Sitka spruce, and works easily to take a good finish. It is not resistant to decay.

USES Spruce finds a variety of uses, as sawn wood, veneer and pulp. It is the world's most important pulp for newsprint because of its bright, white colour. It is sawn for structural use – joists, rafters, studding, etc. – and for interior joinery; poles are split and used for ladder sides. Knot-free Sitka spruce, which has high strength for its weight, is used for the framing of gliders, in boats and for oars and masts. Clear spruce is used for piano sound-boards and the fronts of violins. It is often used as a core-veneer in birch and Douglas fir plywood.

PITCH PINE
Pinus spp.

(×10)

THE TREE Pitch pine, the heaviest commercial softwood, is of special interest in the southern parts of the United States and in the Caribbean region. In the United States, only the heaviest and strongest timber, known locally as longleaf yellow pine, is shipped as pitch pine; other similar, but lighter-weight, wood is classified as southern pine. Caribbean pitch pine is similar to the heavy American wood. Pitch pines are of moderate size, up to 30m in height with a bole 60 to 90cm in diameter. They are extensively planted.

THE WOOD Pitch pine is yellow-brown to red-brown, with conspicuous growth rings marked by a band of dense latewood, and it is often highly resinous. Pitch pine is some 40 per cent heavier than European redwood.

TECHNICAL PROPERTIES The timber dries slowly and with some tendency to split. True pitch pine is noted for its strength, combining hardness, stiffness and good shock resistance. It is fairly hard to work, and resin tends to build up on tools. It is rated moderately durable.

USES Pitch pine was of great importance commercially in the 19th century, used in industrial building, for church and chapel furniture, for school desks, and flooring; today its availability is more limited. It is used particularly for heavy structural purposes – piling, dock work and bridge-building – in ship-building and for chemical vats. In America, southern pine is a major source of plywood and an important pulpwood; lower-grade timber is used for general building purposes.

RADIATA PINE
Pinus radiata

(×10)

THE TREE Radiata pine, though it has a very limited natural distribution in California and the Guadalupe Is off lower California, is of special importance as the most commonly planted softwood. It has been introduced to many warm temperate countries, and the softwood forests in New Zealand, Australia, Chile and, to a lesser extent, South Africa are dominated by radiata pine. Already it is exported from New Zealand. On favourable sites plantation trees grow vigorously, reaching 30m in height and 50cm in diameter in twenty years.

THE WOOD Radiata pine heartwood is pinkish-brown, though much of the timber in vigorously grown young trees is pale sapwood. Growth rings are present but not well marked, so the wood has an even texture, though it is often knotty. It varies

in weight, but, on average, compares closely with European redwood.

TECHNICAL PROPERTIES The timber dries quickly and well and is stable in use. It is strong for its weight and works well, though tending to tear near knots. It is not durable.

USES Use of radiata pine is increasing rapidly as more areas of plantation timber reach felling size. It is replacing imported softwood in Australia, New Zealand and elsewhere as a building timber, for structural purposes and for floors and cladding. It is also used for boxes and crates, brush and broom handles, and better-quality wood is used for furniture. It is peeled for plywood and is used for particleboard and fibre building board. It is an important pulpwood.

YELLOW PINE
Pinus strobus

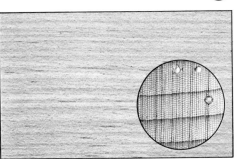

(×10)

THE TREE Yellow pine, or eastern white pine as it is known in America, is one of a number of species with lightweight woods which are known collectively as soft pines. Other commercially important soft pines include American western white and sugar pines and Siberian yellow pine. Yellow pine comes from a tree of medium size, up to 30m in height, with a bole 60 to 90cm in diameter; it occurs naturally in the eastern parts of Canada and the United States. Though it is occasionally seen elsewhere, its scope as a plantation tree is limited by its susceptibility to disease.

THE WOOD Yellow pine is pale-yellow to light-brown, sometimes with a faint pinkish tinge. Growth rings, though present, are not well marked and the wood is noted for its fine, even texture and

straight grain. It is light in weight, some 20 per cent lighter than European redwood. Other soft pines are similar.

TECHNICAL PROPERTIES Yellow pine dries quickly and well and is noted for its small shrinkage on drying and its stability in use. It is low in strength but works very easily, taking an excellent finish. It is not resistant to decay.

USES Yellow pine of good quality commands a very high price and is especially valued for engineers' patterns as it can be cut to give very fine detail and is remarkably stable in use. Other uses are for joinery, panelling, in parts of organs, for drawing boards, wooden rollers, and in ship- and boat-building. In America, lower-grade knotty timber is used for containers and packaging.

Commercial Softwoods

European Redwood

Podocarpus

EUROPEAN REDWOOD
Pinus sylvestris

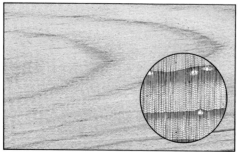

(x10)

THE TREE European redwood, known also as Scots pine, is probably the most commonly used commercial wood. It occurs through much of central Europe and central Asia and is shipped in large quantities from Sweden, Finland, Russia and Poland. It comes from a tree of modest size, 20 to 30m in height and 60cm and occasionally more in diameter. Black, including Corsican, pine of southern Europe, Japanese akamtsu and American red pine are similar.
THE WOOD European redwood is so-called for its reddish-brown heartwood. It has well-marked growth rings, but varies greatly in character from the slowly grown, fine-textured wood of northern Russia to the more vigorously grown, denser wood of southern Sweden and parts of Europe. It is mildly resinous and more or less knotty; it is of medium weight for a softwood, though light in comparison with hardwoods such as oak and beech.
TECHNICAL PROPERTIES Redwood dries easily and well and is stable in use. It has good strength for its weight and works well, taking a good finish. It is not resistant to decay.
USES The timber is graded for use, with the better timber going into joinery and, to a limited extent, furniture, and the general run of production used for construction, especially house-building, and for boxes and crates. It is used, after preservative treatment, for railway sleepers and telegraph and electricity transmission poles. It is an important wood for pulp, especially pulp for wrapping-paper, and is being used increasingly for plywood.

PODOCARPUS
Podocarpus spp.

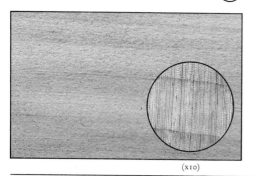

(x10)

THE TREE Species of *Podocarpus* occur in the southern hemisphere and in the warmer parts of the northern hemisphere in central America, East Africa and Asia. There is no single name for their timbers; instead, they are known by different names according to their origin, such as manio from Chile, podo in East Africa, yellowwood and cypress in parts of Central America, yellowwood in South Africa, and in New Zealand by distinctive names, miro, matai, totara and kahikatea, according to the species. They are obtained from trees reaching 30m and sometimes more in height.
THE WOOD *Podocarpus* woods are mostly pale, yellow to yellow-brown, though some from New Zealand are darker, almost reddish-brown in colour. Growth rings are absent or very poorly developed, so that the somewhat featureless wood has a very fine and even texture. The timbers vary from about the weight of whitewood to some 20 per cent heavier than European redwood.
TECHNICAL PROPERTIES The timbers generally dry well, though some have a tendency to split. They have moderate strength but tend to be brittle. They work easily and well and take a very smooth finish. They vary in durability: podo and kahikatea are perishable; totara is very durable.
USES *Podocarpus* species are used, mainly in the countries where they grow, for framing, weather-boarding, flooring and other building purposes. Totara is noted for its resistance to marine borers and is used for dock work, wharves and ship-building; it is also used for chemical vats.

DOUGLAS FIR
Pseudotsuga menziesii

THE TREE Douglas fir, also known as Columbian and Oregon pine though botanically not a pine, is one of the world's outstanding softwoods. It is obtained in British Columbia and on the Pacific coast of the United States from a magnificent tree, commonly 50 to 60m high and exceptionally up to 90m, with a long straight bole 1 to 2m and sometimes more in diameter. It has been introduced and is grown as a plantation tree in many temperate countries, including Britain.
THE WOOD Douglas fir is a pale to medium red-brown, with a conspicuous growth-ring figure, particularly on flat-sawn or rotary-cut surfaces. Typically it is straight-grained, often somewhat resinous and of medium weight for a softwood, marginally heavier than European redwood.
TECHNICAL PROPERTIES Douglas fir dries quickly and well, is noted for its strength, and works readily with hand and machine tools. It is moderately resistant to decay.
USES Douglas fir is the leading structural wood of western North America. It is available in very large sizes, though these are becoming more difficult to obtain, and is used as baulks in piling and for pier, dock and harbour work. It is used in building for roof trusses and laminated beams, and in joinery. It is an important wood for railway sleepers and for chemical vats and tanks; it is used for flooring and decking, but should be quarter-sawn to minimize splintering. It is the world's most important source of plywood, used mainly for structural purposes.

SEQUOIA
Sequoia sempervirens

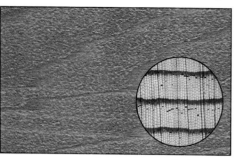

(x10)

THE TREE Sequoia is an American wood, obtained from one of the world's tallest trees, reaching 90m and more in height with a bole up to 3m in diameter, but found only in the coastal regions of California. It is known in America and often marketed as redwood, but should not be confused with European redwood, which is a pine. The closely related giant sequoia, bigtree or wellingtonia, grows to a great age and massive size, but the few remaining stands, also in California, are now protected. Sequoia is often planted as a parkland tree.
THE WOOD Sequoia is a medium to deep red-brown, straight-grained, non-resinous wood. Though often slowly grown, it has a conspicuous growth-ring figure. It is light in weight, about 20 per cent lighter than European redwood.
TECHNICAL PROPERTIES Sequoia dries well and, when dry, is stable in use. Though light in weight, slowly grown timber is firm and moderately strong; it works easily and takes an excellent finish provided that tools are kept sharp. It is noted for its resistance to decay.
USES Only limited quantities of sequoia are available commercially, but it can be obtained in large sizes of virtually knot-free timber. Stable and durable, it is an excellent timber for window and door stock and other joinery purposes. It is used for vats, tanks and silos, and for the slats in water cooling towers. Its light weight makes it unsuitable for heavy structural work, but it is used in farm buildings, greenhouses, garden furniture and for estate work.

Douglas Fir

Sequoia

Southern Cypress

Yew

Western Red Cedar

Hemlock

SOUTHERN CYPRESS
Taxodium distichum

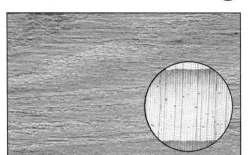
(x10)

THE TREE Southern or swamp cypress is obtained from a tree of the swamps and lowlands of the southeastern United States. The tree reaches 30m or more in height and can grow to a great age, up to 1,000 years. It is of interest in being one of the few deciduous softwoods and is commonly planted in parks in temperate countries as a lakeside or riverside tree.

THE WOOD Southern cypress is unusual for the varied colour of its wood, from pale yellow-brown to dark red-brown and deep chocolate-brown. It has a straight grain and well-marked growth rings giving a conspicuous figure, though it is often slowly grown. It is non-resinous, but is slightly greasy to the touch and has a faintly rancid odour. It is of medium weight for a softwood.

TECHNICAL PROPERTIES Southern cypress dries well though slowly as it is said to contain much water. Its strength properties compare closely with those of European redwood and it works easily to give a good finish with sharp tools. It is noted for its durability.

USES Though once exported in some quantity, Southern cypress is today used mostly in the United States. It is a valuable structural wood where resistance to decay is required, as in dock and bridge work, and for warehouse and factory construction. It is a fine joinery wood, for both outdoor and indoor use, and is considered to be the best wood for chemical vats and tanks. Other uses are for cooperage, in ship- and boat-building, for railway sleepers, and for roofing shingles.

YEW
Taxus baccata

(x10)

THE TREE Yew is a wood of western and central Europe, occurring also in limited parts of western Asia and North Africa. Though it comes from a familiar tree, the wood is of only minor commercial interest for yew is nowhere very common, and the tree reaches only a modest size – 10 to 20m in height – and often has a stem which is short and deeply fluted.

THE WOOD Yew is a beautiful wood, red when first cut, but turning brown on prolonged exposure. Growth rings, though present, are not well marked, often because the wood is very slowly grown; the texture is very fine and even, and an irregular grain often contributes to the decorative appearance of the wood. Yew is one of the heaviest softwoods, about the same weight as pitch pine.

TECHNICAL PROPERTIES Yew dries fairly rapidly and well; it is a strong wood, almost as hard as oak and noted for its resilience and resistance to splitting. It works well, though care is needed in finishing wood with an irregular grain; it turns well and is one of the few softwoods that can be steam-bent. It is durable.

USES Perhaps the most famous use of yew was for archers' bows, especially the English longbow of the Middle Ages. It is a fine furniture wood, though used mainly in craft pieces, and is a traditional wood for the bent parts of Windsor chairs. It is sliced for veneer for panelling and small pieces are turned or used for decorative work. Its durability makes it an excellent estate wood, used for fences and gate-posts.

WESTERN RED CEDAR
Thuja plicata

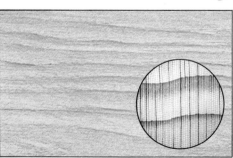
(x10)

THE TREE Western red cedar is obtained in British Columbia and parts of the northwestern United States from a tree which reaches 50m or more in height and 1 to 1.25m in diameter. A similar but commercially less important wood, eastern or northern white cedar, is produced by a smaller tree widely distributed in eastern parts of Canada and the United States.

THE WOOD Western red cedar varies from a pale pinkish-brown to dark-brown; it is non-resinous, but has a faintly pungent smell, and is typically straight-grained with a conspicuous growth-ring figure. It is the lightest-weight softwood in common commercial use; western red cedar is some 25 per cent and white cedar 35 per cent lighter than European redwood.

TECHNICAL PROPERTIES Western red cedar dries quickly and well as thin boards, but thicker stock may be troublesome as it tends to collapse; once dry it is stable in use. A soft, lightweight wood with correspondingly low strength, it is easy to work and takes a good finish provided that tools are kept sharp. It is durable.

USES Western red cedar is one of the most durable softwoods available in quantity, but being light in weight it is suitable only for uses where there is little structural need. It is seen most commonly as vertical cladding and weatherboarding, and is popular for garden buildings and greenhouses. It is used, especially in America, for roofing shingles, and lower-grade timber is used in America for posts and piles.

HEMLOCK
Tsuga spp.

(x10)

THE TREE Hemlock occurs in eastern Asia but North America is the only source of commercial timber. Western hemlock is obtained from southern Alaska to Oregon and is particularly important in British Columbia; it comes from a large tree reaching 50 to 60m in height. Eastern hemlock is a less attractive wood and much less important commercially; it comes from a smaller tree of the eastern United States and adjacent parts of Canada.

THE WOOD Western hemlock is a white wood, though not quite as pale as spruce, as it has a brownish cast and occasional grey streaks. It is non-resinous and straight-grained with well-marked growth rings. It is of medium weight for a softwood. Eastern hemlock is similar but coarser in texture and a little lighter in weight.

TECHNICAL PROPERTIES Hemlock dries rather slowly, but, once dry, is moderately stable in use. It is not as strong as Douglas fir, but is rather better than spruce, with the western timber generally stronger than eastern hemlock. It works easily, taking a good finish. Neither eastern nor western hemlock is resistant to decay.

USES Western hemlock is an important commercial timber shipped to many parts of the world. It is used increasingly as an alternative to Douglas fir for structural purposes, especially in house-building for joists, rafters, studding etc. It is used for boxes and crates, pallets, and clear stock is used for ladder stiles. It is made into plywood and pulp. Eastern hemlock is used usually locally, mainly for lumber and pulp.

Index

Index

Index

Index

Index

Acknowledgements

GENERAL ACKNOWLEDGEMENTS
A great many people, institutions and organizations have given invaluable help and advice during the preparation of this book. The publishers wish to extend their thanks to them all, and in particular to the following:

The Director and staff of the Building Research Establishment, Princes Risborough Laboratory, and in particular Mr R Brodie, Mr DC Bedding (Applications and Services) Mrs C Abbot, Mrs A Cooper, Miss FB Tillott (Library) Mrs A Miles (Properties of Materials) Mr WB Banks (Operations and Services) Mr R Cockcroft (Preservation) Mr JG Savory, Mrs SJ Read (Bio-deterioration); the Director, Conservator and staff of the Forestry Commission, Alice Holt Research Station, and in particular Mr RJ Power, Mr R Gladman, Mr J Wood, Mr R Herring, Mr T Anderson, Mr N Blatchford; the Director, Librarian and staff of the Library of the Royal Botanic Gardens, Kew; the Director and staff of the London College of Furniture, and in particular Mr P Shirtcliffe, Miss P Nasmyth, Mr T Wicking (Dept of Musical Instruments) Mr A Smith; Mr GC Cheeke, American Forestry Institute, Washington DC; The American Library, University of London; Mrs K Bond (Curator) Mrs S Ford, American Museum in Britain, Bath; Mr RE Anderson (Manager, Special Services) American Plywood Association; Mr KS Rolston (Executive Vice-President) American Pulpwood Association; the Librarian of the Ancient Egypt Exploration Society, London; the Institute of Archaeology, London; Miss EM Aslin (Keeper) Bethnal Green Museum, London; Miss C Lattar (Asst Keeper) Birmingham Museum and Art Gallery; British Antarctic Survey; Mr DW Macklin (Press Officer) Council of Forest Industries of British Columbia; Mr RD Barnsdale (Secretary) British Longbow Society; Mr B Hutchinson (Horological Students' Room) British Museum, London; Mr WA Oddy, British Museum Research Laboratory; the Photographic Dept and Librarians (Botany Library) British Museum (Natural History); Prof Dr D Noack, Bundesforschungsanstalt für Forst-und-Holstwirtschaft, Hamburg; California Academy of Sciences, San Francisco; California Redwood Association; Cambridge Botanic Gardens; Commonwealth Forestry Institute, Oxford; Mr N Meeus, Conservatoire Royale de Musique, Brussels; Council for Small Industries in Rural Areas, London; Miss J Collin, Courtauld Institute of Art, London; Crafts Advisory Committee, London; Design Centre, London; Mr J Offerd (Curator) Dodington Carriage Museum, Bristol; Embassies, Tourist Offices and Institutes of Canada, France, Germany, Italy, Japan, Romania, Turkey, UAR, USA, USSR; the Director and staff, Museum of English Rural Life, Reading; Mr L Collins, Dept of Environment, London; Institute of Geological Sciences, London; Mr I Sparkes (Curator) High Wycombe Museum; Historical Monuments Records Office, London; Mr AG Wilkinson (Curator) Henry Moore Museum, Art Gallery of Ontario, Toronto; Mr DM Boston (Curator) and staff, Horniman Museum, London; Kon-Tiki Museet, Oslo; London Library; London School of Oriental and African Studies; Director, staff and Librarians, Museum of Mankind, London; National Maritime Museum, London, and in particular Mrs Etherington; Mr N Dugano, Mr D Maidment, Mr D Sommerin, National Motor Museum, Beaulieu; Mr CR Morschcauser (Technical Director) National Particle Board Association, Maryland; National Trust, and Mr T Graham (Housekeeper) Audley End House; Mr JM Stikvoort (Information Officer) Het Nederlands

Openluft Museum, Arnhem; Norsk Folkemuseum, Oslo; Mr J Scott Taggart, Paper Industries Research Association, Leatherhead; the Curator and staff, Pitt-Rivers Museum, Oxford; the Librarian, Royal Aeronautical Society, London; Royal Botanic Gardens, Kandy; the Librarians, Royal Institute of British Architects; Mr BW Bathe (Ships) Miss S Snowden (Transport) Librarians and staff, Science Museum, London; the Director, Schwarzwalder Freilichtsmuseum Vogtsbaurnhof, Wolfach; Miss CE Stephens (Museum Technician) Miss C Forsyth (Photo Services) Smithsonian Institution, Washington DC; the Director and staff, Timber Research and Development Association, High Wycombe, and in particular Dr OP Hansom, Mrs D Bedding, Mr R Allcorn; Ministry of Trade and Agriculture, Saint Vincent, West Indies; Institute of Tropical Forestry, Puerto Rico; the Director and Librarians, Tropical Products Institute, London; Dept of Wood Science and Technology, University of the Philippines, Laguna; the Director, Librarians and staff, Victoria and Albert Museum, London; Mr F Birkebaek, Vikingeskibshallen, Roskilde; the Director, staff, and Miss H Jackson, Weald and Downland Museum, Singleton; Mr JL Blackwood (Executive Director) Western Forestry Center, Oregon; Mr PM King (Director, Public Affairs) Western Wood Products Association, Washington DC; Mr DJ Bryden (Curator) Whipple Science Museum, Cambridge; Mr C Jagger (Secretary) Worshipful Company of Clockmakers; Mr HC Leslie (Public Information Specialist) US Dept of Agriculture Forest Service; Mr F Davis and Abbey Pattern Works Ltd; the staff, American School in London; Prof S Kapoor and Anthropos Gallery; Mr KE Davis, Mr S Swift and Antique Restorers Ltd; Mr B Arnold; Mr M Biddle; landlord of Boat Inn, Stoke Bruerne; Bluthner Pianos Ltd; Broadwood and Sons Ltd; Mr M Binnings; Mr G Hill, Mr G Corser and Courtaulds Ltd; Mr AJ Robards and Coutts & Co; Mr T Crispin; Mr DB and Miss L Ercolani and Ercol Furniture Ltd; Mr and Mrs A Galliers-Pratt; Mr R Goodearl; Mr B Moser, Mr A Singer and Granada Television Ltd; Rev JLStC Garrington (vicar) Greensted Church; Mrs G Hale; Mr R Hamilton; Information Officer, Hotel d'Hane Steenhuse; Hardy Bros Ltd; Mr M Harris; Mr L Hill; Mr A Hornak; Mr M Lee and JCB Sales Ltd; Mr J Jones; Malden Antiques Ltd; Cmdr JEG McKee; Mercedes Benz Ltd; Mitchell Proctor & Associates; Mr K Potter; Mr C Pierson, Mr A Preston, Mr A Redfern, Mr V Stockton and Price and Pierce (World Pulp) Ltd; Mr J Robins, the vicar, St Wendreda's Church, March; FR Shadbolt and Son Ltd; Miss R Sheradski; Skimaster Ltd; Miss J Foster and Slazengers Ltd; Mr W Souter and WA Souter Ltd; Mr D Stamp; Stuart Surridge & Co Ltd; Mr J Swigoniak; Mr F van Doorninck; Mr D Wason; Mr A Whateley; Wiggins Teape Ltd; Mr Hall (Bursar) Winchester School; Dean and Chapter, Worcester Cathedral.

Our particular thanks are due to: Mr Nicholas Cooper; Mr David Cutler of the Jodrell Laboratory, Royal Botanic Gardens, Kew; Miss Georgina Fuller; Mr Roy King; Mr John Makepeace; Mr Henry Moore; Mr Norman Place; Colonel Humphrey Quill; Mr John L Rawlings; Mr Ronnie Rustin; Mr Noel Taylor; and to the following members of International Paper Company in New York, Washington DC and Longview, Washington: Doug Bartels; Curt R Copenhagen; Arthur D'Arazien; Steven A Mowe; Thomas H Mutchler; Peter Quatz; Milton Wooley. Our thanks are also due to Toni and Peter Schwed at Simon and Schuster, New York.

Acknowledgements

ARTISTS
13 Chris Forsey; 14–15 Peter Morter; 17 Sidney Woods; 18 Chris Forsey, Ian Garrard; 19 Sidney Woods; 20–21 Sidney Woods; 23 Alan Suttie, Errol Bryant; 27 Chris Forsey; 28–29 Charles Pickard; 32 Alan Suttie; 35 Chris Forsey; 40 Chris Forsey; 48 Peter Morter; 50 Alan Suttie; 52 MB Studio; 53 Alan Suttie; 55 Alan Suttie; 58–59 Peter Morter; 59 Chris Forsey; 61 Chris Forsey, Peter Morter, Chris Forsey; 65 Chris Forsey; 67 Ken Ody; 68–69 Peter Morter; 69 Chris Forsey; 71 Chris Forsey; 72–73 main illus.

Peter Morter, subsid. illus. Chris Forsey based on *The Japanese House and Garden* by Tetsuro Yoshida (Architectural Press); 77 Chris Forsey; 81 Chris Forsey; 84–85 Chris Forsey; 86–87 Peter Morter; 88–89 Peter Morter; 89 Ken Ody; 95 Chris Forsey; 99 Chris Forsey; 102–103 Peter Morter; 104–105 Peter Morter; 105 Ken Ody; 109 Ken Ody; 110–111 Sidney Woods; 112–113 John Western; 114–115 Peter Morter; 117 Chris Forsey; 118–119 Sidney Woods; 120–121 Chris Forsey; 124 Chris Forsey; 128–129 Peter Morter; 131 Chris Forsey; 134–135 Brian Delf;

138–139 Brian Delf; 140–141 Brian Delf; 142–143 Peter Hutton; 144–145 Peter Sarson; 156–157 Peter Sarson; 157 Tony Bryan; 158 Peter Morter; 159 David Penney, Peter Morter; 161 Brian Delf; 163 David Ashby; 165 Chris Forsey; 168–169 Sidney Woods; 172–173 Peter Sarson; 174–175 Peter Sarson; 176–177 Peter Sarson; 178–179 Peter Sarson; 180–181 Ron Hayward; 182–183 Peter Morter; 184–185 maps Arka Graphics, illus. Chris Forsey; 186–187 Peter Sarson; 192–193 Peter Morter; 194–195 Peter Morter; 196–197 Peter Morter; 198–199

Peter Morter; 203 Michael Woods; 205 Brian Delf; 208–209 Brian Delf; 209 Chris Forsey; 210–211 Sidney Woods; 212 Michael Woods; 221 Ian Garrard; 226–227 Eugene Fleury; 228–229 Sidney Woods; 230–265 Ian Garrard.

Photographic retouching (colour) JD Studios, (monochrome) Sally Slight; special photogammetry by Products Support Graphics.

Photographers are credited by descending order of the base line of each photograph. Where two or more photographs lie on the same base line, credits read left to right. In some instances, the last-named credit on a page may apply to more than one photograph.

PHOTOGRAPHS
1–11 Michael Freeman; 12–13 Michael Freeman; 13 Mike Busselle; 15 Mike Busselle; 16–17 Mike Busselle; 19 Building Research Establishment Princes Risborough; 20–21 BA Melan and BG Butterfield, published in *The Three-dimensional Structure of Wood* (Chapman and Hall); 22 John Watney, British Museum (Natural History), Reichel/Agence TOP from Colorific; 23 British Museum (Natural History); 24–25 Heather Angel; 26–27 Heather Angel; 28–29 Bruce Coleman/Jane Burton; 30 Tony Carr/Colorific; 31 Ardea Photographics/C McDougal, Paolo Koch, Paolo Koch, Aerofilms, Bruce Coleman/SC Bisserot; 33 Forestry Commission; 34 International Paper Company; 35 USDA Forest Service, Forestry Commission, Forestry Commission, School of Forest Resources North Carolina State University; 36–37 International Paper Company; 38–39 International Paper Company; 40 MJ Bramwell, Arthur d'Arazien; 41 International Paper Company/Arthur Schatz, Forestry Commission, Forestry Commission, Robert Harding Associates; 42–43 International Paper Company; 44–45 International Paper Company except insert far left Arthur d'Arazien; 46–47 International Paper Company; 48 International Paper Company; 49 Koppers, Arthur d'Arazien, Mike Busselle, International Paper Company, Mike Busselle; 50 Forestry Commission; 51 Tiofoto, Mittet Foto/K Hilsen, Paolo Koch, Snark International/E Rousseau; 52 PIRA; 53 Terry Harris; 54 Bruce Coleman/Bill Brooks, Bruce Coleman/CB Frith; 55 Courtaulds Ltd.; 56–57 Shostal Associates; 58 Axel Poignant; 60 Norsk Folkemuseum Oslo, Smithsonian Institution, Schweizerische Verkehrszentrale Zurich, Paolo Koch; 62 Werner Forman Archive; 63 Axel Poignant, Foto

Salmer, Robert Harding Associates, Robert Harding Associates; 64 Barnaby's Picture Library, Paul Dong FJI; 65 Edwin Smith, ZEFA; 66 Michael Holford; 67 Mike Busselle, Mike Busselle, Mike Busselle, Michael Holford; 68 Wayne Andrews; 70–71 Barry Shapiro; 71 Edward Teitelman; 74 AF Kersting; 75 Lauros-Giraudon, Picturepoint, Hoa-Qui, Axel Poignant; 76 Gabriel Wades Ltd., Finnish Plywood Development Association; 77 American Wood Preserving Association, Rainham Timber Engineering Company, gridshell structure made of flat timber lattice designed by Florian Beigel constructed by the students of the North London Polytechnic on principles developed by Professor Frei Otto, Gabriel Wades Ltd./Thomas A Wilkie; 78–79 Western Wood Products Association/Howard Kinney RIA; 78 Western Wood Products Association, Western Wood Products Association/John Stores; 79 California Redwood Association/Ernest Braun, Spectrum Colour Library, Shostal Associates, Western Wood Products Association; 80 Shostal Associates; 81 Shostal Associates, Western Wood Products Association, American Plywood Association, American Plywood Association, American Plywood Association, American Plywood Association, Western Wood Products Association, Western Wood Products Association; 82–83 Mike Busselle; 84 Picturepoint, Mike Busselle, Mike Busselle; 85 Giraudon, Carl Purcell/Colorific; 88 Paolo Koch; 89 Robert Harding Associates, ZEFA; 90 AF Kersting; 91 Cooper-Bridgeman Library, Giraudon, Giraudon, Pictor, Robert Harding Associates; 92 Michael Holford, Angelo Hornak, John Massey Stewart; 93 Sonia Halliday, Sonia Halliday, Scala, Scala; 94–95 Angelo Hornak; 96–97 Mike Busselle; 98 Scala; 99 Michael Holford, the Metropolitan Museum of Art gift of J Pierpont Morgan 1917/Werner Forman Archive, Michael Holford, Michael Holford, Michael Holford, Michael Holford; 100 Susan Griggs/Ian Yeomans, Mike Busselle; 101 Mike Busselle; 106 Edwin Smith; 107 Scala, Michael Holford, Mike

Busselle; 108 Lauros-Giraudon, Mike Busselle; 109 Sally Chappell, Mike Busselle, Mike Busselle, American Museum in Britain/W Morris; 111 Mike Busselle; 116 Mike Busselle, Sotheby's Belgravia; 117 Sotheby's Belgravia, Sotheby's Belgravia, Mike Busselle; 118 Evans Brothers Ltd. from *Antique or Fake* by Charles Hayward, Rijksmuseum Amsterdam, Mike Busselle, Mike Busselle; 119 Mike Busselle; 121 Roger-Viollet, Mike Busselle, Mike Busselle, Cooper-Bridgeman Library; 122 Cooper-Bridgeman Library, Mike Busselle; 123 Angelo Hornak, Cooper-Bridgeman Library, Mike Busselle, Cooper-Bridgeman Library; 124 American Museum in Britain/Derek Balmer; 125 Phillips Son & Neale the Fine Art Auctioneers, the Hispanic Society of America, Mike Busselle, American Museum in Britain/Cooper-Bridgeman Library; 126–127 Terry Harris; 130 Fritz Hansen Denmark; 131 Mike Busselle; 132–133 MJ Bramwell; 136–137 Ricky Hatswell; 146–147 Mike Busselle; 148–149 main photograph and inserts Mike Busselle; 149 Corry Bevington, Corry Bevington, Museum of English Rural Life; 150 Rijksmuseum voor Volkskunde "Het Nederlands Openluchtmuseum", engraving Ronan Picture Library; 151 Mike Busselle; 152–153 Museum of English Rural Life except small insert on 153 HL Edlin; 154–155 Mike Busselle; 157 J Allan Cash; 159 Trustees of the British Museum, Mike Busselle; 160–161 Popperfoto; 162 Dr MJT Lewis, Terry Harris, Terry Harris; 162–163 Terry Harris; 163 Susan Griggs/Adam Woolfitt; 164 Trustees of the British Museum; 165 Crown Copyright National Railway Museum; 166–167 Mike Busselle; 167 Council of the Forest Industries of British Columbia, Council of the Forest Industries of British Columbia, Mike Busselle; 169 Mike Busselle; 170–171 Susan Griggs/Adam Woolfitt; 174 Giraudon; 175 Paul Johnstone; 179 Mauro Pucciarelli, Scala, Mansell Collection; 180 The Director of the Royal Botanic Gardens, The Director of the Royal Botanic Gardens, Trustees of the British Museum; 181 Viking Ship Museum Roskilde; 183 Ricky Hatswell; 188–

189 Mike Busselle except insert bottom left 189 Roger Smith; 190 National Trust; 195 Victoria and Albert Museum, Victoria and Albert Museum, Mike Busselle, Victoria and Albert Museum; 198–199 Mike Busselle; 200–201 Henry Moore/Fischer Fine Art Gallery; 201 Henry Moore; 202 Smithsonian Institution/The Eleanor and Mabel Van Alstyne American Folk Art Collection; 203 American Museum in Britain/Derek Balmer, American Museum in Britain/Derek Balmer, American Museum in Britain/Derek Balmer, Boston Museum of Fine Arts, the collection of the Virginia Museum of Fine Arts, American Museum in Britain/Derek Balmer, Smithsonian Institution/The Eleanor and Mabel Van Alstyne American Folk Art Collection; 204 Trustees of the British Museum; 205 Trustees of the British Museum, Trustees of the British Museum, Victoria and Albert Museum; 206–207 large composite Angelo Hornak, small composites Ricky Hatswell; 208 Associated Press; 209 Sport and General Press; 211 Mike Busselle; 212 Michael Holford, Weidenfeld and Nicolson Ltd./Museo Capitolini Rome; 213 Mike Busselle, Mike Busselle, Mike Busselle, Clive Corless; 214–215 Mike Busselle; 216 Werner Forman Archive, Werner Forman Archive, Mike Busselle; 217 Mike Busselle, Mike Busselle, Mike Busselle, Mike Busselle, Michael Holford, Michael Holford, Michael Holford; 218 Werner Forman Archive, Observer/Transworld; 218–219 Michael MacIntyre; 219 Werner Forman Archive, Werner Forman Archive, Trustees of the British Museum, Trustees of the British Museum, Werner Forman Archive, Observer/Transworld; 220–221 Fotogram; 222 Scala, Trustees of the British Museum, JR Freeman, Werner Forman Archive; 223 Werner Forman Archive/Merseyside County Museums; 224–225 Mike Busselle; 228–229 Building Research Establishment Princes Risborough; 230–265 colour photographs Mike Busselle, black and white photographs Building Research Establishment Princes Risborough.